Stochastic Differential Equations and their Applications

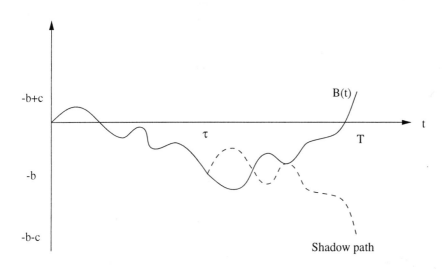

Mathematics possesses not only truth, but supreme beauty – a beauty cold and austere, like that of sculpture, and capable of stern perfection, such as only great art can show.
　　　Bertrand Russell, 3rd Earl Russell (1872-1970): *Principles of Mathematics* (1903)

Dr. XUERONG MAO

Dr Xuerong Mao was born in 1957 in the city of Fuzhou, in the province of Fujian, Peoples Republic of China. After obtaining a Diploma in the Textile University, in Shanghai, he graduated in the Department of Mathematics with a Masters Degree in the Shanghai Textile University. His first teaching appointment was Lecturer, Department of Management, Fuzhou University in the province of Fujian (1982-87).

He came to England in 1987 and was awarded a Doctorate by the Mathematics Institute, University of Warwick, and was then SERC (Science and Research Engineering Council, UK) Post-Doctoral Reseach Fellow 1989-92. Moving to Scotland, he joined the Department of Statistics and Modelling Science, University of Strathclyde, Glasgow as a lecturer in 1992: was promoted to Senior Lecturer in 1995, and was soon made Reader which post he still holds.

Dr Mao is well known and highly respected for his mathematical reputation, and has published 68 papers in mathematical journals in China and the West; and has written two published books, *Stability of Stochastic Differential Equations with Respect to Semimartingales*, (1991) and *Exponential Stability of Stochastic Differential Equations*, (1994).

Recognition came from the American Biographical Institute in 1996 with the Distinguished Leadership Award, and he has been visiting Guest Professor of Huazhong University of Science and Technology, Wuhan, Peoples Republic of China; and he is on the refereeing panel for many journals on stochastics including the Royal Society Proceedings.

Stochastic Differential Equations and their Applications

Xuerong Mao
Department of Statistics and Modelling Science
University of Strathclyde
Glasgow

Horwood Publishing
Chichester

First published in 1997 by
HORWOOD PUBLISHING LIMITED
International Publishers
Coll House, Westergate, Chichester, West Sussex, PO20 6QL England

COPYRIGHT NOTICE
All Rights Reserved. No part of this publication may be reproduced, stored in a retrieval system, or transmitted, in any form or by any means, electronic, mechanical, photocopying, recording, or otherwise, without the permission of Horwood Publishing Limited, Coll House, Westergate, Chichester, West Sussex, PO20 6QL, England

© Xuerong Mao, 1997

British Library Cataloguing in Publication Data
A catalogue record of this book is available from the British Library

ISBN 1-898563-26-8

Printed in Great Britain by Hartnolls, Bodmin, Cornwall

DEDICATION

To the memory of my grandmother Ms Zhang Li Yin
and
my father-in-law Mr Ziang Jie Ming

Preface

Stochastic modelling has come to play an important role in many branches of science and industry where more and more people have encountered stochastic differential equations. There are several excellent books on stochastic differential equations but they are long and difficult, especially for the beginner. There are also a number of book at the introductory level but they do not deal with several important types of stochastic differential equations e.g. stochastic equations of neutral type and backward stochastic differential equations which have been developed recently. There is a need for a book that not only studies the classical theory of stochastic differential equations but also the new developments at an introductory level. It is in this spirit this text is written.

This text will explore stochastic differential equations and their applications. Some important features of this text are as follows:

- This text presents at an introductory level the basic principles of various types of stochastic systems e.g. stochastic differential equations, stochastic functional differential equations, stochastic equations of neutral type and backward stochastic differential equations. The neutral-type and backward equations appear frequently in many branches of science and industry. Although they are are more complicated, this text treats them at an understandable level.

- This text discusses the new developments of Carathedory's and Cauchy–Marayama's approximation schemes in addition to Picard's. The advantage of Cauchy–Marayama's and Carathedory's schemes is that the approximate solutions converge to the accurate solution under a weaker condition than the Lipschitz one, but this is still open for Picard's scheme. These schemes are used to establish the theory of existence and uniqueness of the solution while they also give the procedures to obtain numerical solutions in applications.

- This text demonstrates the manifestations of the general Lyapunov method by showing how this effective technique can be adopted to study entirely differently qualitative and quantitative properties of stochastic systems, e.g.

Preface

asymptotic bounds and exponential stability.

- This text emphasises the analysis of stability in stochastic modelling and illustrates the practical use of stochastic stabilization and destabilization. This is the first text that explains systematically the use of Razumikhin technique in the study of exponential stability for stochastic functional differential equations and the neutral-type equations.

- This text illustrates the practical use of stochastic differential equations through the study of stochastic oscillators, stochastic modelling in finance and stochastic neural networks.

Acknowledgements

I wish to thank Professors G. Gettinby, W.S.C. Gurney and E. Renshaw for their constant support and kind assistance. I am indebted to the University of Strathclyde, the EPSRC, the London Mathematics Society and the Royal Society for their financial support. I also wish to thank Mr. C. Selfridge who reads the manuscript carefully and makes many useful suggestions. Moreover, I should thank my family, in particular, my beloved wife Weihong, for their constant support.

Xuerong Mao
Glasgow May 1997

Contents

Preface

General Notation

1. **Brownian Motions and Stochastic Integrals** — 1
   - 1.1. Introduction — 1
   - 1.2. Basic notations of probability theory — 2
   - 1.3. Stochastic processes — 9
   - 1.4. Brownian motions — 15
   - 1.5. Stochastic integrals — 18
   - 1.6. Itô's formula — 31
   - 1.7. Moment inequalities — 39
   - 1.8. Gronwall-type inequalities — 44
2. **Stochastic Differential Equations** — 47
   - 2.1. Introduction — 47
   - 2.2. Stochastic differential equations — 48
   - 2.3. Existence and uniqueness of solutions — 51
   - 2.4. L^p-estimates — 59
   - 2.5. Almost surely asymptotic estimates — 63
   - 2.6. Caratheodory's approximate solutions — 71
   - 2.7. Cauchy–Maruyama's approximate solutions — 76
   - 2.8. SDE and PDE: Feynman–Kac's formula — 78
   - 2.9. Solutions as Markov processes — 84
3. **Linear Stochastic Differential Equations** — 91
   - 3.1. Introduction — 91
   - 3.2. Stochastic Liouville's formula — 92
   - 3.3. The variation-of-constants formula — 96
   - 3.4. Case studies — 98
   - 3.5. Examples — 101
4. **Stability of Stochastic Differential Equations** — 107
   - 4.1. Introduction — 107
   - 4.2. Stability in probability — 110
   - 4.3. Almost sure exponential stability — 119

Contents

	4.4. Moment exponential stability	127
	4.5. Stochastic stabilization and destabilization	135
	4.6. Further topics	141
5.	**Stochastic Functional Differential Equations**	**147**
	5.1. Introduction	147
	5.2. Existence-and-uniqueness theorems	149
	5.3. Stochastic differential delay equations	155
	5.4. Exponential estimates	158
	5.5. Approximate solutions	162
	5.6. Stability theory—Razumikhin theorems	169
	5.7. Stochastic self-stabilization	188
6.	**Stochastic Equations of Neutral Type**	**199**
	6.1. Introduction	193
	6.2. Neutral stochastic functional differential equations	201
	6.3. Neutral stochastic differential delay solutions	207
	6.4. Moment and pathwise estimates	209
	6.5. L^p-continuity	215
	6.6. Exponential stability	219
7.	**Backward Stochastic Differential Equations**	**233**
	7.1. Introduction	233
	7.2. Martingale representation theorem	234
	7.3. Equations with Lipschitz coefficients	239
	7.4. Equations with non-Lipschitz coefficients	246
	7.5. Regularities	258
	7.6. BSDE and quasilinear PDE	265
8.	**Stochastic Oscillators**	**269**
	8.1. Introduction	269
	8.2. The Cameron–Martin–Girsanov theorem	270
	8.3. Nonlinear stochastic oscillators	272
	8.4. Linear stochastic oscillators	276
	8.5. Energy bounds	286
9.	**Applications to Economics and Finance**	**299**
	9.1. Introduction	299
	9.2. Stochastic modelling in asset prices	299
	9.3. Optimal stopping problems	305
	9.4. Stochastic games	319
10.	**Stochastic Neural Networks**	**327**
	10.1. Introduction	327
	10.2. Stochastic neural networks	327
	10.3. Stochastic neural networks with delays	342
Bibliographical Notes		**353**
References		**355**
Index		**363**

General Notation

Theorem 4.3.2, for example, means Theorem 3.2 (the second theorem in Section 3) in Chapter 4. If this theorem is quoted in Chapter 4, it is written as Theorem 3.2 only.

positive : > 0.
nonpositive : ≤ 0.
negative : < 0.
nonnegative : ≥ 0.
a.s. : almost surely, or P-almost surely, or with probability 1.
$A := B$: A is defined by B or A is denoted by B.
$A(x) \equiv B(x)$: $A(x)$ and $B(x)$ are identically equal, i.e. $A(x) = B(x)$ for all x.
$\emptyset$: the empty set.
I_A : the indicator function of a set A, i.e. $I_A(x) = 1$ if $x \in A$ or otherwise 0.
A^c : the complement of A in Ω, i.e. $A^c = \Omega - A$.
$A \subset B$: $A \cap B^c = \emptyset$.
$A \subset B$ a.s. : $P(A \cap B^c) = 0$.
$\sigma(\mathcal{C})$: the σ-algebra generated by $\mathcal{C}$.
$a \vee b$: the maximum of a and b.
$a \wedge b$: the minimum of a and b.
$f : A \to B$: the mapping f from A to B.
$R = R^1$: the real line.
R_+ : the set of all nonnegative real numbers, i.e. $R_+ = [0, \infty)$.
R^d : the d-dimensional Euclidean space.
$\mathcal{B}^d$: the Borel-σ-algebra on R^d.

General Notation

$\mathcal{B}$: $\mathcal{B} := \mathcal{B}^1$.

$R^{d \times m}$: the space of real $d \times m$-matrices.

$\mathcal{B}^{d \times m}$: the Borel-σ-algebra on $R^{d \times m}$.

C^d : the d-dimensional complex space.

$C^{d \times m}$: the space of complex $d \times m$-matrices.

$|x|$: the Euclidean norm of a vector x.

S_h : $S_h := \{x \in R^d : |x| \leq h\}$.

A^T : the transpose of a vector or matrix A.

(x, y) : the scale product of vectors x and y, i.e. $(x, y) = x^T y$.

trace A : the trace of a square matrix $A = (a_{ij})_{d \times d}$, i.e. trace $A = \sum_{1 \leq i \leq d} a_{ii}$.

$\lambda_{\min}(A)$: the smallest eigenvalue of a matrix A.

$\lambda_{\max}(A)$: the largest eigenvalue of a matrix A.

$|A|$: $|A| = \sqrt{trace(A^T A)}$, i.e. the trace norm of a matrix A.

$||A||$: $||A|| := \sup\{|Ax| : |x| = 1\} = \sqrt{\lambda_{\max}(A^T A)}$, i.e. the operator norm of a matrix A.

δ_{ij} : Dirac's delta function, that is $\delta_{ij} = 1$ if $i = j$ or otherwise 0.

$C(D; R^d)$: the family of continuous R^d-valued functions defined on D.

$C^m(D; R^d)$: the family of continuously m-times differentiable R^d-valued functions defined on D.

$C_0^m(D; R^d)$: the family of functions in $C^m(D; R^d)$ with compact support in D.

$C^{2,1}(D \times R_+; R)$: the family of all real-valued functions $V(x, t)$ defined on $D \times R_+$ which are continuously twice differentiable in $x \in D$ and once differentiable in $t \in R_+$.

∇ : $\nabla = (\frac{\partial}{\partial x_1}, \cdots, \frac{\partial}{\partial x_d})$.

Δ : the Laplace operator, i.e. $\Delta = \sum_{i=1}^d \frac{\partial^2}{\partial x_i^2}$.

V_x : $V_x = \nabla V = (V_{x_1}, \cdots, V_{x_d}) = (\frac{\partial V}{\partial x_1}, \cdots, \frac{\partial V}{\partial x_d})$.

V_{xx} : $V_{xx} = (V_{x_i x_j})_{d \times d} = (\frac{\partial^2 V}{\partial x_i \partial x_j})_{d \times d}$.

$||\xi||_{L^p}$: $||\xi||_{L^p} := (E|\xi|^p)^{1/p}$.

$L^p(\Omega; R^d)$: the family of R^d-valued random variables ξ with $E|\xi|^p < \infty$.

$L^p_{\mathcal{F}_t}(\Omega; R^d)$: the family of R^d-valued $\mathcal{F}_t$-measurable random variables ξ with $E|\xi|^p < \infty$.

$C([-\tau, 0]; R^d)$: the space of all continuous R^d-valued functions φ defined on $[-\tau, 0]$ with a norm $||\varphi|| = \sup_{-\tau \leq \theta \leq 0} |\varphi(\theta)|$.

$L^p_{\mathcal{F}}([-\tau, 0]; R^d)$: the family of all $C([-\tau, 0]; R^d)$-valued random variables ϕ such that $E||\phi||^p < \infty$.

General Notation

$L^p_{\mathcal{F}_t}([-\tau, 0]; R^d)$: the family of all $\mathcal{F}_t$-measurable $C([-\tau, 0]; R^d)$-valued random variables ϕ such that $E\|\phi\|^p < \infty$.

$C^b_{\mathcal{F}_t}([-\tau, 0]; R^d)$: the family of $\mathcal{F}_t$-measurable bounded $C([-\tau, 0]; R^d)$-valued random variables.

$L^p([a, b]; R^d)$: the family of Borel measurable functions $h : [a, b] \to R^d$ such that $\int_a^b |h(t)|^p dt < \infty$.

$\mathcal{L}^p([a, b]; R^d)$: the family of R^d-valued $\mathcal{F}_t$-adapted processes $\{f(t)\}_{a \leq t \leq b}$ such that $\int_a^b |f(t)|^p dt < \infty$ a.s.

$\mathcal{M}^p([a, b]; R^d)$: the family of processes $\{f(t)\}_{a \leq t \leq b}$ in $\mathcal{L}^p([a, b]; R^d)$ such that $E \int_a^b |f(t)|^p dt < \infty$.

$\mathcal{L}^p(R_+; R^d)$: the family of processes $\{f(t)\}_{t \geq 0}$ such that for every $T > 0$, $\{f(t)\}_{0 \leq t \leq T} \subset \mathcal{L}^p([0, T]; R^d)$.

$\mathcal{M}^p(R_+; R^d)$: the family of processes $\{f(t)\}_{t \geq 0}$ such that for every $T > 0$, $\{f(t)\}_{0 \leq t \leq T} \in \mathcal{M}^p([0, T]; R^d)$.

Erf($\cdot$) : the error function given by $\text{Erf}(z) = (2\pi)^{-1/2} \int_0^z e^{-u^2/2} du$.

sign(x) : the sign function, that is sign(x) = +1 if $x \geq 0$ or otherwise -1.

Other notations will be explained where they first appear.

1

Brownian Motions and Stochastic Integrals

1.1 INTRODUCTION

Systems in many branches of science and industry are often perturbed by various types of environmental noise. For example, consider the simple population growth model

$$\frac{dN(t)}{dt} = a(t)N(t) \tag{1.1}$$

with initial value $N(0) = N_0$, where $N(t)$ is the size of the population at time t and $a(t)$ is the relative rate of growth. It might happen that $a(t)$ is not completely known, but subject to some random environmental effects. In other words,

$$a(t) = r(t) + \sigma(t)\text{"noise"},$$

so equation (1.1) becomes

$$\frac{dN(t)}{dt} = r(t)N(t) + \sigma(t)N(t)\text{"noise"}.$$

That is, in form of integration,

$$N(t) = N_0 + \int_0^t r(s)N(s)ds + \int_0^t \sigma(s)N(s)\text{"noise"}ds. \tag{1.2}$$

The questions are: What is the mathematical interpretation for the "noise" term and what is the integration $\int_0^t \sigma(s)N(s)\text{"noise"}ds$?

It turns out that a reasonable mathematical interpretation for the "noise" term is the so-called white noise $\dot{B}(t)$, which is formally regarded as the derivative of a Brownian motion $B(t)$, i.e. $\dot{B}(t) = dB(t)/dt$. So the term "noise"dt can be expressed as $\dot{B}(t)dt = dB(t)$, and

$$\int_0^t \sigma(s)N(s)\text{"noise"}ds = \int_0^t \sigma(s)N(s)dB(s). \tag{1.3}$$

If the Brownian motion $B(t)$ were differentiable, then the integral would have no problem at all. Unfortunately, we shall see that the Brownian motion $B(t)$ is nowhere differentiable hence the integral can not be defined in the ordinary way. On the other hand, if $\sigma(t)N(t)$ is a process of finite variation, one may define the integral by

$$\int_0^t \sigma(s)N(s)dB(s) = \sigma(t)N(t)B(t) - \int_0^t B(s)d[\sigma(s)N(s)].$$

However, if $\sigma(t)N(t)$ is only continuous, or just integrable, this definition does not make sense. To define the integral, we need make use of the stochastic nature of Brownian motion. This integral was first defined by K. Itô in 1949 and is now known as *Itô stochastic integral*. The main aims of this chapter are to introduce the stochastic nature of Brownian motion and to define the stochastic integral with respect to Brownian motion.

To make this book self-contained, we shall briefly review the basic notations of probability theory and stochastic processes. We then give the mathematical definition of Brownian motions and introduce their important properties. Making use of these properties, we proceed to define the stochastic integral with respect to Brownian motion and establish the well-known Itô formula. As the applications of Itô's formula, we establish several moment inequalities e.g. the Burkholder–Davis–Gundy inequality for the stochastic integral as well as the exponential martingale inequality. We shall finally show a number of well-known integral inequalities of Gronwall type.

1.2 BASIC NOTATIONS OF PROBABILITY THEORY

Probability theory deals with mathematical models of trials whose outcomes depend on chance. All the possible outcomes—the elementary events—are grouped together to form a set Ω with typical element $\omega \in \Omega$. Not every subset of Ω is in general an observable or interesting event. So we only group these observable or interesting events together as a family $\mathcal{F}$ of subsets of Ω. For the purpose of probability theory, such a family $\mathcal{F}$ should have the following properties:

(i) $\emptyset \in \mathcal{F}$, where $\emptyset$ denotes the empty set;
(ii) $A \in \mathcal{F} \Rightarrow A^C \in \mathcal{F}$, where $A^C = \Omega - A$ is the complement of A in Ω;
(iii) $\{A_i\}_{i\geq 1} \subset \mathcal{F} \Rightarrow \bigcup_{i=1}^{\infty} A_i \in \mathcal{F}$.

A family $\mathcal{F}$ with these three properties is called a *σ-algebra*. The pair $(\Omega, \mathcal{F})$ is called a *measurable space*, and the elements of $\mathcal{F}$ is henceforth called $\mathcal{F}$-*measurable sets* instead of events. If $\mathcal{C}$ is a family of subsets of Ω, then there exists a smallest σ-algebra $\sigma(\mathcal{C})$ on Ω which contains $\mathcal{C}$. This $\sigma(\mathcal{C})$ is called the *σ-algebra generated by* $\mathcal{C}$. If $\Omega = R^d$ and $\mathcal{C}$ is the family of all open sets in R^d, then $\mathcal{B}^d = \sigma(\mathcal{C})$ is called the *Borel σ-algebra* and the elements of $\mathcal{B}^d$ are called the *Borel sets*.

A real-valued function $X : \Omega \to R$ is said to be $\mathcal{F}$-*measurable* if

$$\{\omega : X(\omega) \leq a\} \in \mathcal{F} \quad \text{for all } a \in R.$$

The function X is also called a real-valued ($\mathcal{F}$-measurable) *random variable*. An R^d-valued function $X(\omega) = (X_1(\omega), \cdots, X_d(\omega))^T$ is said to be $\mathcal{F}$-*measurable* if all the elements X_i are $\mathcal{F}$-measurable. Similarly, a $d \times m$-matrix-valued function $X(\omega) = (X_{ij}(\omega))_{d \times m}$ is said to be $\mathcal{F}$-*measurable* if all the elements X_{ij} are $\mathcal{F}$-measurable. The *indicator function* I_A of a set $A \subset \Omega$ is defined by

$$I_A(\omega) = \begin{cases} 1 & \text{for } \omega \in A, \\ 0 & \text{for } \omega \notin A. \end{cases}$$

The indicator function I_A is $\mathcal{F}$-measurable if and only if A is an $\mathcal{F}$-measurable set, i.e. $A \in \mathcal{F}$. If the measurable space is $(R^d, \mathcal{B}^d)$, a $\mathcal{B}^d$-measurable function is then called a *Borel measurable function*. More generally, let $(\Omega', \mathcal{F}')$ be another measurable space. A mapping $X : \Omega \to \Omega'$ is said to be $(\mathcal{F}, \mathcal{F}')$-*measurable* if

$$\{\omega : X(\omega) \in A'\} \in \mathcal{F} \quad \text{for all } A' \in \mathcal{F}'.$$

The mapping X is then called an Ω'-valued $(\mathcal{F}, \mathcal{F}')$-measurable (or simply, $\mathcal{F}$-measurable) random variable.

Let $X : \Omega \to R^d$ be any function. The *σ-algebra* $\sigma(X)$ *generated by* X is the smallest σ-algebra on Ω containing all the sets $\{\omega : X(w) \in U\}$, $U \subset R^d$ open. That is

$$\sigma(X) = \sigma(\{\omega : X(w) \in U\} : U \subset R^d \text{ open}).$$

Clearly, X will then be $\sigma(X)$-measurable and $\sigma(X)$ is the smallest σ-algebra with this property. If X is $\mathcal{F}$-measurable, then $\sigma(X) \subset \mathcal{F}$, i.e. X generates a sub-σ-algebra of $\mathcal{F}$. If $\{X_i : i \in I\}$ is a collection of R^d-valued functions, define

$$\sigma(X_i : i \in I) = \sigma\left(\bigcup_{i \in I} \sigma(X_i)\right)$$

which is called the *σ-algebra generated by* $\{X_i : i \in I\}$. It is the smallest σ-algebra with respect to which every X_i is measurable. The following result is useful. It is a special case of a result sometimes called the Doob–Dynkin lemma.

Lemma 2.1 *If $X, Y : \Omega \to R^d$ are two given functions, then Y is $\sigma(X)$-measurable if and only if there exists a Borel measurable function $g : R^d \to R^d$ such that $Y = g(X)$.*

A *probability measure* P on a measurable space $(\Omega, \mathcal{F})$ is a function $P : \mathcal{F} \to [0, 1]$ such that
  (i) $P(\Omega) = 1$;
  (ii) for any disjoint sequence $\{A_i\}_{i \geq 1} \subset \mathcal{F}$ (i.e. $A_i \cap A_j = \emptyset$ if $i \neq j$)
$$P\left(\bigcup_{i=1}^{\infty} A_i\right) = \sum_{i=1}^{\infty} P(A_i).$$

The triple $(\Omega, \mathcal{F}, P)$ is called a *probability space*. If $(\Omega, \mathcal{F}, P)$ is a probability space, we set
$$\bar{\mathcal{F}} = \{A \subset \Omega : \exists\, B, C \in \mathcal{F} \text{ such that } B \subset A \subset C,\ P(B) = P(C)\}.$$

Then $\bar{\mathcal{F}}$ is a σ-algebra and is called the *completion* of $\mathcal{F}$. If $\mathcal{F} = \bar{\mathcal{F}}$, the probability space $(\Omega, \mathcal{F}, P)$ is said to be *complete*. If not, one can easily extend P to $\bar{\mathcal{F}}$ by defining $P(A) = P(B) = P(C)$ for $A \in \bar{\mathcal{F}}$, where $B, C \in \mathcal{F}$ with the properties that $B \subset A \subset C$ and $P(B) = P(C)$. Now $(\Omega, \bar{\mathcal{F}}, P)$ is a complete probability space, called the *completion* of $(\Omega, \mathcal{F}, P)$.

In the sequel of this section, we let $(\Omega, \mathcal{F}, P)$ be a probability space. If X is a real-valued random variable and is *integrable* with respect to the probability measure P, then the number
$$EX = \int_{\Omega} X(\omega)\, dP(\omega)$$
is called the *expectation* of X (with respect to P). The number
$$V(X) = E(X - EX)^2$$
is called the *variance* of X (here and in the sequel of this section we assume that all integrals concerned exist). The number $E|X|^p$ ($p > 0$) is called the pth moment of X. If Y is another real-valued random variable,
$$Cov(X, Y) = E[(X - EX)(Y - EY)]$$
is called the *covariance* of X and Y. If $Cov(X, Y) = 0$, X and Y are said to be *uncorrelated*. For an R^d-valued random variable $X = (X_1, \cdots, X_d)^T$, define $EX = (EX_1, \cdots, EX_d)^T$. For a $d \times m$-matrix-valued random variable $X = (X_{ij})_{d \times m}$, define $EX = (EX_{ij})_{d \times m}$. If X and Y are both R^d-valued random variables, the symmetric nonnegative definite $d \times d$ matrix
$$Cov(X, Y) = E[(X - EX)(Y - EY)^T]$$

is called their *covariance matrix*.

Let X be an R^d-valued random variable. Then X induces a probability measure μ_X on the Borel measurable space $(R^d, \mathcal{B}^d)$, defined by

$$\mu_X(B) = P\{\omega : X(\omega) \in B\} \quad \text{for } B \in \mathcal{B}^d,$$

and μ_X is called the *distribution* of X. The expectation of X can now be expressed as

$$EX = \int_{R^d} x d\mu_X(x).$$

More generally, if $g : R^d \to R^m$ is Borel measurable, we then have the following *transformation formula*

$$Eg(X) = \int_{R^d} g(x) d\mu_X(x).$$

For $p \in (0, \infty)$, let $L^p = L^p(\Omega; R^d)$ be the family of R^d-valued random variables X with $E|X|^p < \infty$. In L^1, we have $|EX| \leq E|X|$. Moreover, the following three inequalities are very useful:

(i) **Hölder's inequality**

$$|E(X^T Y)| \leq (E|X|^p)^{1/p} (E|Y|^q)^{1/q}$$

if $p > 1$, $1/p + 1/q = 1$, $X \in L^p$, $Y \in L^q$;

(ii) **Minkovski's inequality**

$$(E|X+Y|^p)^{1/p} \leq (E|X|^p)^{1/p} + (E|Y|^p)^{1/p}$$

if $p > 1, X, Y \in L^p$;

(iii) **Chebyshev's inequality**

$$P\{\omega : |X(\omega)| \geq c\} \leq c^{-p} E|X|^p$$

if $c > 0$, $p > 0$, $X \in L^p$.

A simple application of Hölder's inequality implies

$$(E|X|^r)^{1/r} \leq (E|X|^p)^{1/p}$$

if $0 < r < p < \infty$, $X \in L^p$.

Let X and X_k, $k \geq 1$, be R^d-valued random variables. The following four convergence concepts are very important:

(a) If there exists a P-null set $\Omega_0 \in \mathcal{F}$ such that for every $\omega \notin \Omega_0$, the sequence $\{X_k(\omega)\}$ converges to $X(\omega)$ in the usual sense in R^d, then $\{X_k\}$ is said to converge to X *almost surely* or *with probability 1*, and we write $\lim_{k \to \infty} X_k = X$ a.s.

(b) If for every $\varepsilon > 0$, $P\{\omega : |X_k(\omega) - X(\omega)| > \varepsilon\} \to 0$ as $k \to \infty$, then $\{X_k\}$ is said to converge to X *stochastically* or *in probability*.

(c) If X_k and X belong to L^p and $E|X_k - X|^p \to 0$, then $\{X_k\}$ is said to converge to X *in pth moment* or *in L^p*.

(d) If for every real-valued continuous bounded function g defined on R^d, $\lim_{k \to \infty} Eg(X_k) = Eg(X)$, then $\{X_k\}$ is said to converge to X *in distribution*.

These convergence concepts have the following relationship:

$$\text{convergence in } L^p$$
$$\Downarrow$$
$$\text{a.s. convergence} \Rightarrow \text{convergence in probability}$$
$$\Downarrow$$
$$\text{convergence in distribution}$$

Furthermore, a sequence converges in probability if and only if every subsequence of it contains an almost surely convergent subsequence. A sufficient condition for $\lim_{k \to \infty} X_k = X$ a.s. is the condition

$$\sum_{k=1}^{\infty} E|X_k - X|^p < \infty \quad \text{for some } p > 0.$$

We now state two very important integration convergence theorems.

Theorem 2.2 (Monotonic convergence theorem) *If $\{X_k\}$ is an increasing sequence of nonnegative random variables, then*

$$\lim_{k \to \infty} EX_k = E\left(\lim_{k \to \infty} X_k\right).$$

Theorem 2.3 (Dominated convergence theorem) *Let $p \geq 1$, $\{X_k\} \subset L^p(\Omega; R^d)$ and $Y \in L^p(\Omega; R)$. Assume that $|X_k| \leq Y$ a.s. and $\{X_k\}$ converges to X in probability. Then $X \in L^p(\Omega; R^d)$, $\{X_k\}$ converges to X in L^p, and*

$$\lim_{k \to \infty} EX_k = EX.$$

When Y is bounded, this theorem is also referred as the *bounded convergence theorem*.

Two sets $A, B \in \mathcal{F}$ are said to be *independent* if $P(A \cap B) = P(A)P(B)$. Three sets $A, B, C \in \mathcal{F}$ are said to be *independent* if

$$P(A \cap B) = P(A)P(B), \quad P(A \cap C) = P(A)P(C),$$
$$P(B \cap C) = P(B)P(C) \quad \text{and} \quad P(A \cap B \cap C) = P(A)P(B)P(C).$$

Basic Notations of Probability Theory

Let I be an index set. A collection of sets $\{A_i : i \in I\} \subset \mathcal{F}$ is said to be *independent* if

$$P(A_{i_1} \cap \cdots \cap A_{i_k}) = P(A_{i_1}) \cdots P(A_{i_k})$$

for all possible choices of indices $i_1, \cdots, i_k \in I$. Two sub-σ-algebras $\mathcal{F}_1$ and $\mathcal{F}_2$ of $\mathcal{F}$ are said to be *independent* if

$$P(A_1 \cap A_2) = P(A_1)P(A_2) \quad \text{for all } A_1 \in \mathcal{F}_1, \, A_2 \in \mathcal{F}_2.$$

A collection of sub-σ-algebras $\{\mathcal{F}_i : i \in I\}$ is said to be *independent* if for every possible choice of indices $i_1, \cdots, i_k \in I$,

$$P(A_{i_1} \cap \cdots \cap A_{i_k}) = P(A_{i_1}) \cdots P(A_{i_k})$$

holds for all $A_{i_1} \in \mathcal{F}_{i_1}, \cdots, A_{i_k} \in \mathcal{F}_{i_k}$. A family of random variables $\{X_i : i \in I\}$ (whose ranges may differ for different values of the index) is said to be *independent* if the σ-algebras $\sigma(X_i)$, $i \in I$ generated by them are independent. For example, two random variables $X : \Omega \to R^d$ and $Y : \Omega \to R^m$ are independent if and only if

$$P\{\omega : X(\omega) \in A, \, Y(\omega) \in B\} = P\{\omega : X(\omega) \in A\} \, P\{\omega : Y(\omega) \in B\}$$

holds for all $A \in \mathcal{B}^d$, $B \in \mathcal{B}^m$. If X and Y are two independent real-valued integrable random variables, then XY is also integrable and

$$E(XY) = EX \, EY.$$

If $X, Y \in L^2(\Omega; R)$ are uncorrelated, then

$$V(X+Y) = V(X) + V(Y).$$

If the X and Y are independent, they are uncorrelated. If (X, Y) has a normal distribution, then X and Y are independent if and only if they are uncorrelated.

Let $\{A_k\}$ be a sequence of sets in $\mathcal{F}$. Define the *upper limit of the sets* by

$$\limsup_{k \to \infty} A_k = \{\omega : \omega \in A_k \text{ for infinitely many } k\} = \bigcap_{i=1}^{\infty} \bigcup_{k=i}^{\infty} A_k.$$

Clearly, it belongs to $\mathcal{F}$. With regard to its probability, we have the following well-known *Borel–Cantelli lemma*.

Lemma 2.4 (Borel–Cantelli's lemma)

(1) *If $\{A_k\} \subset \mathcal{F}$ and $\sum_{k=1}^{\infty} P(A_k) < \infty$, then*

$$P\left(\limsup_{k \to \infty} A_k\right) = 0.$$

That is, there exists a set $\Omega_o \in \mathcal{F}$ with $P(\Omega_o) = 1$ and an integer-valued random variable k_o such that for every $w \in \Omega_o$ we have $w \notin A_k$ whenever $k \geq k_o(w)$.

(2) If the sequence $\{A_k\} \subset \mathcal{F}$ is independent and $\sum_{k=1}^{\infty} P(A_k) = \infty$, then

$$P\left(\limsup_{k \to \infty} A_k\right) = 1.$$

That is, there exists a set $\Omega_\theta \in \mathcal{F}$ with $P(\Omega_\theta) = 1$ such that for every $w \in \Omega_\theta$, there exists a sub-sequence $\{A_{k_i}\}$ such that the w belongs to every A_{k_i}.

Let $A, B \in \mathcal{F}$ with $P(B) > 0$. The *conditional probability of A under condition B* is

$$P(A|B) = \frac{P(A \cap B)}{P(B)}.$$

However, we frequently encounter a family of conditions so we need the more general concept of *conditional expectation*. Let $X \in L^1(\Omega; R)$. Let $\mathcal{G} \subset \mathcal{F}$ is a sub-σ-algebra of $\mathcal{F}$ so $(\Omega, \mathcal{G})$ is a measurable space. In general, X is not $\mathcal{G}$-measurable. We now seek an integrable $\mathcal{G}$-measurable random variable Y such that it has the same values as X on the average in the sense that

$$E(I_G Y) = E(I_G X) \quad \text{i.e.} \quad \int_G Y(w) dP(w) = \int_G X(w) dP(w) \quad \text{for all } G \in \mathcal{G}.$$

By the Radon–Nikodym theorem, there exists one such Y, almost surely unique. It is called the *conditional expectation of X under the condition $\mathcal{G}$*, and we write

$$Y = E(X|\mathcal{G}).$$

If $\mathcal{G}$ is the the σ-algebra generated by a random variable Y, we write

$$E(X|\mathcal{G}) = E(X|Y).$$

As an example, consider a collection of sets $\{A_k\} \subset \mathcal{F}$ with

$$\bigcup_k A_k = \Omega, \quad P(A_k) > 0, \quad A_k \cap A_j = \emptyset \text{ if } k \neq j.$$

Let $\mathcal{G} = \sigma(\{A_k\})$, i.e. $\mathcal{G}$ is generated by $\{A_k\}$. Then $E(X|\mathcal{G})$ is a *step function* on Ω given by

$$E(X|\mathcal{G}) = \sum_k \frac{I_{A_k} E(I_{A_k} X)}{P(A_k)}.$$

In other words, if $w \in A_k$,

$$E(X|\mathcal{G})(w) = \frac{E(I_{A_k} X)}{P(A_k)}.$$

It follows from the definition that
$$E(E(X|\mathcal{G})) = E(X)$$
and
$$|E(X|\mathcal{G})| \leq E(|X| \,|\mathcal{G}) \quad a.s.$$

Other important properties of the conditional expectation are as follows (all the equalities and inequalities shown hold almost surely):

(a) $\mathcal{G} = \{\emptyset, \Omega\} \Rightarrow E(X|\mathcal{G}) = EX$;
(b) $X \geq 0 \Rightarrow E(X|\mathcal{G}) \geq 0$;
(c) X is $\mathcal{G}$-measurable $\Rightarrow E(X|\mathcal{G}) = X$;
(d) $X = c = const. \Rightarrow E(X|\mathcal{G}) = c$;
(e) $a, b \in R \Rightarrow E(aX + bY|\mathcal{G}) = aE(X|\mathcal{G}) + bE(Y|\mathcal{G})$;
(f) $X \leq Y \Rightarrow E(X|\mathcal{G}) \leq E(Y|\mathcal{G})$;
(g) X is $\mathcal{G}$-measurable $\Rightarrow E(XY|\mathcal{G}) = XE(Y|\mathcal{G})$,
   in particular, $E(E(X|\mathcal{G}) Y|\mathcal{G}) = E(X|\mathcal{G}) E(Y|\mathcal{G})$;
(h) $\sigma(X), \mathcal{G}$ are independent $\Rightarrow E(X|\mathcal{G}) = EX$,
   in particular, X, Y are independent $\Rightarrow E(X|Y) = EX$;
(i) $\mathcal{G}_1 \subset \mathcal{G}_2 \subset \mathcal{F} \Rightarrow E(E(X|\mathcal{G}_2)|\mathcal{G}_1) = E(X|\mathcal{G}_1)$.

Finally, if $X = (X_1, \cdots, X_d)^T \in L^1(\Omega; R^d)$, its *conditional expectation under* $\mathcal{G}$ is defined as
$$E(X|\mathcal{G}) = (E(X_1|\mathcal{G}), \cdots, E(X_d|\mathcal{G}))^T.$$

1.3 STOCHASTIC PROCESSES

Let $(\Omega, \mathcal{F}, P)$ be a probability space. A *filtration* is a family $\{\mathcal{F}_t\}_{t\geq 0}$ of increasing sub-σ-algebras of $\mathcal{F}$ (i.e. $\mathcal{F}_t \subset \mathcal{F}_s \subset \mathcal{F}$ for all $0 \leq t < s < \infty$). The filtration is said to be *right continuous* if $\mathcal{F}_t = \bigcap_{s>t} \mathcal{F}_s$ for all $t \geq 0$. When the probability space is complete, the filtration is said to satisfy the *usual conditions* if it is right continuous and $\mathcal{F}_0$ contains all P-null sets.

From now on, unless otherwise specified, we shall always work on a given complete probability space $(\Omega, \mathcal{F}, P)$ with a filtration $\{\mathcal{F}_t\}_{t\geq 0}$ satisfying the usual conditions. We also define $\mathcal{F}_\infty = \sigma(\bigcup_{t\geq 0} \mathcal{F}_t)$, i.e. the σ-algebra generated by $\bigcup_{t\geq 0} \mathcal{F}_t$.

A family $\{X_t\}_{t\in I}$ of R^d-valued random variables is called a *stochastic process* with *parameter set* (or *index set*) I and *state space* R^d. The parameter set I is usually (as in this book) the halfline $R_+ = [0, \infty)$, but it may also be an interval $[a, b]$, the nonnegative integers or even subsets of R^d. Note that for each fixed $t \in I$ we have a random variable
$$\Omega \ni \omega \to X_t(\omega) \in R^d.$$

On the other hand, for each fixed $\omega \in \Omega$ we have a function

$$I \ni t \to X_t(\omega) \in R^d$$

which is called a *sample path* of the process, and we shall write $X_\cdot(\omega)$ for the path. Sometimes it is convenient to write $X(t, \omega)$ instead of $X_t(\omega)$, and the stochastic process may be regarded as a function of two variables (t, ω) from $I \times \Omega$ to R^d. Similarly, one can define matrix-valued stochastic processes etc. We often write a stochastic process $\{X_t\}_{t \geq 0}$ as $\{X_t\}$, X_t or $X(t)$.

Let $\{X_t\}_{t \geq 0}$ be an R^d-valued stochastic process. It is said to be *continuous* (resp. *right continuous, left continuous*) if for almost all $\omega \in \Omega$ function $X_t(\omega)$ is continuous (resp. right continuous, left continuous) on $t \geq 0$. It is said to be *cadlag (right continuous and left limit)* if it is right continuous and for almost all $\omega \in \Omega$ the left limit $\lim_{s \uparrow t} X_s(\omega)$ exists and is finite for all $t > 0$. It is said to be *integrable* if for every $t \geq 0$, X_t is an integrable random variable. It is said to be $\{\mathcal{F}_t\}$-*adapted* (or simply, *adapted*) if for every t, X_t is $\mathcal{F}_t$-measurable. It is said to be *measurable* if the stochastic process regarded as a function of two variables (t, ω) from $R_+ \times \Omega$ to R^d is $\mathcal{B}(R_+) \times \mathcal{F}$-measurable, where $\mathcal{B}(R_+)$ is the family of all Borel sub-sets of R_+. The stochastic process is said to be *progressively measurable* or *progressive* if for every $T \geq 0$, $\{X_t\}_{0 \leq t \leq T}$ regarded as a function of (t, ω) from $[0, T] \times \Omega$ to R^d is $\mathcal{B}([0, T]) \times \mathcal{F}_T$-measurable, where $\mathcal{B}([0, T])$ is the family of all Borel sub-sets of $[0, T]$. Let $\mathcal{O}$ (resp. $\mathcal{P}$) denote the smallest σ-algebra on $R_+ \times \Omega$ with respect to which every cadlag adapted process (resp. left continuous process) is a measurable function of (t, ω). A stochastic process is said to be *optional* (resp. *predictable*) if the process regarded as a function of (t, ω) is $\mathcal{O}$-measurable (resp. $\mathcal{P}$-measurable). A real-valued stochastic process $\{A_t\}_{t \geq 0}$ is called an *increasing process* if for almost all $\omega \in \Omega$, $A_t(\omega)$ is nonnegative nondecreasing right continuous on $t \geq 0$. It is called a *process of finite variation* if $A_t = \bar{A}_t - \hat{A}_t$ with $\{\bar{A}_t\}$ and $\{\hat{A}_t\}$ both increasing processes. It is obvious that the processes of finite variation are cadlag. Hence the adapted processes of finite variation are optional.

The relations among the various stochastic processes are summarised below:

continuous adapted	continuous adapted	adapted increasing
$\Downarrow$	$\Downarrow$	$\Downarrow$
left continuous adapted	cadlag adapted	$\Leftarrow$ adapted finite variation
$\Downarrow$	$\Downarrow$	
predictable $\Rightarrow$	optional	
	$\Downarrow$	
	progressive	$\Rightarrow$ adapted
	$\Downarrow$	
	measurable	

Let $\{X_t\}_{t \geq 0}$ be a stochastic process. Another stochastic process $\{Y_t\}_{t \geq 0}$ is called a *version* or *modification* of $\{X_t\}$ if for all $t \geq 0$, $X_t = Y_t$ a.s. (i.e.

$P\{\omega : X_t(\omega) = Y_t(\omega)\} = 1)$. Two stochastic processes $\{X_t\}_{t\geq 0}$ and $\{Y_t\}_{t\geq 0}$ are said to be *indistinguishable* if for almost all $\omega \in \Omega$, $X_t(\omega) = Y_t(\omega)$ for all $t \geq 0$ (i.e. $P\{\omega : X_t(\omega) = Y_t(\omega) \text{ for all } t \geq 0\} = 1$).

A random variable $\tau : \Omega \to [0, \infty]$ (it may take the value ∞) is called an $\{\mathcal{F}_t\}$-*stopping time* (or simply, *stopping time*) if $\{\omega : \tau(\omega) \leq t\} \in \mathcal{F}_t$ for any $t \geq 0$. Let τ and ρ be two stopping times with $\tau \leq \rho$ a.s. We define

$$[[\tau, \rho[[= \{(t, \omega) \in R_+ \times \Omega : \tau(\omega) \leq t < \rho(\omega)\}$$

and call it a *stochastic interval*. Similarly, we can define stochastic intervals $[[\tau, \rho]]$, $]]\tau, \rho]]$ and $]]\tau, \rho[[$. If τ is a stopping time, define

$$\mathcal{F}_\tau = \{A \in \mathcal{F} : A \cap \{\omega : \tau(\omega) \leq t\} \in \mathcal{F}_t \text{ for all } t \geq 0\}$$

which is a sub-σ-algebra of $\mathcal{F}$. If τ and ρ are two stopping times with $\tau \leq \rho$ a.s., then $\mathcal{F}_\tau \subset \mathcal{F}_\rho$. The following two theorems are useful.

Theorem 3.1 *If $\{X_t\}_{t\geq 0}$ is a progressively measurable process and τ is a stopping time, then $X_\tau I_{\{\tau < \infty\}}$ is $\mathcal{F}_\tau$-measurable. In particular, if τ is finite, then X_τ is $\mathcal{F}_\tau$-measurable.*

Theorem 3.2 *Let $\{X_t\}_{t\geq 0}$ be an R^d-valued cadlag $\{\mathcal{F}_t\}$-adapted process, and D an open subset of R^d. Define*

$$\tau = \inf\{t \geq 0 : X_t \notin D\},$$

where we use the convention $\inf \emptyset = \infty$. *Then τ is an $\{\mathcal{F}_t\}$-stopping time, and is called the first exit time from D. Moreover, if ρ is a stopping time, then*

$$\theta = \inf\{t \geq \rho : X_t \notin D\}$$

is also an $\{\mathcal{F}_t\}$-stopping time, and is called the first exit time from D after ρ.

An R^d-valued $\{\mathcal{F}_t\}$-adapted integrable process $\{M_t\}_{t\geq 0}$ is called a *martingale with respect to $\{\mathcal{F}_t\}$* (or simply, *martingale*) if

$$E(M_t | \mathcal{F}_s) = M_s \quad \text{a.s. for all } 0 \leq s < t < \infty.$$

It should be pointed out that every martingale has a cadlag modification since we always assume that the filtration $\{\mathcal{F}_t\}$ is right continuous. Therefore we can always assume that any martingale is cadlag in the sequel. If $X = \{X_t\}_{t\geq 0}$ is a progressively measurable process and τ is a stopping time, then $X^\tau = \{X_{\tau \wedge t}\}_{t\geq 0}$ is called a *stopped process* of X. The following is the well-known Doob martingale stopping theorem.

Theorem 3.3 *Let $\{M_t\}_{t\geq 0}$ be an R^d-valued martingale with respect to $\{\mathcal{F}_t\}$, and let θ, ρ be two finite stopping times. Then*

$$E(M_\theta | \mathcal{F}_\rho) = M_{\theta \wedge \rho} \quad a.s.$$

In particular, if τ is a stopping time, then

$$E(M_{\tau \wedge t}|\mathcal{F}_s) = M_{\tau \wedge s} \quad \text{a.s.}$$

holds for all $0 \leq s < t < \infty$. That is, the stopped process $M^\tau = \{M_{\tau \wedge t}\}$ is still a martingale with respect to the same filtration $\{\mathcal{F}_t\}$.

A stochastic process $X = \{X_t\}_{t \geq 0}$ is called *square-integrable* if $E|X_t|^2 < \infty$ for every $t \geq 0$. If $M = \{M_t\}_{t \geq 0}$ is a real-valued square-integrable continuous martingale, then there exists a unique continuous integrable adapted increasing process denoted by $\{\langle M, M \rangle_t\}$ such that $\{M_t^2 - \langle M, M \rangle_t\}$ is a continuous martingale vanishing at $t = 0$. The process $\{\langle M, M \rangle_t\}$ is called the *quadratic variation* of M. In particular, for any finite stopping time τ,

$$EM_\tau^2 = E\langle M, M \rangle_\tau.$$

If $N = \{N_t\}_{t \geq 0}$ is another real-valued square-integrable continuous martingale, we define

$$\langle M, N \rangle_t = \frac{1}{2}\Big(\langle M+N, M+N \rangle_t - \langle M, M \rangle_t - \langle N, N \rangle_t\Big),$$

and call $\{\langle M, N \rangle_t\}$ the *joint quadratic variation* of M and N. It is useful to know that $\{\langle M, N \rangle_t\}$ is the unique continuous integrable adapted process of finite variation such that $\{M_t N_t - \langle M, N \rangle_t\}$ is a continuous martingale vanishing at $t = 0$. In particular, for any finite stopping time τ,

$$EM_\tau N_\tau = E\langle M, N \rangle_\tau.$$

A right continuous adapted process $M = \{M_t\}_{t \geq 0}$ is called a *local martingale* if there exists a nondecreasing sequence $\{\tau_k\}_{k \geq 1}$ of stopping times with $\tau_k \uparrow \infty$ a.s. such that every $\{M_{\tau_k \wedge t} - M_0\}_{t \geq 0}$ is a martingale. Every martingale is a local martingale (by Theorem 3.3), but the converse is not true. If $M = \{M_t\}_{t \geq 0}$ and $N = \{N_t\}_{t \geq 0}$ are two real-valued continuous local martingales, their *joint quadratic variation* $\{\langle M, N \rangle\}_{t \geq 0}$ is the unique continuous adapted process of finite variation such that $\{M_t N_t - \langle M, N \rangle_t\}_{t \geq 0}$ is a continuous local martingale vanishing at $t = 0$. When $M = N$, $\{\langle M, M \rangle\}_{t \geq 0}$ is called the *quadratic variation* of M. The following result is the useful strong law of large numbers.

Theorem 3.4 (Strong law of large numbers) Let $M = \{M_t\}_{t \geq 0}$ be a real-valued continuous local martingale vanishing at $t = 0$. Then

$$\lim_{t \to \infty} \langle M, M \rangle_t = \infty \quad \text{a.s.} \quad \Rightarrow \quad \lim_{t \to \infty} \frac{M_t}{\langle M, M \rangle_t} = 0 \quad \text{a.s.}$$

and also

$$\limsup_{t \to \infty} \frac{\langle M, M \rangle_t}{t} < \infty \quad \text{a.s.} \quad \Rightarrow \quad \lim_{t \to \infty} \frac{M_t}{t} = 0 \quad \text{a.s.}$$

More generally, if $A = \{A_t\}_{t\geq 0}$ is a continuous adapted increasing process such that
$$\lim_{t\to\infty} A_t = \infty \quad \text{and} \quad \int_0^\infty \frac{d\langle M, M\rangle_t}{(1+A_t)^2} < \infty \quad a.s.$$
then
$$\lim_{t\to\infty} \frac{M_t}{A_t} = 0 \quad a.s.$$

A real-valued $\{\mathcal{F}_t\}$-adapted integrable process $\{M_t\}_{t\geq 0}$ is called a *supermartingale (with respect to $\{\mathcal{F}_t\}$)* if
$$E(M_t|\mathcal{F}_s) \leq M_s \quad a.s. \text{ for all } 0 \leq s < t < \infty.$$
It is called a *submartingale (with respect to $\{\mathcal{F}_t\}$)* if we replace the sign $\leq$ in the last formula with $\geq$. Clearly, $\{M_t\}$ is submartingale if and only if $\{-M_t\}$ is supermartingale. For a real-valued martingale $\{M_t\}$, $\{M_t^+ := \max(M_t, 0)\}$ and $\{M_t^- := \max(0, -M_t)\}$ are submartingales. For a supermartingale (resp. submartingale), EM_t is monotonically decreasing (resp. increasing). Moreover, if $p \geq 1$ and $\{M_t\}$ is an R^d-valued martingale such that $M_t \in L^p(\Omega; R^d)$, then $\{|M_t|^p\}$ is a nonnegative submartingale. Moreover, Doob's stopping theorem 3.3 holds for supermartingales and submartingales as well.

Theorem 3.5 (Doob's martingale convergence theorem)

(i) Let $\{M_t\}_{t\geq 0}$ be a real-valued right-continuous supermartingale. If
$$\sup_{0\leq t<\infty} EM_t^- < \infty,$$
then M_t converges almost surely to a random variable $M_\infty \in L^1(\Omega; R)$. In particular, this holds if M_t is nonnegative.

(ii) Let $\{M_t\}_{t\geq 0}$ be a real-valued right-continuous supermartingale. Then $\{M_t\}_{t\geq 0}$ is uniformly integrable, i.e.
$$\lim_{c\to\infty}\left[\sup_{t\geq 0} E\Big(I_{\{|M_t|\geq c\}}|M_t|\Big)\right] = 0$$
if and only if there exists a random variable $M_\infty \in L^1(\Omega; R)$ such that $M_t \to M_\infty$ a.s. and in L^1 as well.

(iii) Let $X \in L^1(\Omega; R)$. Then
$$E(X|\mathcal{F}_t) \to E(X|\mathcal{F}_\infty) \quad \text{as } t \to \infty$$
a.s. and in L^1 as well.

Theorem 3.6 (Supermartingale inequalities) Let $\{M_t\}_{t\geq 0}$ be a real-valued supermartingale. Let $[a,b]$ be a bounded interval in R_+. Then
$$c\,P\Big\{\omega : \sup_{a\leq t\leq b} M_t(\omega) \geq c\Big\} \leq EM_a + EM_b^-,$$
$$c\,P\Big\{\omega : \inf_{a\leq t\leq b} M_t(\omega) \leq -c\Big\} \leq EM_b^-.$$

hold for all $c > 0$.

For submartingales we have the following well-known Doob inequality.

Theorem 3.7 (Doob's submartingale inequalities) *Let $p > 1$. Let $\{M_t\}_{t \geq 0}$ be a real-valued nonnegative submartingale such that $M_t \in L^p(\Omega; R)$. Let $[a, b]$ be a bounded interval in R_+. Then*

$$E\left(\sup_{a \leq t \leq b} M_t^p\right) \leq \left(\frac{p}{p-1}\right)^p EM_b^p.$$

If we apply these results to an R^d-valued martingale, we obtain the following Doob's martingale inequalities.

Theorem 3.8 (Doob's martingale inequalities) *Let $\{M_t\}_{t \geq 0}$ be an R^d-valued martingale. Let $[a, b]$ be a bounded interval in R_+.*
(i) If $p \geq 1$ and $M_t \in L^p(\Omega; R^d)$, then

$$P\left\{\omega : \sup_{a \leq t \leq b} |M_t(\omega)| \geq c\right\} \leq \frac{E|M_b|^p}{c^p}$$

holds for all $c > 0$.
(ii) If $p > 1$ and $M_t \in L^p(\Omega; R^d)$, then

$$E\left(\sup_{a \leq t \leq b} |M_t|^p\right) \leq \left(\frac{p}{p-1}\right)^p E|M_b|^p.$$

To close this section we state one more useful convergence theorem.

Theorem 3.9 *Let $\{A_t\}_{t \geq 0}$ and $\{U_t\}_{t \geq 0}$ be two continuous adapted increasing processes with $A_0 = U_0 = 0$ a.s. Let $\{M_t\}_{t \geq 0}$ be a real-valued continuous local martingale with $M_0 = 0$ a.s. Let ξ be a nonnegative $\mathcal{F}_0$-measurable random variable. Define*

$$X_t = \xi + A_t - U_t + M_t \quad \text{for } t \geq 0.$$

If X_t is nonnegative, then

$$\left\{\lim_{t \to \infty} A_t < \infty\right\} \subset \left\{\lim_{t \to \infty} X_t \text{ exists and is finite}\right\} \cap \left\{\lim_{t \to \infty} U_t < \infty\right\} \quad a.s.$$

where $B \subset D$ a.s. means $P(B \cap D^c) = 0$. In particular, if $\lim_{t \to \infty} A_t < \infty$ a.s., then for almost all $\omega \in \Omega$

$$\lim_{t \to \infty} X_t(\omega) \text{ exists and is finite, and } \lim_{t \to \infty} U_t(\omega) < \infty.$$

1.4 BROWNIAN MOTIONS

Brownian motion is the name given to the irregular movement of pollen grains, suspended in water, observed by the Scottish botanist Robert Brown in 1828. The motion was later explained by the random collisions with the molecules of water. To describe the motion mathematically it is natural to use the concept of a stochastic process $B_t(\omega)$, interpreted as the position of the pollen grain ω at time t. Let us now give the mathematical definition of Brownian motion.

Definition 4.1 *Let $(\Omega, \mathcal{F}, P)$ be a probability space with a filtration $\{\mathcal{F}_t\}_{t \geq 0}$. A (standard) one-dimensional Brownian motion is a real-valued continuous $\{\mathcal{F}_t\}$-adapted process $\{B_t\}_{t \geq 0}$ with the following properties:*

(i) $B_0 = 0$ a.s.;

(ii) *for $0 \leq s < t < \infty$, the increment $B_t - B_s$ is normally distributed with mean zero and variance $t - s$;*

(iii) *for $0 \leq s < t < \infty$, the increment $B_t - B_s$ is independent of $\mathcal{F}_s$.*

We shall sometimes speak of a Brownian motion $\{B_t\}_{0 \leq t \leq T}$ on $[0, T]$, for some $T > 0$, and the meaning of this terminology is apparent.

If $\{B_t\}_{t \geq 0}$ is a Brownian motion and $0 \leq t_0 < t_1 < \cdots < t_k < \infty$, then the increments $B_{t_i} - B_{t_{i-1}}$, $1 \leq i \leq k$ are independent, and we say that the Brownian motion has *independent increments*. Moreover, the distribution of $B_{t_i} - B_{t_{i-1}}$ depends only on the difference $t_i - t_{i-1}$, and we say that the Brownian motion has *stationary increments*.

The filtration $\{\mathcal{F}_t\}$ is a part of the definition of Brownian motion. However, we sometimes speak of a Brownian motion on a probability space $(\Omega, \mathcal{F}, P)$ without filtration. That is, $\{B_t\}_{t \geq 0}$ is a a real-valued continuous process with properties (i) and (ii) but the property (iii) is replaced by that it has the independent increments. In this case, define $\mathcal{F}_t^B = \sigma(B_s : 0 \leq s \leq t)$ for $t \geq 0$, i.e. $\mathcal{F}_t^B$ is the σ-algebra generated by $\{B_s : 0 \leq s \leq t\}$. We call $\{\mathcal{F}_t^B\}_{t \geq 0}$ the *natural filtration* generated by $\{B_t\}$. Clearly, $\{B_t\}$ is a Brownian motion with respect to the natural filtration $\{\mathcal{F}_t^B\}$. Moreover, if $\{\mathcal{F}_t\}$ is a "larger" filtration in the sense that $\mathcal{F}_t^B \subset \mathcal{F}_t$ for $t \geq 0$, and $B_t - B_s$ is independent of $\mathcal{F}_s$ whenever $0 \leq s < t < \infty$, then $\{B_t\}$ is a Brownian motion with respect to the filtration $\{\mathcal{F}_t\}$.

In the definition we do not require the probability space $(\Omega, \mathcal{F}, P)$ be complete and the filtration $\{\mathcal{F}_t\}$ satisfy the usual conditions. However, it is often necessary to work on a complete probability space with a filtration satisfying the usual conditions. Let $\{B_t\}_{t \geq 0}$ be a Brownian motion defined on a probability space $(\Omega, \mathcal{F}, P)$. Let $(\Omega, \bar{\mathcal{F}}, \bar{P})$ be the completion of $(\Omega, \mathcal{F}, P)$. Clearly, $\{B_t\}$ is a Brownian motion on the complete probability space $(\Omega, \bar{\mathcal{F}}, \bar{P})$. Let $\mathcal{N}$ be the collection of P-null sets, i.e. $\mathcal{N} = \{A \in \bar{\mathcal{F}} : P(A) = 0\}$. For $t \geq 0$, define

$$\bar{\mathcal{F}}_t = \sigma(\mathcal{F}_t^B \cup \mathcal{N}).$$

We called $\{\bar{\mathcal{F}}_t\}$ the *augmentation under P of the natural filtration* $\{\mathcal{F}_t^B\}$ generated by $\{B_t\}$. It is known that the augmentation $\{\bar{\mathcal{F}}_t\}$ is a filtration on $(\Omega, \bar{\mathcal{F}}, P)$ satisfying the usual condition. Moreover, $\{B_t\}$ is a Brownian motion on $(\Omega, \bar{\mathcal{F}}, P)$ with respect to $\{\bar{\mathcal{F}}_t\}$. This shows that given a Brownian motion $\{B_t\}_{t\geq 0}$ on a probability space $(\Omega, \mathcal{F}, P)$, one can construct a complete probability space with a filtration satisfying the usual conditions to work on.

However, throughout this book, unless otherwise specified, we would rather assume that $(\Omega, \mathcal{F}, P)$ is a complete probability space with a filtration $\{\mathcal{F}_t\}$ satisfying the usual conditions, and the one-dimensional Brownian motion $\{B_t\}$ is defined on it. The Brownian motion has many important properties, and some of them are summarized below:

(a) $\{-B_t\}$ is a Brownian motion with respect to the same filtration $\{\mathcal{F}_t\}$.

(b) Let $c > 0$. Define
$$X_t = \frac{B_{ct}}{\sqrt{c}} \quad \text{for } t \geq 0.$$
Then $\{X_t\}$ is a Brownian motion with respect to the filtration $\{\mathcal{F}_{ct}\}$.

(c) $\{B_t\}$ is a continuous square-integrable martingale and its quadratic variation $\langle B, B \rangle_t = t$ for all $t \geq 0$.

(d) The strong law of large numbers states that
$$\lim_{t\to\infty} \frac{B_t}{t} = 0 \quad a.s.$$

(e) For almost every $\omega \in \Omega$, the Brownian sample path $B_\cdot(\omega)$ is nowhere differentiable.

(f) For almost every $\omega \in \Omega$, the Brownian sample path $B_\cdot(\omega)$ is locally Hölder continuous with exponent δ if $\delta \in (0, \frac{1}{2})$. However, for almost every $\omega \in \Omega$, the Brownian sample path $B_\cdot(\omega)$ is nowhere Hölder continuous with exponent $\delta > \frac{1}{2}$.

Besides, we have the following well-known law of the iterated logarithm.

Theorem 4.2 (Law of the Iterated Logarithm, A. Hinčin (1933)) *For almost every $\omega \in \Omega$, we have*

(i) $\limsup\limits_{t\downarrow 0} \dfrac{B_t(\omega)}{\sqrt{2t\log\log(1/t)}} = 1,$ (ii) $\liminf\limits_{t\downarrow 0} \dfrac{B_t(\omega)}{\sqrt{2t\log\log(1/t)}} = -1,$

(iii) $\limsup\limits_{t\to\infty} \dfrac{B_t(\omega)}{\sqrt{2t\log\log t}} = 1,$ (iv) $\liminf\limits_{t\to\infty} \dfrac{B_t(\omega)}{\sqrt{2t\log\log t}} = -1.$

This theorem shows that for any $\varepsilon > 0$ there exists a positive random variable ρ_ε such that for almost every $\omega \in \Omega$, the Brownian sample path $B_\cdot(\omega)$ is within the interval $\pm(1+\varepsilon)\sqrt{2t\log\log t}$ whenever $t \geq \rho_\varepsilon(\omega)$, that is

$$-(1+\varepsilon)\sqrt{2t\log\log t} \leq B_t(\omega) \leq (1+\varepsilon)\sqrt{2t\log\log t} \quad \text{for all } t \geq \rho_\varepsilon(\omega).$$

On the other hand, the bounds $-(1-\varepsilon)\sqrt{2t\log\log t}$ and $(1-\varepsilon)\sqrt{2t\log\log t}$ (for $0 < \varepsilon < 1$) are exceeded in every t-neighbourhood of ∞ for every sample path.

Let us now define a d-dimensional Brownian motion.

Definition 4.3 *A d-dimensional process $\{B_t = (B_t^1, \cdots, B_t^d)\}_{t \geq 0}$ is called a d-dimensional Brownian motion if every $\{B_t^i\}$ is a one-dimensional Brownian motion, and $\{B_t^1\}, \cdots, \{B_t^d\}$ are independent.*

For a d-dimensional Brownian motion, we still have

$$\limsup_{t \to \infty} \frac{|B_t|}{\sqrt{2t\log\log t}} = 1 \quad a.s.$$

This is somewhat surprising because it means that the independent individual components of B_t are not simultaneously of the order $\sqrt{2t\log\log t}$, otherwise $\sqrt{d}$ instead of 1 would have appeared in the right-hand side of the above equality.

It is easy to see that a d-dimensional Brownian motion is a d-dimensional continuous martingale with the joint quadratic variations

$$\langle B^i, B^j \rangle_t = \delta_{ij} t \quad \text{for } 1 \leq i, j \leq d,$$

where δ_{ij} is the Dirac delta function, i.e.

$$\delta_{ij} = \begin{cases} 1 & \text{for } i = i, \\ 0 & \text{for } i \neq j. \end{cases}$$

It turns out that this property characterizes Brownian motion among continuous local martingales. This is described by the following well-known Lévy theorem.

Theorem 4.4 (P. Lévy (1948)) *Let $\{M_t = (M_t^1, \cdots, M_t^d)\}_{t \geq 0}$ be a d-dimensional continuous local martingale with respect to the filtration $\{\mathcal{F}_t\}$ and $M_0 = 0$ a.s. If*

$$\langle M^i, M^j \rangle_t = \delta_{ij} t \quad \text{for } 1 \leq i, j \leq d,$$

then $\{M_t = (M_t^1, \cdots, M_t^d)\}_{t \geq 0}$ is a d-dimensional Brownian motion with respect to $\{\mathcal{F}_t\}$.

As an application of the Lévy theorem, one can show the following useful result.

Theorem 4.5 *Let $M = \{M_t\}_{t \geq 0}$ ba a real-valued continuous local martingale such that $M_0 = 0$ and $\lim_{t \to \infty} \langle M, M \rangle_t = \infty$ a.s. For each $t \geq 0$, define the stopping time*

$$\tau_t = \inf\{s : \langle M, M \rangle_s > t\}.$$

Then $\{M_{\tau_t}\}_{t \geq 0}$ is a Brownian motion with respect to the filtration $\{\mathcal{F}_{\tau_t}\}_{t \geq 0}$.

1.5 STOCHASTIC INTEGRALS

In this section we shall define the stochastic integral

$$\int_0^t f(s)dB_s$$

with respect to an m-dimensional Brownian motion $\{B_t\}$ for a class of $d \times m$-matrix-valued stochastic processes $\{f(t)\}$. Since for almost all $\omega \in \Omega$, the Brownian sample path $B.(\omega)$ is nowhere differentiable, the integral can not be defined in the ordinary way. However, we can define the integral for a large class of stochastic processes by making use of the stochastic nature of Brownian motion. This integral was first defined by K. Itô in 1949 and is now known as *Itô stochastic integral*. We shall now start to define the stochastic integral step by step.

Let $(\Omega, \mathcal{F}, P)$ be a complete probability space with a filtration $\{\mathcal{F}_t\}_{t \geq 0}$ satisfying the usual conditions. Let $B = \{B_t\}_{t \geq 0}$ be a one-dimensional Brownian motion defined on the probability space adapted to the filtration.

Definition 5.1 *Let $0 \leq a < b < \infty$. Denote by $\mathcal{M}^2([a,b]; R)$ the space of all real-valued measurable $\{\mathcal{F}_t\}$-adapted processes $f = \{f(t)\}_{a \leq t \leq b}$ such that*

$$\|f\|_{a,b}^2 = E\int_a^b |f(t)|^2 dt < \infty. \tag{5.1}$$

We identify f and $\bar{f}$ in $\mathcal{M}^2([a,b]; R)$ if $\|f - \bar{f}\|_{a,b}^2 = 0$. In this case we say that f and $\bar{f}$ are equivalent and write $f = \bar{f}$.

Clearly, $\|\cdot\|_{a,b}$ defines a metric on $\mathcal{M}^2([a,b]; R)$ and the space is complete under this metric. Let us point out that for every $f \in \mathcal{M}^2([a,b]; R)$, there is a predictable $\bar{f} \in \mathcal{M}^2([a,b]; R)$ such that $f = \bar{f}$. In fact, f has a progressively measurable modification $\hat{f}$ in $\mathcal{M}^2([a,b]; R)$ and then we may take

$$\bar{f}(t) = \limsup_{h \downarrow 0} \frac{1}{h}\int_{t-h}^t \hat{f}(s)ds.$$

Thus, if necessary, we may assume that $f \in \mathcal{M}^2([a,b]; R)$ is predictable without loss of generality. However, in this book we would rather follow the usual custom of not being very careful about the distinction between the equivalence processes.

For stochastic processes $f \in \mathcal{M}^2([a,b]; R)$ we shall show how to define the Itô integral $\int_a^b f(t)dB_t$. The idea is natural: First define the integral $\int_a^b g(t)dB_t$ for a class of simple processes g. Then we show that each $f \in \mathcal{M}^2([a,b]; R)$ can be approximated by such simple processes g's and we define the limit of $\int_a^b g(t)dB_t$ as the integral $\int_a^b f(t)dB_t$. Let us first introduce the concept of simple processes.

Definition 5.2 A real-valued stochastic process $g = \{g(t)\}_{a \leq t \leq b}$ is called a simple (or step) process if there exists a partition $a = t_0 < t_1 < \cdots < t_k = b$ of $[a,b]$, and bounded random variables ξ_i, $0 \leq i \leq k-1$ such that ξ_i is $\mathcal{F}_{t_i}$-measurable and

$$g(t) = \xi_0 I_{[t_0,\, t_1]}(t) + \sum_{i=1}^{k-1} \xi_i I_{(t_i,\, t_{i+1}]}(t). \tag{5.2}$$

Denote by $\mathcal{M}_0([a,b]; R)$ the family of all such processes.

Clearly, $\mathcal{M}_0([a,b]; R) \subset \mathcal{M}^2([a,b]; R)$. We now give the definition of the Itô integral for such simple processes.

Definition 5.3 (Part 1 of the definition of Itô's integral) For a simple process g with the form of (5.2) in $\mathcal{M}_0([a,b]; R)$, define

$$\int_a^b g(t) dB_t = \sum_{i=0}^{k-1} \xi_i (B_{t_{i+1}} - B_{t_i}) \tag{5.3}$$

and call it the stochastic integral of g with respect to the Brownian motion $\{B_t\}$ or the Itô integral.

Clearly, the stochastic integral $\int_a^b g(t)dB_t$ is $\mathcal{F}_b$-measurable. We shall now show that it belongs to $L^2(\Omega; R)$.

Lemma 5.4 If $g \in \mathcal{M}_0([a,b]; R)$, then

$$E \int_a^b g(t)dB_t = 0, \tag{5.4}$$

$$E \left| \int_a^b g(t)dB_t \right|^2 = E \int_a^b |g(t)|^2 dt. \tag{5.5}$$

Proof. Since ξ_i is $\mathcal{F}_{t_i}$-measurable whereas $B_{t_{i+1}} - B_{t_i}$ is independent of $\mathcal{F}_{t_i}$,

$$E \int_a^b g(t)dB_t = \sum_{i=0}^{k-1} E[\xi_i(B_{t_{i+1}} - B_{t_i})] = \sum_{i=0}^{k-1} E\xi_i \, E(B_{t_{i+1}} - B_{t_i}) = 0.$$

Moreover, note that $B_{t_{j+1}} - B_{t_j}$ is independent of $\xi_i \xi_j (B_{t_{i+1}} - B_{t_i})$ if $i < j$. Thus

$$E \left| \int_a^b g(t)dB_t \right|^2 = \sum_{0 \leq i,j \leq k-1} E[\xi_i \xi_j (B_{t_{i+1}} - B_{t_i})(B_{t_{j+1}} - B_{t_j})]$$

$$= \sum_{i=0}^{k-1} E[\xi_i^2 (B_{t_{i+1}} - B_{t_i})^2] = \sum_{i=0}^{k-1} E\xi_i^2 E(B_{t_{i+1}} - B_{t_i})^2$$

$$= \sum_{i=0}^{k-1} E\xi_i^2 (t_{i+1} - t_i) = E \int_a^b |g(t)|^2 dt.$$

as required.

Lemma 5.5 *Let $g_1, g_2 \in \mathcal{M}_0([a,b]; R)$ and let c_1, c_2 be two real numbers. Then $c_1 g_1 + c_2 g_2 \in \mathcal{M}_0([a,b]; R)$ and*

$$\int_a^b [c_1 g_1(t) + c_2 g_2(t)] dB_t = c_1 \int_a^b g_1(t) dB_t + c_2 \int_a^b g_2(t) dB_t.$$

The proof is left to the reader. We shall now use the properties shown in Lemmas 5.4 and 5.5 to extend the integral definition from simple processes to processes in $\mathcal{M}^2([a,b]; R)$. This is based on the following approximation result.

Lemma 5.6 *For any $f \in \mathcal{M}^2([a,b]; R)$, there exists a sequence $\{g_n\}$ of simple processes such that*

$$\lim_{n \to \infty} E \int_a^b |f(t) - g_n(t)|^2 dt = 0. \tag{5.6}$$

Proof. We divide the whole proof into three steps.

Step 1. We first claim that for any $f \in \mathcal{M}^2([a,b]; R)$, there exists a sequence $\{\varphi_n\}$ of bounded processes in $\mathcal{M}^2([a,b]; R)$ such that

$$\lim_{n \to \infty} E \int_a^b |f(t) - \varphi_n(t)|^2 dt = 0. \tag{5.7}$$

In fact, for each n, put

$$\varphi_n(t) = [-n \vee f(t)] \wedge n.$$

Then (5.7) follows by the dominated convergence theorem (i.e. Theorem 2.3).

Step 2. We next claim that if $\varphi \in \mathcal{M}^2([a,b]; R)$ is bounded, say $|\varphi| \leq C = const.$, then there exists a sequence $\{\phi_n\}$ of bounded continuous processes in $\mathcal{M}^2([a,b]; R)$ such that

$$\lim_{n \to \infty} E \int_a^b |\varphi(t) - \phi_n(t)|^2 dt = 0. \tag{5.8}$$

In fact, for each n, let $\rho_n : R \to R_+$ be a continuous function such that $\rho_n(s) = 0$ for $s \leq -\frac{1}{n}$ and $s \geq 0$ and

$$\int_{-\infty}^\infty \rho_n(s) ds = 1.$$

Define

$$\phi_n(t) = \phi_n(t, \omega) = \int_a^b \rho_n(s-t) \varphi(s, \omega) ds.$$

Then for every ω, $\phi_n(\cdot, \omega)$ is continuous and $|\phi_n(t, \omega)| \leq C$. Also ϕ_n is a measurable $\{\mathcal{F}_t\}$-adapted process. Moreover, for all $\omega \in \Omega$,

$$\lim_{n \to \infty} \int_a^b |\varphi(t, \omega) - \phi_n(t, \omega)|^2 dt = 0.$$

So (5.8) follows by the bounded convergence theorem.

Step 3. We now claim that if $\phi \in \mathcal{M}^2([a, b]; R)$ is bounded and continuous, then there exists a sequence $\{g_n\}$ of simple processes such that

$$\lim_{n \to \infty} E \int_a^b |\phi(t) - g_n(t)|^2 dt = 0. \tag{5.9}$$

In fact, for each n, let

$$g_n(t) = \phi(a) \, I_{[a, \, a+(b-a)/n]}(t)$$
$$+ \sum_{i=1}^{n-1} \phi(a + i(b-a)/n) \, I_{(a+i(b-a)/n, \, a+(i+1)(b-a)/n]}(t).$$

Then $g_n \in \mathcal{M}_0([a, b]; R)$, and for every ω,

$$\lim_{n \to \infty} \int_a^b |\phi(t, \omega) - g_n(t, \omega)|^2 dt = 0.$$

So (5.9) follows by the bounded convergence theorem once again. Finally, the conclusion of the lemma follows clearly from steps 1–3 and the proof is now complete.

We can now explain how to define the Itô integral for a process $f \in \mathcal{M}^2([a, b]; R)$. By Lemma 5.6, there is a sequence $\{g_n\}$ of simple processes such that

$$\lim_{n \to \infty} E \int_a^b |f(t) - g_n(t)|^2 dt = 0.$$

Thus, by Lemmas 5.4 and 5.5,

$$E \left| \int_a^b g_n(t) dB_t - \int_a^b g_m(t) dB_t \right|^2 = E \left| \int_a^b [g_n(t) - g_m(t)] dB_t \right|^2$$
$$= E \int_a^b |g_n(t) - g_m(t)|^2 dt \to 0 \quad \text{as } n, m \to \infty.$$

In other words, $\{\int_a^b g_n(t) dB_t\}$ is a Cauchy sequence in $L^2(\Omega; R)$. So the limit exists and we define the limit as the stochastic integral. This leads to the following definition.

Definition 5.7 (Part 2 of the definition of Itô's integral) Let $f \in \mathcal{M}^2([a, b]; R)$. The Itô integral of f with respect to $\{B_t\}$ is defined by

$$\int_a^b f(t) dB_t = \lim_{n \to \infty} \int_a^b g_n(t) dB_t \quad \text{in } L^2(\Omega; R), \tag{5.10}$$

where $\{g_n\}$ is a sequence of simple processes such that

$$\lim_{n \to \infty} E \int_a^b |f(t) - g_n(t)|^2 dt = 0. \tag{5.11}$$

The above definition is independent of the particular sequence $\{g_n\}$. For if $\{h_n\}$ is another sequence of simple processes converging to f in the sense that

$$\lim_{n\to\infty} E \int_a^b |f(t) - h_n(t)|^2 dt = 0,$$

then the sequence $\{\varphi_n\}$, where $\varphi_{2n-1} = g_n$ and $\varphi_{2n} = h_n$, is also convergent to f in the same sense. Hence, by what we have proved, the sequence $\{\int_a^b \varphi_n(t)dB_t\}$ is convergent in $L^2(\Omega; R)$. It follows that the limits (in L^2) of $\int_a^b g_n(t)dB_t$ and of $\int_a^b h_n(t)dB_t$ are equal almost surely.

The stochastic integral has many nice properties. We first observe the following:

Theorem 5.8 *Let $f, g \in \mathcal{M}^2([a,b]; R)$, and let α, β be two real numbers. Then*

(i) $\int_a^b f(t)dB_t$ *is $\mathcal{F}_b$-measurable;*

(ii) $E \int_a^b f(t)dB_t = 0$;

(iii) $E|\int_a^b f(t)dB_t|^2 = E \int_a^b |f(t)|^2 dt$;

(vi) $\int_a^b [\alpha f(t) + \beta g(t)]dB_t = \alpha \int_a^b f(t)dB_t + \beta \int_a^b g(t)dB_t$.

The proof is left to the reader. The next theorem improves the results (ii) and (iii) of Theorem 5.8

Theorem 5.9 *Let $f \in \mathcal{M}^2([a,b]; R)$. Then*

$$E\left(\int_a^b f(t)dB(t)\Big|\mathcal{F}_a\right) = 0, \tag{5.12}$$

$$E\left(\Big|\int_a^b f(t)dB(t)\Big|^2 \Big|\mathcal{F}_a\right) = E\left(\int_a^b |f(t)|^2 dt \Big|\mathcal{F}_a\right)$$

$$= \int_a^b E(|f(t)|^2|\mathcal{F}_a)dt. \tag{5.13}$$

We need a simple lemma.

Lemma 5.10 *If $f \in \mathcal{M}^2([a,b]; R)$ and ξ is a real-valued bounded $\mathcal{F}_a$-measurable random variable, then $\xi f \in \mathcal{M}^2([a,b]; R)$ and*

$$\int_a^b \xi f(t)dB_t = \xi \int_a^b f(t)dB_t. \tag{5.14}$$

Proof. It is clear that $\xi f \in \mathcal{M}^2([a,b]; R)$. If f is a simple processes, then (5.14) follows from the definition of the stochastic integral. For general $f \in$

$\mathcal{M}^2([a,b]; R)$, let $\{g_n\}$ be a sequence of simple processes satisfying (5.11). Applying (5.14) to each g_n and taking $n \to \infty$, the assertion (5.14) follows.

Proof of Theorem 5.9. By the definition of conditional expectation, (5.12) holds if and only if

$$E\left(I_A \int_a^b f(t)dB(t)\right) = 0$$

for all sets $A \in \mathcal{F}_a$. But by Lemma 5.10 and Theorem 5.8,

$$E\left(I_A \int_a^b f(t)dB(t)\right) = E\int_a^b I_A f(t)dB(t) = 0$$

as required. The proof of (5.13) is similar.

Let $T > 0$ and $f \in \mathcal{M}^2([0,T]; R)$. Clearly, for any $0 \le a < b \le T$, $\{f(t)\}_{a \le t \le b} \in \mathcal{M}^2([a,b]; R)$ so $\int_a^b f(t)dB_t$ is well defined. It is easy to show that

$$\int_a^b f(t)dB_t + \int_b^c f(t)dB_t = \int_a^c f(t)dB_t \tag{5.15}$$

if $0 \le a < b < c \le T$.

Definition 5.11 *Let $f \in \mathcal{M}^2([0,T]; R)$. Define*

$$I(t) = \int_0^t f(s)dB_s \quad \text{for } 0 \le t \le T,$$

where, by definition, $I(0) = \int_0^0 f(s)dB_s = 0$. We call $I(t)$ the indefinite Itô integral of f.

Clearly, $\{I(t)\}$ is $\{\mathcal{F}_t\}$-adapted. We now show the very important martingale property of the indefinite Itô integral.

Theorem 5.12 *If $f \in \mathcal{M}^2([0,T]; R)$, then the indefinite integral $\{I(t)\}_{0 \le t \le T}$ is a square-integrable martingale with respect to the filtration $\{\mathcal{F}_t\}$. In particular,*

$$E\left[\sup_{0 \le t \le T}\left|\int_0^t f(s)dB_s\right|^2\right] \le 4E\int_0^T |f(s)|^2 ds. \tag{5.16}$$

Proof. Clearly, $\{I(t)\}_{0 \le t \le T}$ is square-integrable. To show the martingale property, let $0 \le s < t \le T$. By (5.15) and Theorem 5.9

$$E(I(t)|\mathcal{F}_s) = E(I(s)|\mathcal{F}_s) + E\left(\int_s^t f(r)dB_r|\mathcal{F}_s\right) = I(s)$$

as desired. The inequality (5.16) now follows from Doob's martingale inequality (i.e. Theorem 3.8).

Theorem 5.13 *If $f \in \mathcal{M}^2([0,T]; R)$, then the indefinite integral $\{I(t)\}_{0 \le t \le T}$ has a continuous version.*

Proof. Let $\{g_n\}$ be a sequence of simple processes such that

$$\lim_{n \to \infty} E \int_0^T |f(s) - g_n(s)|^2 ds = 0. \tag{5.17}$$

Note from the definition of the stochastic integral and the continuity of the Brownian motion that the indefinite integrals

$$I_n(t) = \int_0^t g_n(s) dB_s, \qquad 0 \le t \le T$$

are continuous. By Theorem 5.12, $\{I_n(t) - I_m(t)\}$ is a martingale, for each pair of integers n, m. Hence, by Doob's martingale inequality (i.e. Theorem 3.8), for any $\varepsilon > 0$

$$P\left\{\sup_{0 \le t \le T} |I_n(t) - I_m(t)| \ge \varepsilon\right\} \le \frac{1}{\varepsilon^2} E |I_n(T) - I_m(T)|^2$$

$$= \frac{1}{\varepsilon^2} E \int_0^T |g_n(s) - g_m(s)|^2 ds \to 0 \quad \text{as } n, m \to \infty.$$

For each $k = 1, 2, \cdots$, taking $\varepsilon = k^{-2}$ it follows that for some n_k sufficiently large,

$$P\left\{\sup_{0 \le t \le T} |I_{n_k}(t) - I_m(t)| \ge \frac{1}{k^2}\right\} \le \frac{1}{k^2} \quad \text{if } m \ge n_k.$$

One can then choose the n_k in such a way that $n_k \uparrow \infty$ as $k \to \infty$ and

$$P\left\{\sup_{0 \le t \le T} |I_{n_k}(t) - I_{n_{k+1}}(t)| \ge \frac{1}{k^2}\right\} \le \frac{1}{k^2}, \qquad k \ge 1.$$

Since $\sum k^{-2} < \infty$, the Borel–Cantelli lemma (i.e. Lemma 2.4) implies that there exists a set $\Omega_0 \in \mathcal{F}$ with $P(\Omega_0) = 0$ and an integer-valued random variable k_0 such that for every $\omega \in \Omega_0$,

$$\sup_{0 \le t \le T} |I_{n_k}(t, \omega) - I_{n_{k+1}}(t, \omega)| < \frac{1}{k^2} \qquad \text{if } k \ge k_0(\omega).$$

In other words, with probability 1, $\{I_{n_k}(t)\}_{k \ge 1}$ is uniformly convergent in $t \in [0,T]$, and therefore the limit, denoted by $J(t)$, is continuous in $t \in [0,T]$ for almost all $\omega \in \Omega$. Since (5.17) implies

$$\lim_{k \to \infty} I_{n_k}(t) = \int_0^t f(s) dB_s \qquad \text{in } L^2(\Omega, R),$$

it follows that
$$J(t) = \int_0^t f(s)dB_s \qquad a.s.$$
That is, the indefinite integral has a continuous version.

From now on, when we speak of the indefinite integral we always mean a continuous version of it.

Theorem 5.14 Let $f \in \mathcal{M}^2([0,T]; R)$. Then the indefinite integral $I = \{I(t)\}_{0 \le t \le T}$ is a square-integrable continuous martingale and its quadratic variation is given by
$$\langle I, I \rangle_t = \int_0^t |f(s)|^2 ds, \qquad 0 \le t \le T. \tag{5.18}$$

Proof. Obviously we need only to show (5.18). By the definition of the quadratic variation we need to show that $\{I^2(t) - \langle I, I \rangle_t\}$ is a continuous martingale vanishing at $t = 0$. But, obviously $I^2(0) - \langle I, I \rangle_0 = 0$. Moreover, if $0 \le r < t \le T$, by Theorem 5.9,

$$E(I^2(t) - \langle I, I \rangle_t | \mathcal{F}_r)$$
$$= I^2(r) - \langle I, I \rangle_r + 2I(r)E\left(\int_r^t f(s)dB_s \Big| \mathcal{F}_r\right)$$
$$+ E\left(\Big|\int_r^t f(s)dB_s\Big|^2 \Big| \mathcal{F}_r\right) - E\left(\int_r^t |f(s)|^2 ds \Big| \mathcal{F}_r\right)$$
$$= I^2(r) - \langle I, I \rangle_r$$

as desired.

Let us now proceed to define the stochastic integrals with stopping time. We observe that if τ is an $\{\mathcal{F}_t\}$-stopping time, then $\{I_{[\![0,\,\tau]\!]}(t)\}_{t \ge 0}$ is a bounded right continuous $\{\mathcal{F}_t\}$-adapted process. In fact, the boundedness and right continuity are obvious. Moreover, for each $t \ge 0$,

$$\{\omega : I_{[\![0,\,\tau]\!]}(t, \omega) \le r\} = \begin{cases} \emptyset \in \mathcal{F}_t & \text{if } r < 0, \\ \{\omega : \tau(\omega) < t\} \in \mathcal{F}_t & \text{if } 0 \le r < 1, \\ \Omega \in \mathcal{F}_t & \text{if } r \ge 1, \end{cases}$$

that is, $I_{[\![0,\,\tau]\!]}(t)$ is $\mathcal{F}_t$-measurable. Therefore, $\{I_{[\![0,\,\tau]\!]}(t)\}_{t \ge 0}$ is also predictable.

Definition 5.15 Let $f \in \mathcal{M}^2([0,T]; R)$, and let τ be an $\{\mathcal{F}_t\}$-stopping time such that $0 \le \tau \le T$. Then, $\{I_{[\![0,\,\tau]\!]}(t)f(t)\}_{0 \le t \le T} \in \mathcal{M}^2([0,T]; R)$ clearly, and we define
$$\int_0^\tau f(s)dB_s = \int_0^T I_{[\![0,\,\tau]\!]}(s)f(s)dB_s.$$

Furthermore, if ρ is another stopping time with $0 \leq \rho \leq \tau$, we define

$$\int_\rho^\tau f(s)dB_s = \int_0^\tau f(s)dB_s - \int_0^\rho f(s)dB_s.$$

It is easy to see that

$$\int_\rho^\tau f(s)dB_s = \int_0^T I_{]\!]\rho,\ \tau]\!]}(s) f(s) dB_s. \tag{5.19}$$

If applying Theorem 5.8 to this we immediately obtain:

Theorem 5.16 *Let $f \in \mathcal{M}^2([0,T]; R)$, and let ρ, τ be two stopping times such that $0 \leq \rho \leq \tau \leq T$. Then*

$$E \int_\rho^\tau f(s)dB_s = 0,$$

$$E\left|\int_\rho^\tau f(s)dB_s\right|^2 = E \int_\rho^\tau |f(s)|^2 ds.$$

However, the next theorem improves these results and is also a generalization of Theorem 5.9.

Theorem 5.17 *Let $f \in \mathcal{M}^2([0,T]; R)$, and let ρ, τ be two stopping times such that $0 \leq \rho \leq \tau \leq T$. Then*

$$E\left(\int_\rho^\tau f(s)dB_s \Big| \mathcal{F}_\rho\right) = 0, \tag{5.20}$$

$$E\left(\left|\int_\rho^\tau f(s)dB_s\right|^2 \Big| \mathcal{F}_\rho\right) = E\left(\int_\rho^\tau |f(s)|^2 ds \Big| \mathcal{F}_\rho\right). \tag{5.21}$$

We need a useful lemma.

Lemma 5.18 *Let $f \in \mathcal{M}^2([0,T]; R)$, and let τ be a stopping time such that $0 \leq \tau \leq T$. Then*

$$\int_0^\tau f(s)dB_s = I(\tau),$$

where $\{I(t)\}_{0 \leq t \leq T}$ is the indefinite integral of f given by Definition 5.11.

We leave the proof of this lemma to the reader, but prove Theorem 5.17.

Proof of Theorem 5.17. By Theorem 5.14 and the Doob martingale stopping theorem (i.e. Theorem 3.3),

$$E(I(\tau)|\mathcal{F}_\rho) = I(\rho) \tag{5.22}$$

and
$$E(I^2(\tau) - \langle I, I\rangle_\tau | \mathcal{F}_\rho) = I^2(\rho) - \langle I, I\rangle_\rho, \tag{5.23}$$

where $\{\langle I, I\rangle_t\}$ is defined by (5.18). Applying Lemma 5.18 one then sees from (5.22) that
$$E\left(\int_\rho^\tau f(s)dB_s | \mathcal{F}_\rho\right) = E(I(\tau) - I(\rho)|\mathcal{F}_\rho) = 0$$

which is (5.20). Moreover, by (5.22) and (5.23),
$$E(|I(\tau) - I(\rho)|^2 | \mathcal{F}_\rho) = E(I^2(\tau)|\mathcal{F}_\rho) - 2I(\rho)E(I(\tau)|\mathcal{F}_\rho) + I^2(\rho)$$
$$= E(I^2(\tau)|\mathcal{F}_\rho) - I^2(\rho) = E(\langle I, I\rangle_\tau - \langle I, I\rangle_\rho | \mathcal{F}_\rho) = E\left(\int_\rho^\tau |f(s)|^2 ds | \mathcal{F}_\rho\right)$$

which, by Lemma 5.18, is the required (5.21). The proof is complete.

Corollary 5.19 *Let $f, g \in \mathcal{M}^2([0, T]; R)$, and let ρ, τ be two stopping times such that $0 \leq \rho \leq \tau \leq T$. Then*
$$E\left(\int_\rho^\tau f(s)dB_s \int_\rho^\tau g(s)dB_s | \mathcal{F}_\rho\right) = E\left(\int_\rho^\tau f(s)g(s)ds | \mathcal{F}_\rho\right).$$

Proof. By Theorem 5.17,
$$4E\left(\int_\rho^\tau f(s)dB_s \int_\rho^\tau g(s)dB_s | \mathcal{F}_\rho\right)$$
$$= E\left(\left|\int_\rho^\tau (f(s) + g(s))dB_s\right|^2 | \mathcal{F}_\rho\right) - E\left(\left|\int_\rho^\tau (f(s) - g(s))dB_s\right|^2 | \mathcal{F}_\rho\right)$$
$$= E\left(\int_\rho^\tau (f(s) + g(s))^2 ds | \mathcal{F}_\rho\right) - E\left(\int_\rho^\tau (f(s) - g(s))^2 ds | \mathcal{F}_\rho\right)$$
$$= 4E\left(\int_\rho^\tau f(s)g(s)ds | \mathcal{F}_\rho\right)$$

as desired.

We shall now extend the Itô stochastic integral to the multi-dimensional case. Let $\{B_t = (B_t^1, \cdots, B_t^m)^T\}_{t \geq 0}$ be an m-dimensional Brownian motion defined on the complete probability space $(\Omega, \mathcal{F}, P)$ adapted to the filtration $\{\mathcal{F}_t\}$. Let $\mathcal{M}^2([0, T]; R^{d \times m})$ denote the family of all $d \times m$-matrix-valued measurable $\{\mathcal{F}_t\}$-adapted processes $f = \{(f_{ij}(t))_{d \times m}\}_{0 \leq t \leq T}$ such that
$$E\int_0^T |f(s)|^2 dt < \infty.$$

Here, and throughout this book, $|A|$ will denote the trace norm for matrix A, i.e. $|A| = \sqrt{trace(A^T A)}$.

Definition 5.20 Let $f \in \mathcal{M}^2([0,T]; R^{d \times m})$. Using matrix notation, we define the multi-dimensional indefinite Itô integral

$$\int_0^t f(s)dB_s = \int_0^t \begin{pmatrix} f_{11}(s) & \cdots & f_{1m}(s) \\ \vdots & & \vdots \\ f_{d1}(s) & \cdots & f_{dm}(s) \end{pmatrix} \begin{pmatrix} dB_s^1 \\ \vdots \\ dB_s^m \end{pmatrix}$$

to be the d-column-vector-valued process whose i'th component is the following sum of 1-dimensional Itô integrals

$$\sum_{j=1}^m \int_0^t f_{ij}(s)dB_s^j.$$

Clearly, the Itô integral is an R^d-valued continuous martingale with respect to $\{\mathcal{F}_t\}$. Besides, it has the following important properties.

Theorem 5.21 Let $f \in \mathcal{M}^2([0,T]; R^{d \times m})$, and let ρ, τ be two stopping times such that $0 \leq \rho \leq \tau \leq T$. Then

$$E\left(\int_\rho^\tau f(s)dB_s \Big| \mathcal{F}_\rho\right) = 0, \tag{5.24}$$

$$E\left(\left|\int_\rho^\tau f(s)dB_s\right|^2 \Big| \mathcal{F}_\rho\right) = E\left(\int_\rho^\tau |f(s)|^2 ds \Big| \mathcal{F}_\rho\right). \tag{5.25}$$

The assertion (5.24) follows from the definition of multi-dimensional Itô integral and Theorem 5.17, while (5.25) follows from Theorem 5.17 and the following lemma.

Lemma 5.22 Let $\{B_t^1\}_{t \geq 0}$ and $\{B_t^2\}_{t \geq 0}$ be two independent 1-dimensional Brownian motions. Let $f, g \in \mathcal{M}^2([0,T]; R)$, and let ρ, τ be two stopping times such that $0 \leq \rho \leq \tau \leq T$. Then

$$E\left(\int_\rho^\tau f(s)dB_s^1 \int_\rho^\tau g(s)dB_s^2 \Big| \mathcal{F}_\rho\right) = 0. \tag{5.26}$$

Proof. We first claim that if $\varphi, \phi \in \mathcal{M}^2([a,b]; R)$. Then

$$E\left(\int_a^b \varphi(s)dB_s^1 \int_a^b \phi(s)dB_s^2\right) = 0. \tag{5.27}$$

In fact, let φ, ϕ be simple processes with the forms

$$\varphi(t) = \xi_0 I_{[t_0,\, t_1]}(t) + \sum_{i=1}^{k-1} \xi_i I_{(t_i,\, t_{i+1}]}(t)$$

and
$$\phi(t) = \zeta_0 I_{[\bar{t}_0, \bar{t}_1]}(t) + \sum_{j=1}^{m-1} \zeta_i I_{(\bar{t}_j, \bar{t}_{j+1}]}(t).$$

Then
$$E\left(\int_a^b \varphi(s)dB_s^1 \int_a^b \phi(s)dB_s^2\right) = \sum_{i=0}^{k-1}\sum_{j=0}^{m-1} E\left[\xi_i \zeta_j (B_{t_{i+1}}^1 - B_{t_i}^1)(B_{\bar{t}_{j+1}}^2 - B_{\bar{t}_j}^2)\right]$$

But for every pair of i, j, if $t_i \leq \bar{t}_j$, then $B_{\bar{t}_{j+1}}^2 - B_{\bar{t}_j}^2$ is independent of $\xi_i \zeta_j (B_{t_{i+1}}^1 - B_{t_i}^1)$ and hence
$$E\left[\xi_i \zeta_j (B_{t_{i+1}}^1 - B_{t_i}^1)(B_{\bar{t}_{j+1}}^2 - B_{\bar{t}_j}^2)\right] = 0.$$

Similarly, it still holds if $t_i > \bar{t}_j$. In other words, we have shown that (5.27) holds for simple processes φ, ϕ, but the general case follows by the approximation procedure.

We next observe that for any $0 \leq r < t \leq T$
$$E\left(\int_r^t f(s)dB_s^1 \int_r^t g(s)dB_s^2 \middle| \mathcal{F}_r\right) = 0, \tag{5.28}$$

since, by (5.27) and Lemma 5.10, for any $A \in \mathcal{F}_r$
$$E\left(I_A \int_r^t f(s)dB_s^1 \int_r^t g(s)dB_s^2\right) = E\left(\int_r^t I_A f(s)dB_s^1 \int_r^t g(s)dB_s^2\right) = 0.$$

Therefore
$$E\left(\int_0^t f(s)dB_s^1 \int_0^t g(s)dB_s^2 \middle| \mathcal{F}_r\right)$$
$$= \int_0^r f(s)dB_s^1 \int_0^r g(s)dB_s^2 + \int_0^r f(s)dB_s^1 E\left(\int_r^t g(s)dB_s^2 \middle| \mathcal{F}_r\right)$$
$$+ \int_0^r g(s)dB_s^2 E\left(\int_r^t f(s)dB_s^1 \middle| \mathcal{F}_r\right) + E\left(\int_r^t f(s)dB_s^1 \int_r^t g(s)dB_s^2 \middle| \mathcal{F}_r\right)$$
$$= \int_0^r f(s)dB_s^1 \int_0^r g(s)dB_s^2.$$

That is, $\{\int_0^t f(s)dB_s^1 \int_0^t g(s)dB_s^2\}_{0 \leq t \leq T}$ is a martingale with respect to $\{\mathcal{F}_t\}$. Hence, by the Doob martingale stopping theorem,
$$E\left(\int_0^\tau f(s)dB_s^1 \int_0^\tau g(s)dB_s^2 \middle| \mathcal{F}_\rho\right) = \int_0^\rho f(s)dB_s^1 \int_0^\rho g(s)dB_s^2. \tag{5.29}$$

Now the assertion (5.26) follows from (5.29) easily. The proof of the lemma, hence of Theorem 5.21 is now complete.

We shall finally extend the stochastic integral to a larger class of stochastic processes. Let $\mathcal{L}^2(R_+; R^{d\times m})$ denote the family of all $d\times m$-matrix-valued measurable $\{\mathcal{F}_t\}$-adapted processes $f = \{f(t)\}_{t\geq 0}$ such that

$$\int_0^T |f(t)|^2 dt < \infty \quad \text{a.s. for every } T > 0.$$

Let $\mathcal{M}^2(R_+; R^{d\times m})$ denote the family of all processes $f \in \mathcal{L}^2(R_+; R^{d\times m})$ such that

$$E\int_0^T |f(t)|^2 dt < \infty \quad \text{for every } T > 0.$$

Clearly, if $f \in \mathcal{M}^2(R_+; R^{d\times m})$, then $\{f(t)\}_{0\leq t\leq T} \in \mathcal{M}^2([0,T]; R^{d\times m})$ for every $T > 0$. Hence, the indefinite integral $\int_0^t f(s)dB_s$, $t \geq 0$ is well defined, and it is an R^d-valued continuous square-integrable martingale. However, we aim to define the integral for all processes in $\mathcal{L}^2(R_+; R^{d\times m})$. Let $f \in \mathcal{L}^2(R_+; R^{d\times m})$. For each integer $n \geq 1$, define the stopping time

$$\tau_n = n \wedge \inf\{t \geq 0 : \int_0^t |f(s)|^2 ds \geq n\}.$$

Clearly, $\tau_n \uparrow \infty$ a.s. Moreover, $\{f(t)I_{[[0,\tau_n]]}(t)\}_{t\geq 0} \in \mathcal{M}^2(R_+; R^{d\times m})$ so the integral

$$I_n(t) = \int_0^t f(s)I_{[[0,\tau_n]]}(s)dB_s, \quad t \geq 0$$

is well defined. Note that for $1 \leq n \leq m$ and $t \geq 0$,

$$I_m(t \wedge \tau_n) = \int_0^{t\wedge \tau_n} f(s)I_{[[0,\tau_m]]}(s)dB_s = \int_0^t f(s)I_{[[0,\tau_m]]}(s)I_{[[0,\tau_n]]}(s)dB_s$$
$$= \int_0^t f(s)I_{[[0,\tau_n]]}(s)dB_s = I_n(t),$$

which implies

$$I_m(t) = I_n(t), \quad 0 \leq t \leq \tau_n.$$

So we may define the indefinite stochastic integral $\{I(t)\}_{t\geq 0}$ as

$$I(t) = I_n(t) \quad \text{on } 0 \leq t \leq \tau_n. \quad (5.30)$$

Definition 5.23 Let $f = \{f(t)\}_{t\geq 0} \in \mathcal{M}^2(R_+; R^{d\times m})$. The indefinite Itô integral of f with respect to $\{B_t\}$ is the R^d-valued process $\{I(t)\}_{t\geq 0}$ defined by (5.30). As before, we usually write $\int_0^t f(s)dB_s$ instead of $I(t)$.

It is clear that the Itô integral $\int_0^t f(s)dB_s$, $t \geq 0$ is a R^d-valued continuous local martingale.

1.6 ITÔ'S FORMULA

In the previous section we defined the Itô stochastic integrals. However the basic definition of the integrals is not very convenient in evaluating a given integral. This is similar to the situation for classical Lebesgue integrals, where we do not use the basic definition but rather the fundamental theorem of calculus plus the chain rule in the explicit calculations. For example, it is very easy to use the chain rule to calculate $\int_0^t \cos s \, ds = \sin t$ but not so if you use the basic definition. In this section we shall establish the stochastic version of the chain rule for the Itô integrals, which is known as Itô's formula. We shall see in this book that Itô's formula is not only useful for evaluating the Itô integrals but, more importantly, it plays a key role in stochastic analysis.

We shall first establish the one-dimensional Itô formula and then generalize it to the multi-dimensional case. Let $\{B_t\}_{t \geq 0}$ be a one-dimensional Brownian motion defined on the complete probability space $(\Omega, \mathcal{F}, P)$ adapted to the filtration $\{\mathcal{F}_t\}_{t \geq 0}$. Let $\mathcal{L}^1(R_+; R^d)$ denote the family of all R^d-valued measurable $\{\mathcal{F}_t\}$-adapted processes $f = \{f(t)\}_{t \geq 0}$ such that

$$\int_0^T |f(t)| dt < \infty \quad \text{a.s. for every } T > 0.$$

Definition 6.1 *A one-dimensional Itô process is a continuous adapted process $x(t)$ on $t \geq 0$ of the form*

$$x(t) = x(0) + \int_0^t f(s)ds + \int_0^t g(s)dB_s, \quad (6.1)$$

where $f \in \mathcal{L}^1(R_+; R)$ and $g \in \mathcal{L}^2(R_+; R)$. We shall say that $x(t)$ has stochastic differential $dx(t)$ on $t \geq 0$ given by

$$dx(t) = f(t)dt + g(t)dB_t. \quad (6.2)$$

We shall sometimes speak of Itô process $x(t)$ and its stochastic differential $dx(t)$ on $t \in [a,b]$, and the meaning is apparent.

Let $C^{2,1}(R^d \times R_+; R)$ denote the family of all real-valued functions $V(x,t)$ defined on $R^d \times R_+$ such that they are continuously twice differentiable in x and once in t. If $V \in C^{2,1}(R^d \times R_+; R)$, we set

$$V_t = \frac{\partial V}{\partial t}, \quad V_x = \left(\frac{\partial V}{\partial x_1}, \cdots, \frac{\partial V}{\partial x_d}\right),$$

$$V_{xx} = \left(\frac{\partial^2 V}{\partial x_i \partial x_j}\right)_{d \times d} = \begin{pmatrix} \frac{\partial^2 V}{\partial x_1 \partial x_1} & \cdots & \frac{\partial^2 V}{\partial x_1 \partial x_d} \\ \vdots & & \vdots \\ \frac{\partial^2 V}{\partial x_d \partial x_1} & \cdots & \frac{\partial^2 V}{\partial x_d \partial x_d} \end{pmatrix}.$$

Clearly, when $V \in C^{2,1}(R \times R_+; R)$, we have $V_x = \frac{\partial V}{\partial x}$ and $V_{xx} = \frac{\partial^2 V}{\partial x^2}$.

We are now ready to state the first main result in this section.

Theorem 6.2 (The one-dimensional Itô formula) *Let $x(t)$ be an Itô process on $t \geq 0$ with the stochastic differential*

$$dx(t) = f(t)dt + g(t)dB_t,$$

where $f \in \mathcal{L}^1(R_+; R)$ and $g \in \mathcal{L}^2(R_+; R)$. Let $V \in C^{2,1}(R \times R_+; R)$. Then $V(x(t), t)$ is again an Itô process with the stochastic differential given by

$$dV(x(t),t) = \left[V_t(x(t),t) + V_x(x(t),t)f(t) + \frac{1}{2}V_{xx}(x(t),t)g^2(t)\right]dt$$
$$+ V_x(x(t),t)g(t)dB_t \quad \text{a.s.} \qquad (6.3)$$

Proof. The proof is rather technical and we shall divide the whole proof into several steps.

Step 1. We may assume that $x(t)$ is bounded, say by a constant K so the values of $V(x,t)$ for $x \notin [-K, K]$ are irrelevant. Otherwise, for each $n \geq 1$, define the stopping time

$$\tau_n = \inf\{t \geq 0 : |x(t)| \geq n\}.$$

Clearly, $\tau_n \uparrow \infty$ a.s. Also define the stochastic process

$$x_n(t) = [-n \vee x(0)] \wedge n + \int_0^t f(s)I_{[[0,\tau_n]]}(s)ds + \int_0^t g(s)I_{[[0,\tau_n]]}(s)dB_s$$

on $t \geq 0$. Then $|x_n(t)| \leq n$, that is $x_n(t)$ is bounded. Moreover, for every $t \geq 0$ and almost every $\omega \in \Omega$, there exists an integer $n_o = n_o(t, \omega)$ such that

$$x_n(s, \omega) = x(s, \omega) \quad \text{on } 0 \leq s \leq t$$

provided $n \geq n_o$. Therefore, if we can establish (6.3) for $x_n(t)$, that is

$$V(x_n(t), t) - V(x(0), 0)$$
$$= \int_0^t \left[V_t(x_n(s), s) + V_x(x_n(s), s)f(s)I_{[[0,\tau_n]]}(s)\right.$$
$$\left. + \frac{1}{2}V_{xx}(x_n(s), s)g^2(s)I_{[[0,\tau_n]]}(s)\right]ds$$
$$+ \int_0^t V_x(x_n(s), s)g(s)I_{[[0,\tau_n]]}(s)dB_s,$$

then we obtain the desired result upon letting $n \to \infty$.

Step 2. We may assume that $V(x,t)$ is C^2, i.e. it is continuously twice differentiable in both variables (x,t), otherwise we can find a sequence $\{V_n(x,t)\}$ of C^2-functions such that

$$V_n(x,t) \to V(x,t), \qquad \frac{\partial}{\partial t}V_n(x,t) \to V_t(x,t),$$

$$\frac{\partial}{\partial x}V_n(x,t) \to V_x(x,t), \qquad \frac{\partial^2}{\partial x^2}V_n(x,t) \to V_{xx}(x,t)$$

uniformly on every compact subset of $R \times R_+$ (see e.g. Friedman (1975)). If we can show the Itô formula for every V_n, that is

$$V_n(x(t),t) - V_n(x(0),0)$$
$$= \int_0^t \left[\frac{\partial}{\partial t}V_n(x(s),s) + \frac{\partial}{\partial x}V_n(x(s),s)f(s) + \frac{1}{2}\frac{\partial^2}{\partial x^2}V_n(x(s),s)g^2(s)\right]ds$$
$$+ \int_0^t \frac{\partial}{\partial x}V_n(x(s),s)g(s)dB_s,$$

then, letting $n \to \infty$, we obtain the desired result (6.3). By steps 1 and 2, we may assume without loss of generality that $V, V_t, V_{tt}, V_x, V_{tx}$ and V_{xx} are all bounded on $R \times [0,t]$ for every $t \geq 0$.

Step 3. If we can show (6.3) in the case that both f and g are simple processes, then the general case follows by approximation procedure. (The processes in $\mathcal{L}^1(R_+;R)$ can be approximated by simple processes in a similar way as shown in the proof of Lemma 5.6 but we leave the details to the reader.)

Step 4. We now fix $t > 0$ arbitrarily, and assume that $V, V_t, V_{tt}, V_x, V_{tx}, V_{xx}$ are bounded on $R \times [0,t]$, and $f(s), g(s)$ are simple processes on $s \in [0,t]$. Let $\Pi = \{t_0, t_1, \cdots, t_k\}$ be a partition of $[0,t]$ (i.e. $0 = t_0 < t_1 < \cdots < t_k = t$) sufficiently fine that $f(s)$ and $g(s)$ are "random constant" on every $(t_i, t_{i+1}]$ in the sense that

$$f(s) = f_i, \quad g(s) = g_i \quad \text{if } s \in (t_i, t_{i+1}].$$

Using the well-known Taylor expansion formula we get

$$V(x(t),t) - V(x(0),0) = \sum_{i=0}^{k-1}\left[V(x(t_{i+1}),t_{i+1}) - V(x(t_i),t_i)\right]$$

$$= \sum_{i=0}^{k-1} V_t(x(t_i),t_i)\Delta t_i + \sum_{i=0}^{k-1} V_x(x(t_i),t_i)\Delta x_i + \frac{1}{2}\sum_{i=0}^{k-1} V_{tt}(x(t_i),t_i)(\Delta t_i)^2$$

$$+ \sum_{i=0}^{k-1} V_{tx}(x(t_i),t_i)\Delta t_i \Delta x_i + \frac{1}{2}\sum_{i=0}^{k-1} V_{xx}(x(t_i),t_i)(\Delta x_i)^2 + \sum_{i=0}^{k-1} R_i, \qquad (6.4)$$

where

$$\Delta t_i = t_{i+1} - t_i, \quad \Delta x_i = x(t_{i+1}) - x(t_i), \quad R_i = o((\Delta t_i)^2 + (\Delta x_i)^2).$$

Set $|\Pi| = \max_{0 \leq i \leq k-1} \Delta t_i$. It is easy to see that when $|\Pi| \to 0$, with probability 1,

$$\sum_{i=0}^{k-1} V_t(x(t_i), t_i) \Delta t_i \to \int_0^t V_t(x(s), s) ds, \tag{6.5}$$

$$\sum_{i=0}^{k-1} V_x(x(t_i), t_i) \Delta x_i \to \int_0^t V_x(x(s), s) dx(s)$$

$$= \int_0^t V_x(x(s), s) f(s) ds + \int_0^t V_x(x(s), s) g(s) dB_s, \tag{6.6}$$

$$\sum_{i=0}^{k-1} V_{tt}(x(t_i), t_i)(\Delta t_i)^2 \to 0, \quad \text{and} \quad \sum_{i=0}^{k-1} R_i \to 0. \tag{6.7}$$

Note that

$$\sum_{i=0}^{k-1} V_{tx}(x(t_i), t_i) \Delta t_i \Delta x_i$$

$$= \sum_{i=0}^{k-1} V_{tx}(x(t_i), t_i) f_i (\Delta t_i)^2 + \sum_{i=0}^{k-1} V_{tx}(x(t_i), t_i) g_i \Delta t_i \Delta B_i,$$

where $\Delta B_i = B_{t_{i+1}} - B_{t_i}$. When $|\Pi| \to 0$, the first term tends to 0 a.s. while the second term tends to 0 in L^2 since

$$E\Big(\sum_{i=0}^{k-1} V_{tx}(x(t_i), t_i) g_i \Delta t_i \Delta B_i\Big)^2 = \sum_{i=0}^{k-1} E[V_{tx}(x(t_i), t_i) g_i]^2 (\Delta t_i)^3 \to 0.$$

In other words, we have shown (due to the assumption of boundedness) that

$$\sum_{i=0}^{k-1} V_{tx}(x(t_i), t_i) \Delta t_i \Delta x_i \to 0 \quad \text{in } L^2. \tag{6.8}$$

Note also that

$$\sum_{i=0}^{k-1} V_{xx}(x(t_i), t_i)(\Delta x_i)^2$$

$$= \sum_{i=0}^{k-1} V_{xx}(x(t_i), t_i) \big[f_i^2 (\Delta t_i)^2 + 2 f_i g_i \Delta t_i \Delta B_i \big] + \sum_{i=0}^{k-1} V_{xx}(x(t_i), t_i) g_i^2 (\Delta B_i)^2.$$

The first term tends to 0 in L^2 as $|\Pi| \to 0$ in the same reason as before, while we claim the second term tends to $\int_0^t V_{xx}(x(s), s) g^2(s) ds$ in L^2. To show the latter, we set $h(t) = V_{xx}(x(t),t)g^2(t)$, $h_i = V_{xx}(x(t_i),t_i)g_i^2$, and compute

$$E\left(\sum_{i=0}^{k-1} h_i(\Delta B_i)^2 - \sum_{i=0}^{k-1} h_i \Delta t_i\right)^2$$

$$= E\left(\sum_{i=0}^{k-1}\sum_{j=0}^{k-1} h_i h_j [(\Delta B_i)^2 - \Delta t_i][(\Delta B_j)^2 - \Delta t_j]\right)$$

$$= \sum_{i=0}^{k-1} E\left(h_i^2 [(\Delta B_i)^2 - \Delta t_i]^2\right)$$

$$= \sum_{i=0}^{k-1} E h_i^2 \, E\left[(\Delta B_i)^4 - 2(\Delta B_i)^2 \Delta t_i + (\Delta t_i)^2\right]$$

$$= \sum_{i=0}^{k-1} E h_i^2 \left[3(\Delta t_i)^2 - 2(\Delta t_i)^2 + (\Delta t_i)^2\right]$$

$$= 2 \sum_{i=0}^{k-1} E h_i^2 (\Delta t_i)^2 \to 0,$$

where we have used the known fact that $E(\Delta B_i)^{2n} = (2n)!(\Delta t_i)^n/(2^n n!)$. Thus

$$\sum_{i=0}^{k-1} h_i(\Delta B_i)^2 \to \int_0^t h(s)ds \quad \text{in } L^2.$$

In other words, we have already shown that

$$\sum_{i=0}^{k-1} V_{xx}(x(t_i),t_i)(\Delta x_i)^2 \to \int_0^t V_{xx}(x(s),s)g^2(s)ds \quad \text{in } L^2. \quad (6.9)$$

Substituting (6.5)–(6.9) into (6.4) we obtain that

$$V(x(t),t) - V(x(0),0)$$
$$= \int_0^t \left[V_t(x(s),s) + V_x(x(s),s)f(s) + \frac{1}{2}V_{xx}(x(s),s)g^2(s)\right]ds$$
$$+ \int_0^t V_x(x(s),s)g(s)dB_s \quad a.s.$$

which is the required (6.3). The proof is now complete.

We shall now extend the 1-dimensional Itô formula to the multi-dimensional case. Let $B(t) = (B_1(t), \cdots, B_m(t))^T$, $t \geq 0$ be an m-dimensional Brownian motion defined on the complete probability space $(\Omega, \mathcal{F}, P)$ adapted to the filtration $\{\mathcal{F}_t\}_{t \geq 0}$.

Definition 6.3 A d-dimensional Itô process is an R^d-valued continuous adapted process $x(t) = (x_1(t), \cdots, x_d(t))^T$ on $t \geq 0$ of the form

$$x(t) = x(0) + \int_0^t f(s)ds + \int_0^t g(s)dB(s),$$

where $f = (f_1, \cdots, f_d)^T \in \mathcal{L}^1(R_+; R^d)$ and $g = (g_{ij})_{d \times m} \in \mathcal{L}^2(R_+; R^{d \times m})$. We shall say that $x(t)$ has stochastic differential $dx(t)$ on $t \geq 0$ given by

$$dx(t) = f(t)dt + g(t)dB(t).$$

Theorem 6.4 (The multi-dimensional Itô formula) Let $x(t)$ be a d-dimensional Itô process on $t \geq 0$ with the stochastic differential

$$dx(t) = f(t)dt + g(t)dB(t),$$

where $f \in \mathcal{L}^1(R_+; R^d)$ and $g \in \mathcal{L}^2(R_+; R^{d \times m})$. Let $V \in C^{2,1}(R^d \times R_+; R)$. Then $V(x(t), t)$ is again an Itô process with the stochastic differential given by

$$dV(x(t), t) = \Big[V_t(x(t), t) + V_x(x(t), t)f(t)$$
$$+ \frac{1}{2}\text{trace}\big(g^T(t)V_{xx}(x(t), t)g(t)\big)\Big]dt + V_x(x(t), t)g(t)dB(t) \quad \text{a.s.} \quad (6.10)$$

The proof is similar to the one-dimensional case so is omitted. Let us now introduce formally a multiplication table:

$$dtdt = 0, \quad dB_i dt = 0,$$
$$dB_i dB_i = dt, \quad dB_i dB_j = 0 \quad \text{if } i \neq j.$$

Then, for example,

$$dx_i(t)dx_j(t) = \sum_{k=1}^m g_{ik}(t)g_{jk}(t)dt. \tag{6.11}$$

Moreover, the Itô formula can be written as

$$dV(x(t), t) = V_t(x(t), t)dt + V_x(x(t), t)dx(t)$$
$$+ \frac{1}{2}dx^T(t)V_{xx}(x(t), t)dx(t). \tag{6.12}$$

Note that if $x(t)$ were continuously differentiable in t, then (by the classical calculus formula for total derivatives) the term $\frac{1}{2}dx^T(t)V_{xx}(x(t), t)dx(t)$ would not appear. For example, let $V(x, t) = x_1 x_2$, then (6.11) and (6.12) yield

$$d[x_1(t)x_2(t)] = x_1(t)dx_2(t) + x_2(t)dx_1(t) + dx_1 dx_2 \tag{6.13}$$
$$= x_1(t)dx_2(t) + x_2(t)dx_1(t) + \sum_{k=1}^m g_{1k}(t)g_{2k}(t)dt,$$

which is different from the classical formula of integration by parts $d(uv) = vdu + udv$ if both u, v are differentiable. However we do have the stochastic version of integration by parts formula which is similar to the classical one.

Theorem 6.5 (Integration by parts formula) Let $x(t)$, $t \geq 0$ be a 1-dimensional Itô process with the stochastic differential

$$dx(t) = f(t)dt + g(t)dB(t),$$

where $f \in \mathcal{L}^1(R_+; R)$ and $g \in \mathcal{L}^2(R_+; R^{1\times m})$. Let $y(t)$, $t \geq 0$ be a real-valued continuous adapted process of finite variation. Then

$$d[x(t)y(t)] = y(t)dx(t) + x(t)dy(t), \tag{6.14}$$

that is

$$x(t)y(t) - x(0)y(0) = \int_0^t y(s)[f(s)ds + g(s)dB(s)] + \int_0^t x(s)dy(s), \tag{6.15}$$

where the last integral is the Lebesgue-Stieltjes integral.

Let us now give a number of examples to illustrate the use of Itô's formula in evaluating the stochastic integrals.

Example 6.6 Let $B(t)$ be a 1-dimensional Brownian motion. To compute the stochastic integral

$$\int_0^t B(s)dB(s),$$

we apply the Itô formula to $B^2(t)$ (i.e. let $V(x,t) = x^2$ and $x(t) = B(t)$), and get

$$d(B^2(t)) = 2B(t)dB(t) + dt.$$

That is

$$B^2(t) = 2\int_0^t B(s)dB(s) + t,$$

which implies that

$$\int_0^t B(s)dB(s) = \frac{1}{2}[B^2(t) - t].$$

Example 6.7 Let $B(t)$ be a 1-dimensional Brownian motion. To compute the stochastic integral

$$\int_0^t e^{-s/2+B(s)}dB(s),$$

we let $V(x,t) = e^{-t/2+x}$ and $x(t) = B(t)$, and then, by the Itô formula, we obtain

$$d\left[e^{-t/2+B(t)}\right] = -\frac{1}{2}e^{-t/2+B(t)}dt + e^{-t/2+B(t)}dB(t) + \frac{1}{2}e^{-t/2+B(t)}dt$$
$$= e^{-t/2+B(t)}dB(t).$$

That yields

$$\int_0^t e^{-s/2+B(s)} dB(s) = e^{-t/2+B(t)} - 1.$$

Example 6.8 Let $B(t)$ be a 1-dimensional Brownian motion. What is the integration of the Brownian sample path over the time interval $[0,t]$, i.e. $\int_0^t B(s)ds$? The integration by parts formula yields

$$d[tB(t)] = B(t)dt + tdB(t).$$

Therefore

$$\int_0^t B(s)ds = tB(t) - \int_0^t s dB(s).$$

On the other hand, we may apply Itô's formula to $B^3(t)$ to obtain

$$dB^3(t) = 3B^2(t)dB(t) + 3B(t)dt,$$

which gives the alternative

$$\int_0^t B(s)ds = \frac{1}{3}B^3(t) - \int_0^t B^2(s)dB(s).$$

Example 6.9 Let $B(t)$ be an m-dimensional Brownian motion. Let $V : R^m \to R$ be C^2. Then Itô's formula implies

$$V(B(t)) = V(0) + \frac{1}{2}\int_0^t \Delta V(B(s))ds + \int_0^t V_x(B(s))dB(s),$$

where $\Delta = \sum_{i=1}^m \frac{\partial^2}{\partial x_i^2}$ is the Laplace operator. In particular, let V be a quadratic function, i.e. $V(x) = x^T Q x$, where Q is an $m \times m$ matrix. Then

$$B^T(t)QB(t) = trace(Q)t + \int_0^t B^T(s)(Q+Q^T)dB(s).$$

Example 6.10 Let $x(t)$ be a d-dimensional Itô process as given by Definition 6.3. Let Q be a $d \times d$ matrix. Then

$$x^T(t)Qx(t) - x^T(0)Qx(0)$$
$$= \int_0^t \left(x^T(s)(Q+Q^T)f(s) + \frac{1}{2}trace[g^T(s)(Q+Q^T)g(s)]\right)ds$$
$$+ \int_0^t x^T(s)(Q+Q^T)g(s)dB(s).$$

1.7 MOMENT INEQUALITIES

In this section we shall apply Itô's formula to establish several very important moment inequalities for stochastic integrals as well as the exponential martingale inequality. These will demonstrate the powerfulness of the Itô formula.

Throughout this section, we let $B(t) = (B_1(t), \cdots, B_m(t))^T$, $t \geq 0$ be an m-dimensional Brownian motion defined on the complete probability space $(\Omega, \mathcal{F}, P)$ adapted to the filtration $\{\mathcal{F}_t\}_{t \geq 0}$.

Theorem 7.1 Let $p \geq 2$. Let $g \in \mathcal{M}^2([0,T]; R^{d \times m})$ such that
$$E \int_0^T |g(s)|^p ds < \infty.$$
Then
$$E \left| \int_0^T g(s) dB(s) \right|^p \leq \left(\frac{p(p-1)}{2} \right)^{\frac{p}{2}} T^{\frac{p-2}{2}} E \int_0^T |g(s)|^p ds. \tag{7.1}$$
In particular, for $p = 2$, there is equality.

Proof. For $p = 2$ the required result follows from Theorem 5.21 so we only need to show the theorem for the case of $p > 2$. For $0 \leq t \leq T$, set
$$x(t) = \int_0^t g(s) dB(s).$$
By Itô's formula and Theorem 5.21,
$$E|x(t)|^p$$
$$= \frac{p}{2} E \int_0^t \left(|x(s)|^{p-2} |g(s)|^2 + (p-2)|x(s)|^{p-4} |x^T(s) g(s)|^2 \right) ds \tag{7.2}$$
$$\leq \frac{p(p-1)}{2} E \int_0^t |x(s)|^{p-2} |g(s)|^2 ds. \tag{7.3}$$
Using Hölder's inequality one then sees that
$$E|x(t)|^p \leq \frac{p(p-1)}{2} \left(E \int_0^t |x(s)|^p ds \right)^{\frac{p-2}{p}} \left(E \int_0^t |g(s)|^p ds \right)^{\frac{2}{p}}$$
$$= \frac{p(p-1)}{2} \left(\int_0^t E|x(s)|^p ds \right)^{\frac{p-2}{p}} \left(E \int_0^t |g(s)|^p ds \right)^{\frac{2}{p}}.$$
Note from (7.2) that $E|x(t)|^p$ is nondecreasing in t. It then follows
$$E|x(t)|^p \leq \frac{p(p-1)}{2} \left[tE|x(t)|^p \right]^{\frac{p-2}{p}} \left(E \int_0^t |g(s)|^p ds \right)^{\frac{2}{p}}.$$

This yields
$$E|x(t)|^p \le \left(\frac{p(p-1)}{2}\right)^{\frac{p}{2}} t^{\frac{p-2}{2}} E\int_0^t |g(s)|^p ds,$$
and the required (7.1) follows by replacing t with T.

Theorem 7.2 *Under the same assumptions as Theorem 7.1,*
$$E\left(\sup_{0\le t\le T}\left|\int_0^t g(s)dB(s)\right|^p\right) \le \left(\frac{p^3}{2(p-1)}\right)^{p/2} T^{\frac{p-2}{2}} E\int_0^T |g(s)|^p ds. \quad (7.4)$$

Proof. Recall that the stochastic integral $\int_0^t g(s)dB(s)$ is an R^d-valued continuous martingale. Hence, by the Doob martingale inequality (i.e. Theorem 3.8), we have
$$E\left(\sup_{0\le t\le T}\left|\int_0^t g(s)dB(s)\right|^p\right) \le \left(\frac{p}{p-1}\right)^p E\left|\int_0^T g(s)dB(s)\right|^p.$$

In view of Theorem 7.1, we then obtain the desired (7.4).

The following theorem is known as the Burkholder–Davis–Gundy inequality.

Theorem 7.3 *Let* $g \in \mathcal{L}^2(R_+; R^{d\times m})$. *Define, for* $t \ge 0$,
$$x(t) = \int_0^t g(s)dB(s) \quad \text{and} \quad A(t) = \int_0^t |g(s)|^2 ds.$$

Then for every $p > 0$, *there exist universal positive constants* c_p, C_p *(depending only on* p), *such that*
$$c_p E|A(t)|^{\frac{p}{2}} \le E\left(\sup_{0\le s\le t} |x(s)|^p\right) \le C_p E|A(t)|^{\frac{p}{2}} \quad (7.5)$$

for all $t \ge 0$. *In particular, one may take*

$$\begin{array}{lll} c_p = (p/2)^p, & C_p = (32/p)^{p/2} & \text{if } 0 < p < 2; \\ c_p = 1, & C_p = 4 & \text{if } p = 2; \\ c_p = (2p)^{-p/2}, & C_p = [p^{p+1}/2(p-1)^{p-1}]^{p/2} & \text{if } p > 2. \end{array}$$

Proof. We may assume without loss of generality that both $x(t)$ and $A(t)$ are bounded. Otherwise, for each integer $n \ge 1$, define the stopping time
$$\tau_n = \inf\{t \ge 0 : |x(t)| \vee A(t) \ge n\}.$$

If we can show (7.5) for the stopped processes $x(t \wedge \tau_n)$ and $A(t \wedge \tau_n)$, then the general case follows upon letting $n \to \infty$. Besides, for convenience, we set $x^*(t) = \sup_{0\le s\le t} |x(s)|$.

Moment Inequalities

Case 1: $p = 2$. The required (7.5) follows from Theorem 5.21 and the Doob martingale inequality immediately.

Case 2: $p > 2$. It follows from (7.3) that

$$E|x(t)|^p \leq \frac{p(p-1)}{2} E\left[|x^*(t)|^{p-2} A(t)\right]$$

$$\leq \frac{p(p-1)}{2} \left[E|x^*(t)|^p\right]^{\frac{p-2}{p}} \left[E|A(t)|^{\frac{p}{2}}\right]^{\frac{2}{p}}, \qquad (7.6)$$

where the Hölder inequality has been used. But, by the Doob martingale inequality,

$$E|x^*(t)|^p \leq \left(\frac{p}{p-1}\right)^p E|x(t)|^p.$$

Substituting this into (7.6) yields

$$E|x^*(t)|^p \leq \left(\frac{p^{(p+1)}}{2(p-1)^{p-1}}\right)^{\frac{p}{2}} E|A(t)|^{\frac{p}{2}}$$

which is the right-hand-side inequality of (7.5). To show the left-hand-side one, we set

$$y(t) = \int_0^t |A(s)|^{\frac{p-2}{4}} g(s) dB(s).$$

Then

$$E|y(t)|^2 = E \int_0^t |A(s)|^{\frac{p-2}{2}} |g(s)|^2 ds$$

$$= E \int_0^t |A(s)|^{\frac{p-2}{2}} dA(s) = \frac{2}{p} E|A(t)|^{\frac{p}{2}}. \qquad (7.7)$$

On the other hand, the integration by parts formula yields

$$x(t)|A(t)|^{\frac{p-2}{4}} = \int_0^t |A(s)|^{\frac{p-2}{4}} dx(s) + \int_0^t x(s) d\left(|A(s)|^{\frac{p-2}{4}}\right)$$

$$= y(t) + \int_0^t x(s) d\left(|A(s)|^{\frac{p-2}{4}}\right).$$

Thus

$$|y(t)| \leq |x(t)||A(t)|^{\frac{p-2}{4}} + \int_0^t |x(s)| d\left(|A(s)|^{\frac{p-2}{4}}\right) \leq 2x^*(t)|A(t)|^{\frac{p-2}{4}}.$$

Substituting this into (7.7) one sees that

$$\frac{2}{p} E|A(t)|^{\frac{p}{2}} \leq 4E\left[|x^*(t)|^2 |A(t)|^{\frac{p-2}{2}}\right] \leq 4\left[E|x^*(t)|^p\right]^{\frac{2}{p}} \left[E|A(t)|^{\frac{p}{2}}\right]^{\frac{p-2}{p}}.$$

This implies
$$\frac{1}{(2p)^{p/2}} E|A(t)|^{\frac{p}{2}} \le E|x^*(t)|^p$$
as desired.

Case 3: $0 < p < 2$. Fix $\varepsilon > 0$ arbitrarily and define
$$\eta(t) = \int_0^t [\varepsilon + A(s)]^{\frac{p-2}{4}} g(s) dB(s) \quad \text{and} \quad \eta^*(t) = \sup_{0 \le s \le t} |\eta(s)|.$$

Then
$$E|\eta(t)|^2 = E \int_0^t [\varepsilon + A(s)]^{\frac{p-2}{2}} dA(s) \le \frac{2}{p} E[\varepsilon + A(t)]^{\frac{p}{2}}. \tag{7.8}$$

On the other hand, the integration by parts formula gives
$$\eta(t)[\varepsilon + A(t)]^{\frac{2-p}{4}} = \int_0^t g(s) dB(s) + \int_0^t \eta(s) d\left([\varepsilon + A(s)]^{\frac{2-p}{4}}\right)$$
$$= x(t) + \int_0^t \eta(s) d\left([\varepsilon + A(s)]^{\frac{2-p}{4}}\right).$$

Thus
$$|x(t)| \le |\eta(t)|[\varepsilon + A(t)]^{\frac{2-p}{4}} + \int_0^t |\eta(s)| d\left([\varepsilon + A(s)]^{\frac{2-p}{4}}\right)$$
$$\le 2\eta^*(t)[\varepsilon + A(t)]^{\frac{2-p}{4}}.$$

Since this holds for all $t \ge 0$ and the right-hand side is nondecreasing, we must have
$$E|x^*(t)|^p \le 2^p E\left[|\eta^*(t)|^p [\varepsilon + A(t)]^{\frac{p(2-p)}{4}}\right]$$
$$\le 2^p \left[E|\eta^*(t)|^2\right]^{\frac{p}{2}} \left[E[\varepsilon + A(t)]^{\frac{p}{2}}\right]^{\frac{2-p}{2}}. \tag{7.9}$$

But, by Doob's martingale inequality and (7.8),
$$E|\eta^*(t)|^2 \le 4E|\eta(t)|^2 \le \frac{8}{p} E[\varepsilon + A(t)]^{\frac{p}{2}}.$$

Substituting this into (7.9) one sees that
$$E|x^*(t)|^p \le \left(\frac{32}{p}\right)^{\frac{p}{2}} E[\varepsilon + A(t)]^{\frac{p}{2}}.$$

Letting $\varepsilon \to 0$ we obtain the right-hand-side inequality of (7.5). To show the left-hand-side one, we write, for any fixed $\varepsilon > 0$,
$$|A(t)|^{\frac{p}{2}} = \left(|A(t)|^{\frac{p}{2}} [\varepsilon + x^*(t)]^{\frac{-p(2-p)}{2}}\right) [\varepsilon + x^*(t)]^{\frac{p(2-p)}{2}}.$$

Then, applying Hölder's inequality, one sees that

$$E|A(t)|^{\frac{p}{2}} \leq \left[E\big(A(t)[\varepsilon + x^*(t)]^{p-2}\big)\right]^{\frac{p}{2}} \left(E[\varepsilon + x^*(t)]^p\right)^{\frac{2-p}{2}}. \qquad (7.10)$$

Define

$$\xi(t) = \int_0^t [\varepsilon + x^*(s)]^{\frac{p-2}{2}} g(s) dB(s).$$

Then

$$E|\xi(t)|^2 = E \int_0^t [\varepsilon + x^*(s)]^{p-2} dA(s) \geq E\Big([\varepsilon + x^*(t)]^{p-2} A(t)\Big). \qquad (7.11)$$

On the other hand, the integration by parts formula gives

$$x(t)[\varepsilon + x^*(t)]^{\frac{p-2}{2}} = \xi(t) + \int_0^t x(s) d\Big([\varepsilon + x^*(s)]^{\frac{p-2}{2}}\Big)$$

$$= \xi(t) + \frac{p-2}{2} \int_0^t x(s)[\varepsilon + x^*(s)]^{\frac{p-4}{2}} d[\varepsilon + x^*(s)].$$

Thus

$$|\xi(t)| \leq x^*(t)[\varepsilon + x^*(t)]^{\frac{p-2}{2}} + \frac{2-p}{2} \int_0^t x^*(s)[\varepsilon + x^*(s)]^{\frac{p-4}{2}} d[\varepsilon + x^*(s)]$$

$$\leq [\varepsilon + x^*(t)]^{\frac{p}{2}} + \frac{2-p}{2} \int_0^t [\varepsilon + x^*(s)]^{\frac{p-2}{2}} d[\varepsilon + x^*(s)]$$

$$\leq \frac{2}{p}[\varepsilon + x^*(t)]^{\frac{p}{2}}.$$

This, together with (7.11), implies

$$E\Big([\varepsilon + x^*(t)]^{p-2} A(t)\Big) \leq \left(\frac{2}{p}\right)^2 E[\varepsilon + x^*(t)]^p.$$

Substituting this into (7.10) we get that

$$E|A(t)|^{\frac{p}{2}} \leq \left(\frac{2}{p}\right)^p E[\varepsilon + x^*(t)]^p.$$

Finally, letting $\varepsilon \to 0$ we have

$$\left(\frac{p}{2}\right)^p E|A(t)|^{\frac{p}{2}} \leq E|x^*(t)|^p$$

as required. The proof is now complete.

The following theorem is known as the exponential martingale inequality which will play an important role in this book.

Theorem 7.4 Let $g = (g_1, \cdots, g_m) \in \mathcal{L}^2(R_+; R^{1 \times m})$, and let T, α, β be any positive numbers. Then

$$P\left\{\sup_{0 \le t \le T}\left[\int_0^t g(s)dB(s) - \frac{\alpha}{2}\int_0^t |g(s)|^2 ds\right] > \beta\right\} \le e^{-\alpha\beta}. \tag{7.12}$$

Proof. For every integer $n \ge 1$, define the stopping time

$$\tau_n = \inf\left\{t \ge 0 : \left|\int_0^t g(s)dB(s)\right| + \int_0^t |g(s)|^2 ds \ge n\right\},$$

and the Itô process

$$x_n(t) = \alpha \int_0^t g(s)I_{[[0,\tau_n]]}(s)dB(s) - \frac{\alpha^2}{2}\int_0^t |g(s)|^2 I_{[[0,\tau_n]]}(s)ds.$$

Clearly, $x_n(t)$ is bounded and $\tau_n \uparrow \infty$ a.s. Applying the Itô's formula to $\exp[x_n(t)]$ we obtain that

$$\exp[x_n(t)] = 1 + \int_0^t \exp[x_n(s)]dx_n(s) + \frac{\alpha^2}{2}\int_0^t \exp[x_n(s)]|g(s)|^2 I_{[[0,\tau_n]]}(s)ds$$

$$= 1 + \alpha \int_0^t \exp[x_n(s)]g(s)I_{[[0,\tau_n]]}(s)dB(s).$$

In view of Theorem 5.21, one sees that $\exp[x_n(t)]$ is a nonnegative martingale on $t \ge 0$ with $E(\exp[x_n(t)]) = 1$. Hence, by Theorem 3.8, we get that

$$P\left\{\sup_{0 \le t \le T} \exp[x_n(t)] \ge e^{\alpha\beta}\right\} \le e^{-\alpha\beta} E(\exp[x_n(T)]) = e^{-\alpha\beta}.$$

That is,

$$P\left\{\sup_{0 \le t \le T}\left[\int_0^t g(s)I_{[[0,\tau_n]]}(s)dB(s) - \frac{\alpha}{2}\int_0^t |g(s)|^2 I_{[[0,\tau_n]]}(s)ds\right] > \beta\right\} \le e^{-\alpha\beta}.$$

Now the required (7.12) follows by letting $n \to \infty$ and the proof is therefore complete.

1.8 GRONWALL-TYPE INEQUALITIES

The integral inequalities of Gronwall type have been widely applied in the theory of ordinary differential equations and stochastic differential equations to prove the results on existence, uniqueness, boundedness, comparison, continuous

Gronwall-Type Inequalities

dependence, perturbation and stability etc. Naturally, Gronwall-type inequalities will play an important role in this book. For the convenience of the reader, we establish a number of well-known inequalities of this type in this section.

Theorem 8.1 (Gronwall's inequality) *Let $T > 0$ and $c \geq 0$. Let $u(\cdot)$ be a Borel measurable bounded nonnegative function on $[0, T]$, and let $v(\cdot)$ be a nonnegative integrable function on $[0, T]$. If*

$$u(t) \leq c + \int_0^t v(s)u(s)ds \quad \text{for all } 0 \leq t \leq T, \tag{8.1}$$

then

$$u(t) \leq c \exp\left(\int_0^t v(s)ds\right) \quad \text{for all } 0 \leq t \leq T. \tag{8.2}$$

Proof. Without loss of generality we may assume that $c > 0$. Set

$$z(t) = c + \int_0^t v(s)u(s)ds \quad \text{for } 0 \leq t \leq T.$$

Then $u(t) \leq z(t)$. Moreover, by the chain rule of classical calculus, we have

$$\log(z(t)) = \log(c) + \int_0^t \frac{v(s)u(s)}{z(s)}ds \leq \log(c) + \int_0^t v(s)ds.$$

This implies

$$z(t) \leq c \exp\left(\int_0^t v(s)ds\right) \quad \text{for } 0 \leq t \leq T,$$

and the required inequality (8.2) follows since $u(t) \leq z(t)$.

Theorem 8.2 (Bihari's inequality) *Let $T > 0$ and $c > 0$. Let $K : R_+ \to R_+$ be a continuous nondecreasing function such that $K(t) > 0$ for all $t > 0$. Let $u(\cdot)$ be a Borel measurable bounded nonnegative function on $[0, T]$, and let $v(\cdot)$ be a nonnegative integrable function on $[0, T]$. If*

$$u(t) \leq c + \int_0^t v(s)K(u(s))ds \quad \text{for all } 0 \leq t \leq T, \tag{8.3}$$

then

$$u(t) \leq G^{-1}\left(G(c) + \int_0^t v(s)ds\right) \tag{8.4}$$

holds for all such $t \in [0, T]$ that

$$G(c) + \int_0^t v(s)ds \in Dom(G^{-1}), \tag{8.5}$$

where
$$G(r) = \int_1^r \frac{ds}{K(s)} \quad \text{on } r > 0,$$
and G^{-1} is the inverse function of G.

Proof. Set
$$z(t) = c + \int_0^t v(s)K(u(s))ds \quad \text{for } 0 \le t \le T.$$
Then $u(t) \le z(t)$. By the chain rule of classical calculus, one can derive that
$$G(z(t)) = G(c) + \int_0^t \frac{v(s)K(u(s))}{K(z(s))}ds \le G(c) + \int_0^t v(s)ds \quad (8.6)$$
for all $t \in [0,T]$. Hence, for $t \in [0,T]$ satisfying (8.5) one sees from (8.6) that
$$z(t) \le G^{-1}\left(G(c) + \int_0^t v(s)ds\right),$$
and the desired inequality (8.4) follows since $u(t) \le z(t)$.

Theorem 8.3 *Let $T > 0$, $\alpha \in [0,1)$ and $c > 0$. Let $u(\cdot)$ be a Borel measurable bounded nonnegative function on $[0,T]$, and let $v(\cdot)$ be a nonnegative integrable function on $[0,T]$. If*
$$u(t) \le c + \int_0^t v(s)[u(s)]^\alpha ds \quad \text{for all } 0 \le t \le T, \quad (8.7)$$
then
$$u(t) \le \left(c^{1-\alpha} + (1-\alpha)\int_0^t v(s)ds\right)^{\frac{1}{1-\alpha}} \quad (8.8)$$
holds for all $t \in [0,T]$.

Proof. Without loss of generality, we may assume $c > 0$. Set
$$z(t) = c + \int_0^t v(s)[u(s)]^\alpha ds \quad \text{for } 0 \le t \le T.$$
Then $u(t) \le z(t)$ and $z(t) > 0$. By the fundamental differential formula, one can show that
$$[z(t)]^{1-\alpha} = c^{1-\alpha} + (1-\alpha)\int_0^t \frac{v(s)[u(s)]^\alpha}{[z(s)]^\alpha}ds$$
$$\le c^{1-\alpha} + (1-\alpha)\int_0^t v(s)ds$$
for all $t \in [0,T]$, and the required inequality (8.8) follows immediately.

2

Stochastic Differential Equations

2.1 INTRODUCTION

In section 1.1 we introduced, as an example, the simple stochastic population growth model

$$N(t) = N_0 + \int_0^t r(s)N(s)ds + \int_0^t \sigma(s)N(s)dB(s),$$

or, in differential form,

$$dN(t) = r(t)N(t)dt + \sigma(t)N(t)dB(t) \quad \text{on } t \geq 0 \qquad (1.1)$$

with initial value $N(0) = N_0$. In this chapter we shall turn to find the solution to the equation. In general, we shall investigate the solution to a non-linear stochastic differential equation

$$dx(t) = f(x(t),t)dt + g(x(t),t)dB(t) \quad \text{on } t \in [t_0, T] \qquad (1.2)$$

with initial value $x(t_0) = x_0$, where $0 \leq t_0 < T < \infty$. The questions are:
- What is the solution?
- Are there the existence-and-uniqueness theorems for such solution?
- What are the properties of the solution?
- How can the solution be obtained in practice?

In this chapter we shall answer these questions one by one. Besides, as an important application of stochastic differential equations, we shall establish the well-known Feynman-Kac formula which gives the stochastic representation for the solution to a linear parabolic partial differential equation in terms of the solution to the corresponding stochastic differential equation.

2.2 STOCHASTIC DIFFERENTIAL EQUATIONS

Let $(\Omega, \mathcal{F}, P)$ be a complete probability space with a filtration $\{\mathcal{F}_t\}_{t \geq 0}$ satisfying the usual conditions. Throughout this chapter, unless otherwise specified, we let $B(t) = (B_1(t), \cdots, B_m(t))^T$, $t \geq 0$ be an m-dimensional Brownian motion defined on the space. Let $0 \leq t_0 < T < \infty$. Let x_0 be an $\mathcal{F}_{t_0}$-measurable R^d-valued random variable such that $E|x_0|^2 < \infty$. Let $f : R^d \times [t_0, T] \to R^d$ and $g : R^d \times [t_0, T] \to R^{d \times m}$ be both Borel measurable. Consider the d-dimensional stochastic differential equation of Itô type

$$dx(t) = f(x(t), t)dt + g(x(t), t)dB(t) \quad \text{on } t_0 \leq t \leq T \qquad (2.1)$$

with initial value $x(t_0) = x_0$. By the definition of stochastic differential, this equation is equivalent to the following stochastic integral equation

$$x(t) = x_0 + \int_{t_0}^{t} f(x(s), s)ds + \int_{t_0}^{t} g(x(s), s)dB(s) \quad \text{on } t_0 \leq t \leq T. \qquad (2.2)$$

Let us first give the definition of the solution.

Definition 2.1 *An R^d-valued stochastic process $\{x(t)\}_{t_0 \leq t \leq T}$ is called a solution of equation (2.1) if it has the following properties:*

(i) *$\{x(t)\}$ is continuous and $\mathcal{F}_t$-adapted;*

(ii) *$\{f(x(t), t)\} \in \mathcal{L}^1([t_0, T]; R^d)$ and $\{g(x(t), t)\} \in \mathcal{L}^2([t_0, T]; R^{d \times m})$;*

(iii) *equation (2.2) holds for every $t \in [t_0, T]$ with probability 1.*

A solution $\{x(t)\}$ is said to be unique if any other solution $\{\bar{x}(t)\}$ is indistinguishable from $\{x(t)\}$, that is

$$P\{x(t) = \bar{x}(t) \text{ for all } t_0 \leq t \leq T\} = 1.$$

Remark 2.2 (a) Denote the solution of equation (2.1) by $x(t; t_0, x_0)$. Note from equation (2.2) that for any $s \in [t_0, T]$,

$$x(t) = x(s) + \int_{s}^{t} f(x(r), r)dr + \int_{s}^{t} g(x(r), r)dB(r) \quad \text{on } s \leq t \leq T. \qquad (2.3)$$

But, this is a stochastic differential equation on $[s, T]$ with initial value $x(s) = x(s; t_0, x_0)$, whose solution is denoted by $x(t; s, x(s; t_0, x_0))$. Therefore, we see that the solution of equation (2.1) satisfies the following *flow* or *semigroup* property

$$x(t; t_0, x_0) = x(t; s, x(s; t_0, x_0)), \quad t_0 \leq s \leq t \leq T.$$

(b) The coefficients f and g can depend on ω in a general manner as long as they are adapted. For further details, please see Gihman & Skorohod (1972).

(c) In this book we require the initial value x_0 be L^2, but in general it is enough for x_0 to be a random variable as long as it is $\mathcal{F}_{t_0}$-measurable. For further details, please see Gihman & Skorohod (1972).

We shall now give some examples of stochastic differential equations.

Example 2.3 Let $B(t)$, $t \geq 0$ be a one-dimensional Brownian motion. Define the two-dimensional stochastic process

$$x(t) = (x_1(t), x_2(t))^T = (\cos(B(t)), \sin(B(t)))^T \quad \text{on } t \geq 0. \quad (2.4)$$

The process $x(t)$ is called *Brownian motion on the unit circle*. We now show that $x(t)$ satisfies a linear stochastic differential equation. By Itô's formula,

$$\begin{cases} dx_1(t) = -\sin(B(t))dB(t) - \dfrac{1}{2}\cos(B(t))dt = -\dfrac{1}{2}x_1(t)dt - x_2(t)dB(t), \\ dx_2(t) = \cos(B(t))dB(t) - \dfrac{1}{2}\sin(B(t))dt = -\dfrac{1}{2}x_2(t)dt + x_1(t)dB(t). \end{cases}$$

That is, in matrix notation,

$$dx(t) = -\frac{1}{2}x(t)dt + Kx(t)dB(t), \quad \text{where } K = \begin{pmatrix} 0 & -1 \\ 1 & 0 \end{pmatrix}. \quad (2.5)$$

Example 2.4 The charge $Q(t)$ at time t at a fixed point in an electrical circuit satisfies the second order differential equation

$$L\ddot{Q}(t) + R\dot{Q}(t) + \frac{1}{C}Q(t) = F(t), \quad (2.6)$$

where L is the inductance, R the resistance, C the capacitance and $F(t)$ the potential source. Suppose that the potential source is subject to the environmental noise and is described by $F(t) = G(t) + \alpha \dot{B}(t)$, where $\dot{B}(t)$ is a 1-dimensional white noise (i.e. $B(t)$ is a Brownian motion) and α is the intensity of the noise. Then equation (2.6) becomes

$$L\ddot{Q}(t) + R\dot{Q}(t) + \frac{1}{C}Q(t) = G(t) + \alpha \dot{B}(t). \quad (2.7)$$

Introduce the 2-dimensional process $x(t) = (x_1(t), x_2(t))^T = (Q(t), \dot{Q}(t))^T$. Then equation (2.7) can be expressed as an Itô equation

$$\begin{cases} dx_1(t) = x_2(t)dt, \\ dx_2(t) = \dfrac{1}{L}\left(-\dfrac{1}{C}x_1(t) - Rx_2(t) + G(t)\right)dt + \dfrac{\alpha}{L}dB(t). \end{cases}$$

That is,
$$dx(t) = [Ax(t) + H(t)]dt + K dB(t), \quad (2.8)$$
where
$$A = \begin{pmatrix} 0 & 1 \\ -1/CL & -R/L \end{pmatrix}, \quad H(t) = \begin{pmatrix} 0 \\ G(t)/L \end{pmatrix}, \quad K = \begin{pmatrix} 0 \\ \alpha/L \end{pmatrix}.$$

Example 2.5 More generally, consider a dth-order differential equation with white noise of the form
$$y^{(d)}(t) = F(y(t), \cdots, y^{(d-1)}(t), t) + G(y(t), \cdots, y^{(d-1)}(t), t)\dot{B}(t), \quad (2.9)$$
where $F : R^d \times R_+ \to R$, $G : R^d \times R_+ \to R^{1 \times m}$, and $\dot{B}(t)$ is an m-dimensional white noise, i.e. $B(t)$ is an m-dimensional Brownian motion. Introducing the R^d-valued stochastic process $x(t) = (x_1(t), \cdots, x_d(t))^T = (y(t), \cdots, y^{(d-1)}(t))^T$, we can then convert equation (2.9) into a d-dimensional Itô equation
$$dx(t) = \begin{pmatrix} x_2(t) \\ \vdots \\ x_d(t) \\ F(x(t), t) \end{pmatrix} dt + \begin{pmatrix} 0 \\ \vdots \\ 0 \\ G(x(t), t) \end{pmatrix} dB(t). \quad (2.10)$$

Example 2.6 If $g(x, t) \equiv 0$, equation (2.1) becomes the ordinary differential equation
$$\dot{x}(t) = f(x(t), t) \quad \text{on } t \in [t_0, T] \quad (2.11)$$
with initial value $x(t_0) = x_0$. In this case, the random influence can only show up in the initial value x_0. As a special case, consider the one-dimensional equation
$$\dot{x}(t) = 3[x(t)]^{2/3} \quad \text{on } t \in [t_0, T] \quad (2.12)$$
with initial value $x(t_0) = 1_A$, where $A \in \mathcal{F}_{t_0}$. It is easy to verify that for any $0 < \alpha < T - t_0$, the stochastic process
$$x(t) = x(t, \omega) = \begin{cases} (t - t_0 + 1)^3 & \text{for } t_0 \leq t \leq T, \ \omega \in A, \\ 0 & \text{for } t_0 \leq t \leq t_0 + \alpha, \ \omega \notin A, \\ (t - t_0 - \alpha)^3 & \text{for } t_0 + \alpha < t \leq T, \ \omega \notin A \end{cases}$$
is a solution to equation (2.12). In other words, equation (2.12) has infinitely many solutions. As another special case, consider the one-dimensional equation
$$\dot{x}(t) = [x(t)]^2 \quad \text{on } t \in [t_0, T] \quad (2.13)$$
with initial value $x(t_0) = x_0$, which is a random variable taking values larger than $1/[T-t_0]$. It is easy to verify that equation (2.13) has the (unique) solution
$$x(t) = \left[\frac{1}{x_0} - (t - t_0) \right]^{-1} \quad \text{only on } t_0 \leq t < t_0 + \frac{1}{x_0}(<T),$$

but there is no solution defined for all $t \in [t_0, T]$ in this case.

Example 2.7 Let $B(t)$ be a one-dimensional Brownian motion. Girsanov (1962) has shown that the one-dimensional Itô equation

$$x(t) = \int_{t_0}^{t} |x(s)|^\alpha dB(s)$$

has a unique solution when $\alpha \geq 1/2$, but it has infinitely many solutions when $0 < \alpha < 1/2$.

2.3 EXISTENCE AND UNIQUENESS OF SOLUTIONS

Examples 2.6 and 2.7 show that an Itô equation may not have a unique solution defined on the whole interval $[t_0, T]$. Let us now turn to find the conditions that guarantee the existence and uniqueness of the solution to equation (2.1).

Theorem 3.1 *Assume that there exist two positive constants $\bar{K}$ and K such that*

(i) (Lipschitz condition) for all $x, y \in R^d$ and $t \in [t_0, T]$

$$|f(x,t) - f(y,t)|^2 \bigvee |g(x,t) - g(y,t)|^2 \leq \bar{K}|x-y|^2; \quad (3.1)$$

(ii) (Linear growth condition) for all $(x, t) \in R^d \times [t_0, T]$

$$|f(x,t)|^2 \bigvee |g(x,t)|^2 \leq K(1 + |x|^2). \quad (3.2)$$

Then there exists a unique solution $x(t)$ to equation (2.1) and the solution belongs to $\mathcal{M}^2([t_0, T]; R^d)$.

We first prepare a lemma.

Lemma 3.2 *Assume that the linear growth condition (3.2) holds. If $x(t)$ is a solution of equation (2.1), then*

$$E\left(\sup_{t_0 \leq t \leq T} |x(t)|^2\right) \leq (1 + 3E|x_0|^2)e^{3K(T-t_0)(T-t_0+4)}. \quad (3.3)$$

In particular, $x(t)$ belongs to $\mathcal{M}^2([t_0, T]; R^d)$.

Proof. For every integer $n \geq 1$, define the stopping time

$$\tau_n = T \wedge \inf\{t \in [t_0, T] : |x(t)| \geq n\}.$$

Clearly, $\tau_n \uparrow T$ a.s. Set $x_n(t) = x(t \wedge \tau_n)$ for $t \in [t_0, T]$. Then $x_n(t)$ satisfies the equation

$$x_n(t) = x_0 + \int_{t_0}^t f(x_n(s), s) I_{[[t_0, \tau_n]]}(s) ds + \int_{t_0}^t g(x_n(s), s) I_{[[t_0, \tau_n]]}(s) dB(s).$$

Using the elementary inequality $|a + b + c|^2 \leq 3(|a|^2 + |b|^2 + |c|^2)$, the Hölder inequality and condition (3.2), one can show that

$$|x_n(t)|^2 \leq 3|x_0|^2 + 3K(t - t_0) \int_{t_0}^t (1 + |x_n(s)|^2) ds$$

$$+ 3 \left| \int_{t_0}^t g(x_n(s), s) I_{[[t_0, \tau_n]]}(s) dB(s) \right|^2.$$

Hence, by Theorem 1.7.2 and condition (3.2) one can further show that

$$E\left(\sup_{t_0 \leq s \leq t} |x_n(s)|^2\right) \leq 3E|x_0|^2 + 3K(T - t_0) \int_{t_0}^t (1 + E|x_n(s)|^2) ds$$

$$+ 12 E \int_{t_0}^t |g(x_n(s), s)|^2 I_{[[t_0, \tau_n]]}(s) ds$$

$$\leq 3E|x_0|^2 + 3K(T - t_0 + 4) \int_{t_0}^t (1 + E|x_n(s)|^2) ds.$$

Consequently

$$1 + E\left(\sup_{t_0 \leq s \leq t} |x_n(s)|^2\right)$$

$$\leq 1 + 3E|x_0|^2 + 3K(T - t_0 + 4) \int_{t_0}^t \left[1 + E\left(\sup_{t_0 \leq r \leq s} |x_n(r)|^2\right)\right] ds.$$

Now the Gronwall inequality (i.e. Theorem 1.8.1) yields that

$$1 + E\left(\sup_{t_0 \leq t \leq T} |x_n(t)|^2\right) \leq (1 + 3E|x_0|^2) e^{3K(T-t_0)(T-t_0+4)}.$$

Thus

$$E\left(\sup_{t_0 \leq t \leq \tau_n} |x(t)|^2\right) \leq (1 + 3E|x_0|^2) e^{3K(T-t_0)(T-t_0+4)}.$$

Finally the required inequality (3.3) follows by letting $n \to \infty$.

Proof of Theorem 3.1. <u>Uniqueness</u>. Let $x(t)$ and $\bar{x}(t)$ be two solutions of equation (2.1). By Lemma 3.2, both of them belong to $\mathcal{M}^2([t_0, T]; R^d)$. Note that

$$x(t) - \bar{x}(t) = \int_{t_0}^t [f(x(s), s) - f(\bar{x}(s), s)] ds + \int_{t_0}^t [g(x(s), s) - g(\bar{x}(s), s)] dB(s).$$

Using the Hölder inequality, Theorem 1.7.2 and the Lipschitz condition (3.1) one can show in the same way as the proof of Lemma 3.2 that

$$E\left(\sup_{t_0\leq s\leq t}|x(s)-\bar{x}(s)|^2\right) \leq 2\bar{K}(T+4)\int_{t_0}^{t} E\left(\sup_{t_0\leq r\leq s}|x(r)-\bar{x}(r)|^2\right)ds.$$

The Gronwall inequality then yields that

$$E\left(\sup_{t_0\leq t\leq T}|x(t)-\bar{x}(t)|^2\right) = 0.$$

Hence, $x(t) = \bar{x}(t)$ for all $t_0 \leq t \leq T$ almost surely. The uniqueness has been proved.

<u>Existence.</u> Set $x_0(t) \equiv x_0$, and for $n = 1, 2, \cdots$, define the Picard iterations

$$x_n(t) = x_0 + \int_{t_0}^{t} f(x_{n-1}(s), s)ds + \int_{t_0}^{t} g(x_{n-1}(s), s)dB(s) \quad (3.4)$$

for $t \in [t_0, T]$. Obviously, $x_0(\cdot) \in \mathcal{M}^2([t_0, T]; R^d)$. Moreover, it is easy to see by induction that $x_n(\cdot) \in \mathcal{M}^2([t_0, T]; R^d)$, because we have from (3.4) that

$$E|x_n(t)|^2 \leq c_1 + 3K(T+1)\int_{t_0}^{t} E|x_{n-1}(s)|^2 ds, \quad (3.5)$$

where $c_1 = 3E|x_0|^2 + 3KT(T+1)$. It also follows from (3.5) that for any $k \geq 1$,

$$\max_{1\leq n\leq k} E|x_n(t)|^2 \leq c_1 + 3K(T+1)\int_{t_0}^{t} \max_{1\leq n\leq k} E|x_{n-1}(s)|^2 ds$$

$$\leq c_1 + 3K(T+1)\int_{t_0}^{t} \left(E|x_0|^2 + \max_{1\leq n\leq k} E|x_n(s)|^2\right)ds$$

$$\leq c_2 + 3K(T+1)\int_{t_0}^{t} \max_{1\leq n\leq k} E|x_n(s)|^2 ds,$$

where $c_2 = c_1 + 3KT(T+1)E|x_0|^2$. The Gronwall inequality implies

$$\max_{1\leq n\leq k} E|x_n(t)|^2 \leq c_2 e^{3KT(T+1)}.$$

Since k is arbitrary, we must have

$$E|x_n(t)|^2 \leq c_2 e^{3KT(T+1)} \quad \text{for all } t_0 \leq t \leq T, \, n \geq 1. \quad (3.6)$$

Next, we note that

$$|x_1(t) - x_0(t)|^2 = |x_1(t) - x_0|^2$$

$$\leq 2\left|\int_{t_0}^{t} f(x_0, s)ds\right|^2 + 2\left|\int_{t_0}^{t} g(x_0, s)dB(s)\right|^2.$$

Taking the expectation and using (3.2), we get

$$E|x_1(t) - x_0(t)|^2$$
$$\leq 2K(t-t_0)^2(1+E|x_0|^2) + 2K(t-t_0)(1+E|x_0|^2) \leq C, \quad (3.7)$$

where $C = 2K(T - t_0 + 1)(T - t_0)(1 + E|x_0|^2)$. We now claim that for $n \geq 0$,

$$E|x_{n+1}(t) - x_n(t)|^2 \leq \frac{C[M(t-t_0)]^n}{n!} \quad \text{for } t_0 \leq t \leq T, \quad (3.8)$$

where $M = 2\bar{K}(T - t_0 + 1)$. We shall show this by induction. In view of (3.7) we see that (3.8) holds when $n = 0$. Under the inductive assumption that (3.8) holds for some $n \geq 0$, we shall show that (3.8) still holds for $n + 1$. Note that

$$|x_{n+2}(t) - x_{n+1}(t)|^2 \leq 2\left|\int_{t_0}^t [f(x_{n+1}(s),s) - f(x_n(s),s)]ds\right|^2$$
$$+ 2\left|\int_{t_0}^t [g(x_{n+1}(s),s) - g(x_n(s),s)]dB(s)\right|^2. \quad (3.9)$$

Taking the expectation and using (3.1) as well as the inductive assumption, we derive that

$$E|x_{n+2}(t) - x_{n+1}(t)|^2 \leq 2\bar{K}(t - t_0 + 1)E\int_{t_0}^t |x_{n+1}(s) - x_n(s)|^2 ds$$
$$\leq M \int_{t_0}^t E|x_{n+1}(s) - x_n(s)|^2 ds$$
$$\leq M \int_{t_0}^t \frac{C[M(s-t_0)]^n}{n!} ds = \frac{C[M(t-t_0)]^{n+1}}{(n+1)!}.$$

That is, (3.8) holds for $n + 1$. Hence, by induction, (3.8) holds for all $n \geq 0$. Furthermore, replacing n in (3.9) with $n - 1$ we see that

$$\sup_{t_0 \leq t \leq T} |x_{n+1}(t) - x_n(t)|^2 \leq 2\bar{K}(T - t_0) \int_{t_0}^T |x_n(s) - x_{n-1}(s)|^2 ds$$
$$+ 2 \sup_{t_0 \leq t \leq T} \left|\int_{t_0}^T [g(x_n(s),s) - g(x_{n-1}(s),s)]dB(s)\right|^2.$$

Taking the expectation and using Theorem 1.7.2 and (3.8), we find that

$$E\left(\sup_{t_0 \leq t \leq T} |x_{n+1}(t) - x_n(t)|^2\right) \leq 2\bar{K}(T - t_0 + 4) \int_{t_0}^T E|x_n(s) - x_{n-1}(s)|^2 ds$$
$$\leq 4M \int_{t_0}^T \frac{C[M(s-t_0)]^{n-1}}{(n-1)!} ds = \frac{4C[M(T-t_0)]^n}{n!}.$$

Hence
$$P\left\{\sup_{t_0 \leq t \leq T} |x_{n+1}(t) - x_n(t)| > \frac{1}{2^n}\right\} \leq \frac{4C[4M(T-t_0)]^n}{n!}.$$

Since $\sum_{n=0}^{\infty} 4C[4M(T-t_0)]^n/n! < \infty$, the Borel–Cantelli lemma yields that for almost all $\omega \in \Omega$ there exists a positive integer $n_0 = n_0(\omega)$ such that
$$\sup_{t_0 \leq t \leq T} |x_{n+1}(t) - x_n(t)| \leq \frac{1}{2^n} \qquad \text{whenever } n \geq n_0.$$

It follows that, with probability 1, the partial sums
$$x_0(t) + \sum_{i=0}^{n-1}[x_{i+1}(t) - x_i(t)] = x_n(t)$$

are convergent uniformly in $t \in [0,T]$. Denote the limit by $x(t)$. Clearly, $x(t)$ is continuous and $\mathcal{F}_t$-adapted. On the other hand, one sees from (3.8) that for every t, $\{x_n(t)\}_{n \geq 1}$ is a Cauchy sequence in L^2 as well. Hence we also have $x_n(t) \to x(t)$ in L^2. Letting $n \to \infty$ in (3.6) gives
$$E|x(t)|^2 \leq c_2 e^{3KT(T+1)} \qquad \text{for all } t_0 \leq t \leq T.$$

Therefore, $x(\cdot) \in \mathcal{M}^2([t_0,T]; R^d)$. It remains to show that $x(t)$ satisfies equation (2.2). Note that
$$E\left|\int_{t_0}^t f(x_n(s),s)ds - \int_{t_0}^t f(x(s),s)ds\right|^2$$
$$+ E\left|\int_{t_0}^t g(x_n(s),s)dB(s) - \int_{t_0}^t g(x(s),s)dB(s)\right|^2$$
$$\leq \bar{K}(T-t_0+1) \int_{t_0}^T E|x_n(s) - x(s)|^2 ds \to 0 \qquad \text{as } n \to \infty.$$

Hence we can let $n \to \infty$ in (3.4) to obtain that
$$x(t) = x_0 + \int_{t_0}^t f(x(s),s)ds + \int_{t_0}^t g(x(s),s)dB(s) \qquad \text{on } t_0 \leq t \leq T$$

as desired. The proof is now complete.

In the proof above we have shown that the Picard iterations $x_n(t)$ converge to the unique solution $x(t)$ of equation (2.1). The following theorem gives an estimate on how fast the convergence is.

Theorem 3.3 *Let the assumptions of Theorem 3.1 hold. Let $x(t)$ be the unique solution of equation (2.1) and $x_n(t)$ be the Picard iterations defined by (3.4). Then*
$$E\left(\sup_{t_0 \leq t \leq T} |x_n(t) - x(t)|^2\right) \leq \frac{8C[M(T-t_0)]^n}{n!} e^{8M(T-t_0)} \qquad (3.10)$$

for all $n \geq 1$, where C and M are the same as defined in the proof of Theorem 3.1, that is $C = 2K(T - t_0 + 1)(T - t_0)(1 + E|x_0|^2)$ and $M = 2\bar{K}(T - t_0 + 1)$.

Proof. From

$$x_n(t) - x(t) = \int_{t_0}^{t} [f(x_{n-1}(s), s) - f(x(s), s)]ds$$
$$+ \int_{t_0}^{t} [g(x_{n-1}(s), s) - g(x(s), s)]dB(s),$$

we can derive that

$$E\left(\sup_{t_0 \leq s \leq t} |x_n(s) - x(s)|^2\right) \leq 2\bar{K}(T - t_0 + 4) \int_{t_0}^{t} E|x_{n-1}(s) - x(s)|^2 ds$$
$$\leq 8M \int_{t_0}^{t} E|x_n(s) - x_{n-1}(s)|^2 ds + 8M \int_{t_0}^{t} E|x_n(s) - x(s)|^2 ds.$$

Substituting (3.8) into this yields that

$$E\left(\sup_{t_0 \leq s \leq t} |x_n(s) - x(s)|^2\right)$$
$$\leq 8M \int_{t_0}^{T} \frac{C[M(s - t_0)]^{n-1}}{(n-1)!} ds + 8M \int_{t_0}^{t} E|x_n(s) - x(s)|^2 ds$$
$$\leq \frac{8C[M(T - t_0)]^n}{n!} + 8M \int_{t_0}^{t} E\left(\sup_{t_0 \leq r \leq s} |x_n(r) - x(r)|^2\right) ds.$$

Consequently, the required inequality (3.10) follows by applying the Gronwall inequality. The proof is complete.

This theorem shows that one can use the Picard iteration procedure to obtain the approximate solutions of equation (2.1), and (3.10) gives the estimate for the error of the approximation. We shall discuss other approximation procedures later.

The Lipschitz condition (3.1) means that the coefficients $f(x, t)$ and $g(x, t)$ do not change faster than a linear function of x as change in x. This implies in particular the continuity of $f(x, t)$ and $g(x, t)$ in x for all $t \in [t_0, T]$. Thus, functions that are discontinuous with respect to x are excluded as the coefficients. Besides, functions like $\sin x^2$ do not satisfy the Lipschitz condition. These indicate that the Lipschitz condition is too restrictive. The next theorem is a generalization of Theorem 3.1 in which this (uniform) Lipschitz condition is replaced by the local Lipschitz condition.

Theorem 3.4 *Assume that the linear growth condition (3.2) holds, but the Lipschitz condition (3.1) is replaced with the following local Lipschitz condition:*

For every integer $n \geq 1$, there exists a positive constant K_n such that, for all $t \in [t_0, T]$ and all $x, y \in R^d$ with $|x| \vee |y| \leq n$,

$$|f(x,t) - f(y,t)|^2 \vee |g(x,t) - g(y,t)|^2 \leq K_n |x-y|^2. \qquad (3.11)$$

Then there exists a unique solution $x(t)$ to equation (2.1) in $\mathcal{M}^2([t_0, T]; R^d)$.

Proof. This theorem is proved by a truncation procedure. We only outline the proof but leave the details to the reader. For each $n \geq 1$, define the truncation function

$$f_n(x,t) = \begin{cases} f(x,t) & \text{if } |x| \leq n, \\ f(nx/|x|, t) & \text{if } |x| > n, \end{cases}$$

and $g_n(x,t)$ similarly. Then f_n and g_n satisfy the Lipschitz condition (3.1) and the linear growth condition (3.2). Hence by Theorem 3.1, there is a unique solution $x_n(\cdot)$ in $\mathcal{M}^2([t_0, T]; R^d)$ to the equation

$$x_n(t) = x_0 + \int_{t_0}^{t} f_n(x_n(s), s) dt + \int_{t_0}^{t} g_n(x_n(s), s) dB(s), \quad t \in [t_0, T]. \qquad (3.12)$$

Define the stopping time

$$\tau_n = T \wedge \inf\{t \in [t_0, T] : |x_n(t)| \geq n\}.$$

We can show that

$$x_n(t) = x_{n+1}(t) \quad \text{if } t_0 \leq t \leq \tau_n. \qquad (3.13)$$

This implies that τ_n is increasing. We can then use the linear growth condition to show that for almost all $\omega \in \Omega$, there exists an integer $n_0 = n_0(\omega)$ such that $\tau_n = T$ whenever $n \geq n_0$. Now define $x(t)$ by

$$x(t) = x_{n_0}(t), \quad t \in [t_0, T].$$

By (3.13), $x(t \wedge \tau_n) = x_n(t \wedge \tau_n)$, and it therefore follows from (3.12) that

$$x(t \wedge \tau_n) = x_0 + \int_{t_0}^{t \wedge \tau_n} f_n(x(s), s) ds + \int_{t_0}^{t \wedge \tau_n} g_n(x(s), s) dB(s)$$

$$= x_0 + \int_{t_0}^{t \wedge \tau_n} f(x(s), s) ds + \int_{t_0}^{t \wedge \tau_n} g(x(s), s) dB(s).$$

Letting $n \to \infty$ we see that $x(t)$ is a solution of equation (2.1), which, by Lemma 3.2, belongs to $\mathcal{M}^2([t_0, T]; R^d)$. The uniqueness can be proved via a stopping procedure.

The local Lipschitz condition allows us to include many functions as the coefficients $f(x,t)$ and $g(x,t)$ e.g. functions that have continuous partial derivatives of first order with respect to x on $R^d \times [t_0, T]$. However, the linear growth condition still excludes some important functions like $-|x|^2 x$ as the coefficients. The following result improves the situation.

Theorem 3.5 *Assume that the local Lipschitz condition (3.11) holds, but the linear growth condition (3.2) is replaced with the following monotone condition: There exists a positive constant K such that for all $(x,t) \in R^d \times [t_0, T]$*

$$x^T f(x,t) + \frac{1}{2}|g(x,t)|^2 \leq K(1+|x|^2). \qquad (3.14)$$

Then there exists a unique solution $x(t)$ to equation (2.1) in $\mathcal{M}^2([t_0,T]; R^d)$.

This theorem can be proved in a similar way as Theorem 3.4—the local Lipschitz condition guarantees that the solution exists in $[t_0, \tau_\infty]$, where $\tau_\infty = \lim_{n\to\infty} \tau_n$, but the monotone condition instead of the linear growth condition guarantees that $\tau_\infty = T$ i.e. the solution exists on the whole interval $[t_0, T]$. We leave the details to the reader. It should be stressed that if the linear growth condition (3.2) holds then the monotone condition (3.14) is satisfied, but conversely not. For example, consider the one-dimensional stochastic differential equation

$$dx(t) = [x(t) - x^3(t)]dt + x^2(t)dB(t) \quad \text{on } t \in [t_0, T]. \qquad (3.15)$$

Here $B(t)$ is a one-dimensional Brownian motion. Clearly, the local Lipschitz condition is satisfied but the linear growth condition is not. On the other hand, note that

$$x[x - x^3] + \frac{1}{2}x^4 \leq x^2 < 1 + x^2.$$

That is, the monotone condition is fulfilled. Hence Theorem 3.4 guarantees that equation (3.15) has a unique solution. An even more general condition than the monotone one is the Has'minskii condition which is described by using a Lyapunov-like function. For the details please see Has'minskii (1980).

In this book we often discuss a stochastic differential equation on $[t_0, \infty)$, that is

$$dx(t) = f(x(t), t)dt + g(x(t), t)dB(t) \quad \text{on } t \in [t_0, \infty) \qquad (3.16)$$

with initial value $x(t_0) = x_0$. If the assumptions of the existence-and-uniqueness theorem hold on every finite subinterval $[t_0, T]$ of $[t_0, \infty)$, then equation (3.16) has a unique solution $x(t)$ on the entire interval $[t_0, \infty)$. Such a solution is called a *global* solution. For the convenience of the reader, we state a theorem.

Theorem 3.6 *Assume that for every real number $T > t_0$ and integer $n \geq 1$, there exists a positive constant $K_{T,n}$ such that for all $t \in [t_0, T]$ and all $x, y \in R^d$ with $|x| \vee |y| \leq n$,*

$$|f(x,t) - f(y,t)|^2 \bigvee |g(x,t) - g(y,t)|^2 \leq K_{T,n}|x-y|^2.$$

Assume also that for every $T > t_0$, there exists a positive constant K_T such that for all $(x,t) \in R^d \times [t_0, T]$,

$$x^T f(x,t) + \frac{1}{2}|g(x,t)|^2 \leq K_T(1+|x|^2).$$

Then there exists a unique global solution $x(t)$ to equation (3.16) and the solution belongs to $\mathcal{M}^2([t_0, \infty); R^d)$.

2.4 L^p-ESTIMATES

In this section, we assume that $x(t)$, $t_0 \leq t \leq T$ is the unique solution of equation (2.1) with initial value $x(t_0) = x_0$, and we shall investigate the pth moment of the solution.

Theorem 4.1 Let $p \geq 2$ and $x_0 \in L^p(\Omega; R^d)$. Assume that there exists a constant $\alpha > 0$ such that for all $(x,t) \in R^d \times [t_0, T]$,

$$x^T f(x,t) + \frac{p-1}{2}|g(x,t)|^2 \leq \alpha(1+|x|^2). \tag{4.1}$$

Then

$$E|x(t)|^p \leq 2^{\frac{p-2}{2}}(1+E|x_0|^p)e^{p\alpha(t-t_0)} \quad \text{for all } t \in [t_0, T]. \tag{4.2}$$

Proof. By Itô's formula and condition (4.1) we can derive that for $t \in [t_0, T]$,

$$[1+|x(t)|^2]^{\frac{p}{2}} = [1+|x_0|^2]^{\frac{p}{2}} + p\int_{t_0}^t [1+|x(s)|^2]^{\frac{p-2}{2}} x^T(s)f(x(s),s)ds$$

$$+ \frac{p}{2}\int_{t_0}^t [1+|x(s)|^2]^{\frac{p-2}{2}} |g(x(s),s)|^2 ds$$

$$+ \frac{p(p-2)}{2}\int_{t_0}^t [1+|x(s)|^2]^{\frac{p-4}{2}} |x^T(s)g(x(s),s)|^2 ds$$

$$+ p\int_{t_0}^t [1+|x(s)|^2]^{\frac{p-2}{2}} x^T(s)g(x(s),s)dB(s)$$

$$\leq 2^{\frac{p-2}{2}}(1+|x_0|^p) + p\int_{t_0}^t [1+|x(s)|^2]^{\frac{p-2}{2}}$$

$$\times \left(x^T(s)f(x(s),s) + \frac{p-1}{2}|g(x(s),s)|^2\right)ds$$

$$+ p\int_{t_0}^t [1+|x(s)|^2]^{\frac{p-2}{2}} x^T(s)g(x(s),s)dB(s)$$

$$\leq 2^{\frac{p-2}{2}}(1+|x_0|^p) + p\alpha\int_{t_0}^t [1+|x(s)|^2]^{\frac{p}{2}} ds$$

$$+ p\int_{t_0}^t [1+|x(s)|^2]^{\frac{p-2}{2}} x^T(s)g(x(s),s)dB(s). \tag{4.3}$$

For every integer $n \geq 1$, define the stopping time
$$\tau_n = T \wedge \inf\{t \in [t_0, T] : |x(t)| \geq n\}.$$

Clearly, $\tau_n \uparrow T$ a.s. Moreover, it follows from (4.3) and the property of Itô's integral that

$$E\left([1+|x(t \wedge \tau_n)|^2]^{\frac{p}{2}}\right)$$
$$\leq 2^{\frac{p-2}{2}}(1+E|x_0|^p) + p\alpha E \int_{t_0}^{t \wedge \tau_n} [1+|x(s)|^2]^{\frac{p}{2}} ds$$
$$\leq 2^{\frac{p-2}{2}}(1+E|x_0|^p) + p\alpha \int_{t_0}^{t} E\left([1+|x(s \wedge \tau_n)|^2]^{\frac{p}{2}}\right) ds$$

The Gronwall inequality yields
$$E\left([1+|x(t \wedge \tau_n)|^2]^{\frac{p}{2}}\right) \leq 2^{\frac{p-2}{2}}(1+E|x_0|^p)e^{p\alpha(t-t_0)}.$$

Letting $n \to \infty$ yields
$$E\left([1+|x(t)|^2]^{\frac{p}{2}}\right) \leq 2^{\frac{p-2}{2}}(1+E|x_0|^p)e^{p\alpha(t-t_0)}, \tag{4.4}$$

and the desired inequality (4.2) follows.

We now verify that if the linear growth condition (3.2) is fulfilled, then (4.1) is satisfied with $\alpha = \sqrt{K} + K(p-1)/2$. In fact, using (3.2) and the elementary inequality $2ab \leq a^2 + b^2$ one can derive that for any $\varepsilon > 0$,

$$2x^T f(x,t) \leq 2|x||f(x,t)| = 2(\sqrt{\varepsilon}|x|)(|f(x,t)|/\sqrt{\varepsilon})$$
$$\leq \varepsilon|x|^2 + \frac{1}{\varepsilon}|f(x,t)|^2 \leq \varepsilon|x|^2 + \frac{K}{\varepsilon}(1+|x|^2).$$

Letting $\varepsilon = \sqrt{K}$ yields
$$x^T f(x,t) \leq \sqrt{K}(1+|x|^2).$$

Consequently
$$x^T f(x,t) + \frac{p-1}{2}|g(x,t)|^2 \leq \left[\sqrt{K} + \frac{K(p-1)}{2}\right](1+|x|^2).$$

We therefore obtain the following useful corollary.

Corollary 4.2 *Let $p \geq 2$ and $x_0 \in L^p(\Omega; R^d)$. Assume that the linear growth condition (3.2) holds. Then inequality (4.2) holds with $\alpha = \sqrt{K} + K(p-1)/2$.*

We now apply these results to show one of the important properties of the solution.

Theorem 4.3 *Let $p \geq 2$ and $x_0 \in L^p(\Omega; R^d)$. Assume that the linear growth condition (3.2) holds. Then*

$$E|x(t) - x(s)|^p \leq C(t-s)^{\frac{p}{2}} \quad \text{for all } t_0 \leq s < t \leq T, \tag{4.5}$$

where

$$C = 2^{p-2}(1 + E|x_0|^p)e^{p\alpha(T-t_0)}\left([2(T-t_0)]^{\frac{p}{2}} + [p(p-1)]^{\frac{p}{2}}\right)$$

and $\alpha = \sqrt{K} + K(p-1)/2$. In particular, the pth moment of the solution is continuous on $[t_0, T]$.

Proof. By the elementary inequality $|a+b|^p \leq 2^{p-1}(|a|^p + |b|^p)$, it is easy to see that

$$E|x(t) - x(s)|^p \leq 2^{p-1}E\left|\int_s^t f(x(r), r)dr\right|^p + 2^{p-1}E\left|\int_s^t g(x(r), r)dB(r)\right|^p.$$

Using the Hölder inequality, Theorem 1.7.1 and the linear growth condition, one can then derive that

$$E|x(t) - x(s)|^p \leq [2(t-s)]^{p-1}E\int_s^t |f(x(r), r)|^p dr$$

$$+ \frac{1}{2}[2p(p-1)]^{\frac{p}{2}}(t-s)^{\frac{p-2}{2}}E\int_s^t |g(x(r), r)|^p dr$$

$$\leq c_1(t-s)^{\frac{p-2}{2}}\int_s^t E(1 + |x(r)|^2)^{\frac{p}{2}},$$

where $c_1 = 2^{\frac{p-2}{2}}K^{\frac{p}{2}}\left([2(T-t_0)]^{\frac{p}{2}} + [p(p-1)]^{\frac{p}{2}}\right)$. Applying (4.4) one sees that

$$E|x(t) - x(s)|^p \leq c_1(t-s)^{\frac{p-2}{2}}\int_s^t 2^{\frac{p-2}{2}}(1 + E|x_0|^p)e^{p\alpha(r-t_0)}dr$$

$$\leq c_1 2^{\frac{p-2}{2}}(1 + E|x_0|^p)e^{p\alpha(T-t_0)}(t-s)^{\frac{p}{2}},$$

which is the required inequality (4.5).

Theorem 4.4 *Let $p \geq 2$ and $x_0 \in L^p(\Omega; R^d)$. Assume that the linear growth condition (3.2) holds. Then*

$$E\left(\sup_{t_0 \leq s \leq t} |x(s)|^p\right) \leq (1 + 3^{p-1}E|x_0|^p)e^{\beta(t-t_0)} \tag{4.6}$$

for all $t_0 \leq t \leq T$, where

$$\beta = \frac{1}{6}(18K)^{\frac{p}{2}}(T-t_0)^{\frac{p-2}{2}}\left[(T-t_0)^{\frac{p}{2}} + \left(\frac{p^3}{2(p-1)}\right)^{\frac{p}{2}}\right].$$

Proof. Using the Hölder inequality, Theorem 1.7.2 and condition (3.2) one can derive that

$$E\left(\sup_{t_0\le s\le t}|x(s)|^p\right) \le 3^{p-1}E|x_0|^p + 3^{p-1}E\left(\int_{t_0}^t |f(x(s),s)|ds\right)^p$$

$$+ 3^{p-1}E\left(\sup_{t_0\le s\le t}\left|\int_{t_0}^s g(x(r),r)dB(r)\right|\right)^p$$

$$\le 3^{p-1}E|x_0|^p + [3(t-t_0)]^{p-1}E\int_{t_0}^t |f(x(s),s)|^p ds$$

$$+ 3^{p-1}\left(\frac{p^3}{2(p-1)}\right)^{\frac{p}{2}}(t-t_0)^{\frac{p-2}{2}}E\int_{t_0}^t |g(x(s),s)|^p ds$$

$$\le 3^{p-1}E|x_0|^p + \beta\int_{t_0}^t (1+E|x(s)|^p)ds.$$

Hence

$$1 + E\left(\sup_{t_0\le s\le t}|x(s)|^p\right) \le 1 + 3^{p-1}E|x_0|^p + \beta\int_{t_0}^t\left[1 + E\left(\sup_{t_0\le r\le s}|x(r)|^p\right)\right]ds.$$

By the Gronwall inequality one sees that

$$1 + E\left(\sup_{t_0\le s\le t}|x(s)|^p\right) \le (1 + 3^{p-1}E|x_0|^p)e^{\beta(t-t_0)},$$

and the required inequality (4.6) follows.

Let us now turn to consider the case of $0 < p < 2$. This is rather easy if we note that the Hölder inequality implies

$$E|x(t)|^p \le [E|x(t)|^2]^{\frac{p}{2}}. \tag{4.7}$$

In other words, the estimate for $E|x(t)|^p$ can be done via the estimate for the second moment. For example, we have the following corollaries.

Corollary 4.5 *Let $0 < p < 2$ and $x_0 \in L^2(\Omega; R^d)$. Assume that there exists a constant $\alpha > 0$ such that for all $(x,t) \in R^d \times [t_0,T]$,*

$$x^T f(x,t) + \frac{1}{2}|g(x,t)|^2 \le \alpha(1+|x|^2). \tag{4.8}$$

Then

$$E|x(t)|^p \le (1+E|x_0|^2)^{\frac{p}{2}}e^{p\alpha(t-t_0)} \quad \text{for all } t \in [t_0,T]. \tag{4.9}$$

Corollary 4.6 Let $0 < p < 2$ and $x_0 \in L^2(\Omega; R^d)$. Assume that the linear growth condition (3.2) holds. Then

$$E|x(t) - x(s)|^p \leq C^{\frac{p}{2}}(t-s)^{\frac{p}{2}} \quad \text{for all } t_0 \leq s < t \leq T, \tag{4.10}$$

where

$$C = 2(1 + E|x_0|^2)(T - t_0 + 1)\exp\left[(2\sqrt{K} + K)(T - t_0)\right].$$

In particular, the pth moment of the solution is continuous on $[t_0, T]$.

2.5 ALMOST SURELY ASYMPTOTIC ESTIMATES

Let us now consider the d-dimensional stochastic differential equation

$$dx(t) = f(x(t), t)dt + g(x(t), t)dB(t) \quad \text{on } t \in [t_0, \infty) \tag{5.1}$$

with initial value $x(t_0) = x_0 \in L^2(\Omega; R^d)$. Assume that the equation has a unique global solution $x(t)$ on $[t_0, \infty)$. Besides, we shall impose the monotone condition: There is a positive constant α such that, for all $(x, t) \in R^d \times [t_0, \infty)$,

$$x^T f(x, t) + \frac{1}{2}|g(x, t)|^2 \leq \alpha(1 + |x|^2). \tag{5.2}$$

Let $0 < p \leq 2$. In view of Theorem 4.1 and Corollary 4.5, we know that the pth moment of the solution satisfies

$$E|x(t)|^p \leq (1 + E|x_0|^2)^{\frac{p}{2}} e^{p\alpha(t-t_0)} \quad \text{for all } t \geq t_0.$$

This means that the pth moment will grow at most exponentially with exponent $p\alpha$. This can also be expressed as

$$\limsup_{t \to \infty} \frac{1}{t} \log(E|x(t)|^p) \leq p\alpha. \tag{5.3}$$

The left-hand side of (5.3) is called the *pth moment Lyapunov exponent* (for $p > 2$ too), and (5.3) shows that the pth moment Lyapunov exponent should not be greater than $p\alpha$. In this section, we shall establish the asymptotic estimate for the solution almost surely. More precisely, we shall estimate

$$\limsup_{t \to \infty} \frac{1}{t} \log |x(t)| \tag{5.4}$$

almost surely, which is called the *sample Lyapunov exponent*, or simply *Lyapunov exponent*.

Theorem 5.1 *Under the monotone condition (5.2), the sample Lyapunov exponent of the solution of equation (5.1) should not be greater than α, that is*

$$\limsup_{t \to \infty} \frac{1}{t} \log |x(t)| \leq \alpha \quad \text{a.s.} \tag{5.5}$$

Proof. By Itô's formula and the monotone condition (5.2),

$$\log(1 + |x(t)|^2) = \log(1 + |x_0|^2)$$
$$+ \int_{t_0}^t \frac{1}{1 + |x(s)|^2} \left(2x^T(s) f(x(s), s) + |g(x(s), s)|^2 \right) ds$$
$$- 2 \int_{t_0}^t \frac{|x^T(s) g(x(s), s)|^2}{[1 + |x(s)|^2]^2} ds + M(t)$$
$$\leq \log(1 + |x_0|^2) + 2\alpha(t - t_0) - 2 \int_{t_0}^t \frac{|x^T(s) g(x(s), s)|^2}{[1 + |x(s)|^2]^2} ds + M(t), \quad (5.6)$$

where

$$M(t) = 2 \int_{t_0}^t \frac{x^T(s) g(x(s), s)}{1 + |x(s)|^2} dB(s). \quad (5.7)$$

On the other hand, for every integer $n \geq t_0$, using the exponential martingale inequality (i.e. Theorem 1.7.4) one sees that

$$P\left\{ \sup_{t_0 \leq t \leq n} \left[M(t) - 2 \int_{t_0}^t \frac{|x^T(s) g(x(s), s)|^2}{[1 + |x(s)|^2]^2} ds \right] > 2 \log n \right\} \leq \frac{1}{n^2}.$$

An application of the Borel–Cantelli lemma then yields that for almost all $\omega \in \Omega$ there is a random integer $n_0 = n_0(\omega) \geq t_0 + 1$ such that

$$\sup_{t_0 \leq t \leq n} \left[M(t) - 2 \int_{t_0}^t \frac{|x^T(s) g(x(s), s)|^2}{[1 + |x(s)|^2]^2} ds \right] \leq 2 \log n \quad \text{if } n \geq n_0.$$

That is,

$$M(t) \leq 2 \log n + 2 \int_{t_0}^t \frac{|x^T(s) g(x(s), s)|^2}{[1 + |x(s)|^2]^2} ds \quad (5.8)$$

for all $t_0 \leq t \leq n$, $n \geq n_0$ almost surely. Substituting (5.8) into (5.6) deduces that

$$\log(1 + |x(t)|^2) \leq \log(1 + |x_0|^2) + 2\alpha(t - t_0) + 2 \log n$$

for all $t_0 \leq t \leq n$, $n \geq n_0$ almost surely. Therefore, for almost all $\omega \in \Omega$, if $n \geq n_0$, $n - 1 \leq t \leq n$,

$$\frac{1}{t} \log(1 + |x(t)|^2) \leq \frac{1}{n-1} \Big[\log(1 + |x_0|^2) + 2\alpha(n - t_0) + 2 \log n \Big].$$

This implies

$$\limsup_{t \to \infty} \frac{1}{t} \log |x(t)| \leq \limsup_{t \to \infty} \frac{1}{2t} \log(1 + |x(t)|^2)$$
$$\leq \limsup_{n \to \infty} \frac{1}{2(n-1)} \Big[\log(1 + |x_0|^2) + 2\alpha(n - t_0) + 2 \log n \Big] = \alpha.$$

almost surely. The proof is complete.

Let us now recall the linear growth condition: There exists a $K > 0$ such that for all $(x,t) \in R^d \times [t_0, \infty)$

$$|f(x,t)|^2 \bigvee |g(x,t)|^2 \leq K(1+|x|^2). \tag{5.9}$$

As shown on page 60 (just before the statement of Corollary 4.2), we know that (5.9) implies (5.2) with $\alpha = \sqrt{K} + K/2$. So we have the following useful corollary.

Corollary 5.2 *Under the linear growth condition (5.9), the solution of equation (5.1) has the property*

$$\limsup_{t \to \infty} \frac{1}{t} \log |x(t)| \leq \sqrt{K} + \frac{K}{2} \quad a.s. \tag{5.10}$$

On the other hand, we would like to point out that this corollary can also be proved straightforward. Note that under the linear growth condition (5.9), the continuous local martingale $M(t)$, $t \geq t_0$ (in fact it is a martingale) defined by (5.7) has its quadratic variation (cf. Theorem 1.5.14)

$$\langle M, M \rangle_t = 4 \int_{t_0}^t \frac{|x^T(s)g(x(s),s)|^2}{[1+|x(s)|^2]^2} ds \leq 4K \int_{t_0}^t \frac{|x^T(s)|^2}{1+|x(s)|^2} ds \leq 4K(t-t_0).$$

Hence

$$\limsup_{t \to \infty} \frac{\langle M, M \rangle_t}{t} \leq 4K \quad a.s.$$

and then, by Theorem 1.3.4,

$$\limsup_{t \to \infty} \frac{M(t)}{t} = 0 \quad a.s. \tag{5.11}$$

Now (5.10) follows from (5.6) immediately. However, the monotone condition (5.2) does not necessarily guarantee (5.11) so the more careful arguments carried out in the proof of Theorem 5.1 are quite necessary.

In the sequel of this section, we shall consider a special case of equation (5.1), i.e. the equation of the form

$$dx(t) = f(x(t),t)dt + \sigma dB(t) \quad \text{on } t \in [t_0, \infty) \tag{5.12}$$

with initial value $x(t_0) = x_0 \in L^2(\Omega; R^d)$, where $\sigma = (\sigma_{ij})_{d \times m}$ is a constant matrix. Such a stochastic differential equation appears frequently when a system described by an ordinary differential equation $\dot{x}(t) = f(x(t),t)$ is subject to environmental noises that are independent of the state $x(t)$. We shall impose a condition on $f(x,t)$. That is, there exists a pair of positive constants γ and ρ such that

$$x^T f(x,t) \leq \gamma |x|^2 + \rho \quad \text{for all } (x,t) \in R^d \times [t_0, \infty). \tag{5.13}$$

Note that
$$x^T f(x,t) + \frac{1}{2}|\sigma|^2 \leq \left[\gamma \vee \left(\rho + \frac{|\sigma|^2}{2}\right)\right](1+|x|^2).$$

By Theorem 5.1 one deduces that the solution of equation (5.12) has the property

$$\limsup_{t\to\infty} \frac{1}{t} \log |x(t)| \leq \gamma \vee \left(\rho + \frac{|\sigma|^2}{2}\right) \quad a.s. \quad (5.14)$$

However, with a bit of extra effort, we can obtain much stronger results.

Theorem 5.3 *Let (5.13) hold. Then the solution of equation (5.12) has the property*

$$\lim_{t\to\infty} \frac{|x(t)|}{e^{\gamma t}\sqrt{\log\log t}} = 0 \quad a.s. \quad (5.15)$$

Before the proof, let us emphasis that the conclusion is independent of ρ and σ. Besides, (5.15) implies that for almost all $\omega \in \Omega$,

$$|x(t)| \leq e^{\gamma t}\sqrt{\log\log t} \quad \text{provided } t \text{ is sufficiently large.}$$

We therefore see that

$$\limsup_{t\to\infty} \frac{1}{t}\log|x(t)| \leq \gamma \quad a.s.$$

which is better than (5.14).

Proof. By Itô's formula and condition (5.13), one can derive that

$$e^{-2\gamma t}|x(t)|^2 = e^{-2\gamma t_0}|x_0|^2 + M(t)$$
$$+ \int_{t_0}^t e^{-2\gamma s}\left[-2\gamma|x(s)|^2 + 2x^T(s)f(x(s),s) + |\sigma|^2\right]ds$$
$$\leq e^{-2\gamma t_0}\left[|x_0|^2 + \frac{1}{2\gamma}(2\rho + |\sigma|^2)\right] + M(t), \quad (5.16)$$

where

$$M(t) = 2\int_{t_0}^t e^{-2\gamma s}x^T(s)\sigma dB(s).$$

Assign $p > 1$ arbitrarily. Let $\bar{n}$ be an integer sufficiently large for $2^{\bar{n}^p} > t_0$. For each integer $n \geq \bar{n}$, by Theorem 1.7.4, one sees that

$$P\left\{\sup_{t_0 \leq t \leq 2^{n^p}}\left[M(t) - 2\int_{t_0}^t e^{-4\gamma s}|x^T(s)\sigma|^2 ds\right] > 2\log n\right\} \leq \frac{1}{n^2}.$$

The Borel–Cantelli lemma then yields that for almost all $\omega \in \Omega$, there is a random integer $\hat{n} = \hat{n}(\omega) \geq \bar{n}+1$ such that

$$M(t) \leq 2\log n + 2|\sigma|^2 \int_{t_0}^t e^{-4\gamma s}|x(s)|^2 ds, \quad t_0 \leq t \leq 2^{n^p}$$

whenever $n \geq \hat{n}$. Substituting this into (5.16) gives that

$$e^{-2\gamma t}|x(t)|^2 \leq e^{-2\gamma t_0}\left[|x_0|^2 + \frac{1}{2\gamma}(2\rho + |\sigma|^2)\right] + 2\log n$$
$$+ 2|\sigma|^2 \int_{t_0}^t e^{-2\gamma s}\left[e^{-2\gamma s}|x(s)|^2\right]ds,$$

which then implies, by the Gronwall inequality, that

$$e^{-2\gamma t}|x(t)|^2 \leq \left(e^{-2\gamma t_0}\left[|x_0|^2 + \frac{1}{2\gamma}(2\rho + |\sigma|^2)\right] + 2\log n\right)\exp\left(\frac{|\sigma|^2}{\gamma}\right)$$

for all $t_0 \leq t \leq 2^{n^p}$, $n \geq \hat{n}$ almost surely. Therefore, for almost all $\omega \in \Omega$, if $2^{(n-1)^p} \leq t \leq 2^{n^p}$, $n \geq \hat{n}$,

$$\frac{|x(t)|^2}{e^{2\gamma t}\log\log t} \leq \left(e^{-2\gamma t_0}\left[|x_0|^2 + \frac{1}{2\gamma}(2\rho + |\sigma|^2)\right] + 2\log n\right)$$
$$\times \exp\left(\frac{|\sigma|^2}{\gamma}\right)[p\log(n-1) + \log\log 2]^{-1}.$$

It then follows that

$$\limsup_{t \to \infty} \frac{|x(t)|^2}{e^{2\gamma t}\log\log t} \leq \frac{2}{p}\exp\left(\frac{|\sigma|^2}{\gamma}\right) \quad a.s.$$

Since $p > 1$ is arbitrary, we must have that

$$\lim_{t \to \infty} \frac{|x(t)|}{e^{\gamma t}\sqrt{\log\log t}} = 0 \quad a.s.$$

which is the desired conclusion.

We now strengthen condition (5.13) by letting $\gamma = 0$ to see how the solution behaves asymptotically. Before we state a new result, let us emphasis that although we only use the trace norm $|A| = \sqrt{trace(A^T A)}$ for a matrix A so far, we shall use in the sequel of this book the another norm, i.e. the operator norm $\|A\| = \sup\{|Ax| : |x| = 1\}$. The reader should distinguish these two different (though equivalent) norms and note that $\|A\| \leq |A|$.

Theorem 5.4 *Assume that there exists a positive constant ρ such that*

$$x^T f(x, t) \leq \rho \quad \text{for all } (x, t) \in R^d \times [t_0, \infty). \quad (5.17)$$

Then the solution of equation (5.12) has the property

$$\limsup_{t \to \infty} \frac{|x(t)|}{\sqrt{2t\log\log t}} \leq \|\sigma\|\sqrt{e} \quad a.s. \quad (5.18)$$

Proof. Using Itô's formula and hypothesis (5.17) one can show that for $t \geq t_0$

$$|x(t)|^2 \leq |x_0|^2 + (2\rho + |\sigma|^2)(t - t_0) + M(t), \qquad (5.19)$$

where

$$M(t) = 2 \int_{t_0}^t x^T(s)\sigma dB(s).$$

Assign $\beta > 0$ and $\theta > 1$ arbitrarily. For every integer n sufficiently large for $\theta^n > t_0$, one can apply Theorem 1.7.4 to show that

$$P\left\{\sup_{t_0 \leq t \leq \theta^n}\left[M(t) - 2\beta\theta^{-n}\int_{t_0}^t |x^T(s)\sigma|^2 ds\right] > \beta^{-1}\theta^{n+1}\log n\right\} \leq \frac{1}{n^\theta}.$$

The Borel–Cantelli lemma then yields that for almost all $\omega \in \Omega$, there is a random integer $n_0 = n_0(\omega)$ sufficiently large such that

$$M(t) \leq \beta^{-1}\theta^{n+1}\log n + 2\beta||\sigma||^2\theta^{-n}\int_{t_0}^t |x(s)|^2 ds, \qquad t_0 \leq t \leq \theta^n.$$

Substituting this into (5.19) yields that for almost all $\omega \in \Omega$,

$$|x(t)|^2 \leq |x_0|^2 + (2\rho + |\sigma|^2)(t - t_0) + \beta^{-1}\theta^{n+1}\log n$$
$$+ 2\beta||\sigma||^2\theta^{-n}\int_{t_0}^t |x(s)|^2 ds$$

for all $t_0 \leq t \leq \theta^n$, $n \geq n_0$, which implies

$$|x(t)|^2 \leq \left[|x_0|^2 + (2\rho + |\sigma|^2)(\theta^n - t_0) + \beta^{-1}\theta^{n+1}\log n\right]e^{2\beta||\sigma||^2}.$$

In particular, for almost all $\omega \in \Omega$, if $\theta^{n-1} \leq t \leq \theta^n$, $n \geq n_0$,

$$\frac{|x(t)|^2}{2t\log\log t} \leq \left[|x_0|^2 + (2\rho + |\sigma|^2)(\theta^n - t_0) + \beta^{-1}\theta^{n+1}\log n\right]$$
$$\times e^{2\beta||\sigma||^2}\left(2\theta^{n-1}[\log(n-1) + \log\log\theta]\right)^{-1}.$$

Consequently

$$\limsup_{t \to \infty} \frac{|x(t)|^2}{2t\log\log t} \leq \frac{\theta^2}{2\beta}e^{2\beta||\sigma||^2} \qquad a.s.$$

Finally, letting $\theta \to 1$ and choosing $\beta = (2||\sigma||^2)^{-1}$ we obtain that

$$\limsup_{t \to \infty} \frac{|x(t)|^2}{2t\log\log t} \leq e||\sigma||^2 \qquad a.s.$$

and the required conclusion (5.18) follows immediately. The proof is complete.

Theorem 5.5 *Assume that there exists a pair of positive constants γ and ρ such that*

$$x^T f(x,t) \leq -\gamma |x|^2 + \rho \quad \text{for all } (x,t) \in R^d \times [t_0, \infty). \quad (5.20).$$

Then the solution of equation (5.12) has the property

$$\limsup_{t \to \infty} \frac{|x(t)|}{\sqrt{\log t}} \leq \|\sigma\| \sqrt{\frac{e}{\gamma}} \quad a.s. \quad (5.21)$$

Proof. Assign $\delta > 0$, $\beta > 0$ and $\theta > 1$ arbitrarily. In the same way as the proof of Theorem 5.3, one can show that for almost all $\omega \in \Omega$, there is a random integer $n_0 = n_0(\omega)$ sufficiently large such that

$$e^{2\gamma t}|x(t)|^2 \leq e^{2\gamma t_0}|x_0|^2 + \frac{e^{2\gamma t}}{2\gamma}(2\rho + |\sigma|^2) + \beta^{-1}\theta e^{2\gamma n\delta}\log n$$

$$+ 2\beta|\sigma|^2 e^{-2\gamma n\delta} \int_{t_0}^{t} e^{2\gamma s}\left[e^{2\gamma s}|x(s)|^2\right]ds,$$

for all $t_0 \leq t \leq n\delta$, $n \geq n_0$, which implies

$$e^{2\gamma t}|x(t)|^2 \leq \left[e^{2\gamma t_0}|x_0|^2 + \frac{e^{2\gamma t}}{2\gamma}(2\rho + |\sigma|^2) + \beta^{-1}\theta e^{2\gamma n\delta}\log n\right]\exp\left(\frac{\beta|\sigma|^2}{\gamma}\right).$$

Therefore, for almost all $\omega \in \Omega$, if $(n-1)\delta \leq t \leq n\delta$, $n \geq n_0$,

$$\frac{|x(t)|^2}{\log t} \leq \left[|x_0|^2 + \frac{1}{2\gamma}(2\rho + |\sigma|^2) + \beta^{-1}\theta e^{2\gamma\delta}\log n\right]$$

$$\times \exp\left(\frac{\beta|\sigma|^2}{\gamma}\right)[\log(n-1) + \log\delta]^{-1}.$$

So

$$\limsup_{t \to \infty} \frac{|x(t)|^2}{\log t} \leq \beta^{-1}\theta e^{2\gamma\delta}\exp\left(\frac{\beta|\sigma|^2}{\gamma}\right) \quad a.s.$$

Finally, letting $\theta \to 1$, $\delta \to 0$ and choosing $\beta = \gamma/\|\sigma\|^2$ we obtain that

$$\limsup_{t \to \infty} \frac{|x(t)|}{\sqrt{\log t}} \leq \|\sigma\|\sqrt{\frac{e}{\gamma}} \quad a.s.$$

as required. The proof is complete.

It is not difficult to see that Theorems 5.3–5.5 can be extended to equation (5.1) as long as the coefficient $g(x,t)$ is bounded. More precisely, if there exists a $K > 0$ such that

$$\|g(x,t)\| \leq K \quad \text{for all } (x,t) \in R^d \times [t_0, \infty),$$

then Theorems 5.3–5.5 still hold for the solution of equation (5.1) and, of course, the corresponding $\|\sigma\|$ should be replaced by K. We leave the details to the reader.

To close this section, let us discuss two special cases of equation (5.12) in order to show that the estimates obtained above are quite sharp. First of all, let $m = d$, $f(x,t) \equiv 0$ and σ be the $d \times d$ identity matrix. By Theorem 5.4, we have that

$$\limsup_{t \to \infty} \frac{|x(t)|}{\sqrt{2t \log \log t}} \leq \sqrt{e} \quad a.s. \tag{5.22}$$

On the other hand, in this case, the equation has the explicit solution $x(t) = x_0 + B(t) - B(t_0)$. Applying the law of the iterated logarithm for d-dimensional Brownian motion (cf. page 17) we see that the left-hand side of (5.22) equals to 1. In other words, even in this very special case, Theorem 5.4 still gives reasonably sharp estimate. Next, we let $d = m = 1$, $t_0 = 0$, $f(x,t) = -\gamma x$, σ and γ be both positive constants. This is, we consider the one-dimensional equation

$$dx(t) = -\gamma x(t) dt + \sigma dB(t) \quad \text{on } t \geq 0, \tag{5.23}$$

where $B(t)$ is the one-dimensional Brownian motion. The integration by parts formula yields that

$$e^{\gamma t} x(t) = x(0) + M(t),$$

where $M(t) = \sigma \int_0^t e^{\gamma s} dB(s)$ is a continuous martingale with the quadratic variation $\langle M, M \rangle_t = \sigma^2 (e^{2\gamma t} - 1)/2\gamma := \mu(t)$. It is easy to show that the inverse function of $\mu(t)$ is $\mu^{-1}(t) = \log(2\gamma t/\sigma^2 + 1)/2\gamma$ and, by Theorem 1.4.4, $\{M(\mu^{-1}(t))\}_{t \geq 0}$ is a Brownian motion. Hence, by the law of the iterated logarithm (i.e. Theorem 1.4.2),

$$\limsup_{t \to \infty} \frac{|M(\mu^{-1}(t))|}{\sqrt{2t \log \log t}} = 1 \quad a.s.$$

which implies

$$\limsup_{t \to \infty} \frac{|M(t)|}{\sqrt{2\mu(t) \log \log \mu(t)}} = \limsup_{t \to \infty} \frac{|M(t)|}{e^{\gamma t} \sqrt{(\sigma^2/\gamma) \log t}} = 1 \quad a.s.$$

Therefore

$$\limsup_{t \to \infty} \frac{|x(t)|}{\sqrt{\log t}} = \limsup_{t \to \infty} \frac{|M(t)|}{e^{\gamma t} \sqrt{\log t}} = \frac{\sigma}{\sqrt{\gamma}} \quad a.s. \tag{5.24}$$

On the other hand, by Theorem 5.5, we can estimate that the left-hand side of (5.24) is less or equal to $\sigma \sqrt{e/\gamma}$, which is reasonably closed to the above accurate value $\sigma/\sqrt{\gamma}$.

2.6 CARATHEODORY'S APPROXIMATE SOLUTIONS

In the previous sections we have established the existence-and-uniqueness theorems and discussed the properties of the solution for the stochastic differential equation

$$dx(t) = f(x(t),t)dt + g(x(t),t)dB(t), \qquad t \in [t_0, T] \qquad (6.1)$$

with initial value $x(t_0) = x_0 \in L^2$. However, the Lipschitz condition etc. only guarantee the existence and uniqueness of the solution and, in general, the solution does not have an explicit expression except the linear case which will be discussed in Chapter 3 below. In practice, we therefore often seek the approximate solution rather than the accurate solution.

In Section 2.3 we use the Picard iteration procedure to establish the theorem on the existence and uniqueness of the solution. As the by-product, we also obtain the Picard approximate solution for the equation, and Theorem 3.3 gives an estimate on the difference, called the *error*, between the approximate and the accurate solution. In practice, given the error $\varepsilon > 0$, one can determine n for the left-hand side of (3.10) to be less than ε, and then compute $x_0(t), x_1(t), \cdots, x_n(t)$ by the Picard iteration (3.4). According to Theorem 3.3, we have

$$E\left(\sup_{t_0 \leq t \leq T} |x_n(t) - x(t)|^2\right) < \varepsilon.$$

So we can use $x_n(t)$ as the approximate solution to equation (6.1). The disadvantage of the Picard approximation is that one needs to compute $x_0(t), x_1(t), \cdots, x_{n-1}(t)$ in order to compute $x_n(t)$, and this will involve a lot of calculations on stochastic integrals. More efficient ways in this direction are Caratheodory's approximation procedure and Cauchy–Maruyama's. We shall discuss the former in this section and latter in the next section.

Let us now give the definition of Caratheodory's approximate solutions. For every integer $n \geq 1$, define $x_n(t) = x_0$ for $t_0 - 1 \leq t \leq t_0$ and

$$x_n(t) = x_0 + \int_{t_0}^{t} f(x_n(s - 1/n), s)ds + \int_{t_0}^{t} g(x_n(s - 1/n), s)dB(s) \qquad (6.2)$$

for $t_0 < t \leq T$. Note that for $t_0 \leq t \leq t_0 + 1/n$, $x_n(t)$ can be computed by

$$x_n(t) = x_0 + \int_{t_0}^{t} f(x_0, s)ds + \int_{t_0}^{t} g(x_0, s)dB(s);$$

then for $t_0 + 1/n < t \leq t_0 + 2/n$,

$$x_n(t) = x_n(t_0 + 1/n) + \int_{t_0+1/n}^{t} f(x_n(s - 1/n), s)ds$$
$$+ \int_{t_0+1/n}^{t} g(x_n(s - 1/n), s)dB(s)$$

and so on. In other words, $x_n(t)$ can be computed step-by-step on the intervals $[t_0, t_0 + 1/n]$, $(t_0 + 1/n, t_0 + 2/n]$, $\cdots$. We need to prepare two lemmas in order to establish the main results.

Lemma 6.1 *Under the linear growth condition (3.2), for all $n \geq 1$,*

$$\sup_{t_0 \leq t \leq T} E|x_n(t)|^2 \leq C_1 := (1 + 3E|x_0|^2)e^{3K(T-t_0)(T-t_0+1)}. \tag{6.3}$$

Proof. Fix $n \geq 1$ arbitrarily. It is easy to see from the definition of $x_n(t)$ and condition (3.2) that $\{x_n(t)\}_{t_0 \leq t \leq T} \in \mathcal{M}^2([t_0, T]; R^d)$. Note from (6.2) that for $t_0 \leq t \leq T$,

$$|x_n(t)|^2 \leq 3|x_0|^2 + 3\left|\int_{t_0}^t f(x_n(s - 1/n), s)ds\right|^2$$

$$+ 3\left|\int_{t_0}^t g(x_n(s - 1/n), s)dB(s)\right|^2.$$

Using the Hölder inequality, Theorem 1.5.21 as well as condition (6.2) one can then derive that

$$E|x_n(t)|^2 \leq 3E|x_0|^2 + 3(t - t_0)E\int_{t_0}^t |f(x_n(s - 1/n), s)|^2 ds$$

$$+ 3E\int_{t_0}^t |g(x_n(s - 1/n), s)|^2 ds$$

$$\leq 3E|x_0|^2 + 3K(T - t_0 + 1)\int_{t_0}^t [1 + E|x_n(s - 1/n)|^2]ds$$

$$\leq 3E|x_0|^2 + 3K(T - t_0 + 1)\int_{t_0}^t \left[1 + \sup_{t_0 \leq r \leq s} E|x_n(r)|^2\right]ds$$

for all $t_0 \leq t \leq T$. Consequently

$$1 + \sup_{t_0 \leq r \leq t} E|x_n(r)|^2$$

$$\leq 1 + 3E|x_0|^2 + 3K(T - t_0 + 1)\int_{t_0}^t \left[1 + \sup_{t_0 \leq r \leq s} E|x_n(r)|^2\right]ds.$$

The Gronwall inequality implies

$$1 + \sup_{t_0 \leq r \leq t} E|x_n(r)|^2 \leq (1 + 3E|x_0|^2)e^{3K(t-t_0)(T-t_0+1)}$$

for all $t_0 \leq t \leq T$. In particular, the required (6.3) follows when $t = T$.

Lemma 6.2 *Under the linear growth condition (3.2), for all $n \geq 1$ and $t_0 \leq s < t \leq T$ with $t - s \leq 1$,*

$$E|x_n(t) - x_n(s)|^2 \leq C_2(t - s), \tag{6.4}$$

where $C_2 = 4K(1+C_1)$ and C_1 is defined in Lemma 6.1.

Proof. Note that

$$x_n(t) - x_n(s) = \int_s^t f(x_n(r-1/n), r) dr + \int_s^t g(x_n(r-1/n), r) dB(r).$$

Hence, by Lemma 6.1,

$$E|x_n(t) - x_n(s)|^2$$
$$\leq 2E\left|\int_s^t f(x_n(r-1/n), r) dr\right|^2 + 2E\left|\int_s^t g(x_n(r-1/n), r) dB(r)\right|^2$$
$$\leq 2K(t-s+1)\int_s^t [1 + E|x_n(r-1/n)|^2] dr$$
$$\leq 4K(1+C_1)(t-s)$$

as required.

We can now state the main result.

Theorem 6.3 *Assume that the Lipschitz condition (3.1) and the linear growth condition (3.2) hold. Let $x(t)$ be the unique solution of equation (6.1). Then, for $n \geq 1$,*

$$E\left(\sup_{t_0 \leq t \leq T} |x_n(t) - x(t)|^2\right) \leq \frac{C_3}{n}, \qquad (6.5)$$

where $C_3 = 4C_2 \bar{K}(T-t_0)(T-t_0+4)\exp[4\bar{K}(T-t_0)(T-t_0+4)]$ and C_2 is defined in Lemma 6.2.

Proof. It is not difficult to derive that

$$E\left(\sup_{t_0 \leq r \leq t} |x_n(r) - x(r)|^2\right) \leq 2\bar{K}(T-t_0+4)\int_{t_0}^t E|x_n(s-1/n) - x(s)|^2 ds$$
$$\leq 4\bar{K}(T-t_0+4)\int_{t_0}^t \left[E|x_n(s) - x_n(s-1/n)|^2 + E|x_n(s) - x(s)|^2\right] ds.$$

But, by Lemma 6.2, $E|x_n(s) - x_n(s-1/n)|^2 \leq C_2/n$ if $s \geq t_0 + 1/n$, otherwise if $t_0 \leq s < t_0 + 1/n$, $E|x_n(s) - x_n(s-1/n)|^2 = E|x_n(s) - x_n(t_0)|^2 \leq C_2(s-t_0)$ which is less than C_2/n. Therefore, it follows from the above inequality that

$$E\left(\sup_{t_0 \leq r \leq t} |x_n(r) - x(r)|^2\right) \leq \frac{4}{n}C_2\bar{K}(T-t_0)(T-t_0+4)$$
$$+ 4\bar{K}(T-t_0+4)\int_{t_0}^t E\left(\sup_{t_0 \leq r \leq s} |x_n(r) - x(r)|^2\right) ds.$$

Finally, the required inequality (6.5) follows by applying the Gronwall inequality. The proof is complete.

In practice, given the error $\varepsilon > 0$, one can let n be an integer larger than C_3/ε and then compute $x_n(t)$ over the intervals $[t_0, t_0 + 1/n]$, $(t_0 + 1/n, t_0 + 2/n], \cdots$, step by step. Theorem 6.3 guarantees that this $x_n(t)$ is closed enough to the accurate solution $x(t)$ in the sense

$$E\left(\sup_{t_0 \le t \le T} |x_n(t) - x(t)|^2\right) < \varepsilon.$$

Comparing Picard's approximation, we see the advantage of Caratheodory's approximation that we do not need to compute $x_1(t), \cdots, x_{n-1}(t)$ but compute $x_n(t)$ directly.

In the proof of Theorem 6.3 we have made use of the fact that equation (6.1) has a unique solution under conditions (3.1) and (3.2), and the proof therefore becomes relatively easier. On the other hand, it is possible to show, without using this fact, that Caratheodory's approximation sequence $\{x_n(t)\}$ is Cauchy in L^2 hence converges to a limit, say $x(t)$; and then show that $x(t)$ is the unique solution to equation (6.1), and (6.5) holds. In other words, we can completely use the Caratheodory approximation procedure to establish the existence-and-uniqueness theorem. The details can be found in the author's previous book Mao (1994a).

Moreover, under quite general conditions, we are still able to show that the Caratheodory approximate solutions converge to the unique solution of equation (6.1). This is described as follows.

Theorem 6.4 *Let $f(x,t)$ and $g(x,t)$ be continuous. Let x_0 be a bounded R^d-valued $\mathcal{F}_{t_0}$-measurable random variable. Let the linear growth condition (3.2) hold. Assume that the equation (6.1) has a unique[1] solution $x(t)$. Then the Caratheodory approximate solutions $x_n(t)$ converge to $x(t)$ in the sense that*

$$\lim_{n \to \infty} E\left(\sup_{t_0 \le t \le T} |x_n(t) - x(t)|^2\right) = 0. \tag{6.6}$$

The proof is omitted here but can be found in Mao (1994b). We shall now use this theorem to establish one useful result.

Theorem 6.5 *Let $f(x,t)$ and $g(x,t)$ be continuous. Let x_0 be a bounded R^d-valued $\mathcal{F}_{t_0}$-measurable random variable. Assume that there exists a continuous increasing concave function $\kappa : R_+ \to R_+$ such that*

$$\int_{0+} \frac{du}{\kappa(u)} = \infty, \tag{6.7}$$

and for all $x, y \in R^d$, $t_0 \le t \le T$

$$|f(x,t) - f(y,t)|^2 \bigvee |g(x,t) - g(y,t)|^2 \le \kappa(|x-y|^2). \tag{6.8}$$

---

[1] More precisely we mean the pathwise uniqueness here. Please see Remark 7.5 in the next section for the definition of pathwise uniqueness.

Then the equation (6.1) has a unique solution $x(t)$. Moreover, the Caratheodory approximate solutions $x_n(t)$ converge to $x(t)$ in the sense of (6.6).

Proof. We leave the proof of existence to the reader (cf. Yamada (1981)). To show the uniqueness, let $x(t)$ and $\bar{x}(t)$ be two solutions to equation (6.1). By (6.8), it is easy to show that

$$E\left(\sup_{t_0 \leq r \leq t} |x(r) - \bar{x}(r)|^2\right) \leq 2(T - t_0 + 4) \int_{t_0}^{t} E\kappa(|x(s) - \bar{x}(s)|^2) ds.$$

Since $\kappa(\cdot)$ is concave, by the well-known Jensen inequality, we have

$$E\kappa(|x(s) - \bar{x}(s)|^2) \leq \kappa(E|x(s) - \bar{x}(s)|^2) \leq \kappa\left[E\left(\sup_{t_0 \leq r \leq s} |x(r) - \bar{x}(r)|^2\right)\right].$$

Consequently, for any $\varepsilon > 0$,

$$E\left(\sup_{t_0 \leq r \leq t} |x(r) - \bar{x}(r)|^2\right)$$

$$\leq \varepsilon + 2(T - t_0 + 4) \int_{t_0}^{t} \kappa\left[E\left(\sup_{t_0 \leq r \leq s} |x(r) - \bar{x}(r)|^2\right)\right] ds \qquad (6.9)$$

for all $t_0 \leq t \leq T$. Define

$$G(r) = \int_{1}^{r} \frac{du}{\kappa(u)} \quad \text{on } r > 0,$$

and let $G^{-1}(\cdot)$ be the inverse function of $G(\cdot)$. By condition (6.7), one sees that $\lim_{\varepsilon \downarrow 0} G(\varepsilon) = -\infty$ and $Dom(G^{-1}) = (-\infty, G(\infty))$. Therefore, by the Bihari inequality (i.e. Theorem 1.8.2), one deduces from (6.9) that, for all sufficiently small $\varepsilon > 0$,

$$E\left(\sup_{t_0 \leq r \leq T} |x(r) - \bar{x}(r)|^2\right) \leq G^{-1}\left[G(\varepsilon) + 2(T - t_0 + 4)(T - t_0)\right].$$

Letting $\varepsilon \to 0$ gives

$$E\left(\sup_{t_0 \leq r \leq T} |x(r) - \bar{x}(r)|^2\right) = 0.$$

Hence, $x(t) = \bar{x}(t)$ for all $t_0 \leq t \leq T$ almost surely. The uniqueness has been proved. To show (6.6), we only need to verify the linear growth condition according to Theorem 6.4. Since $\kappa(\cdot)$ is concave and increasing, there must exist a positive number a such that

$$\kappa(u) \leq a(1 + u) \quad \text{on } u \geq 0.$$

Besides, let $b = \sup_{t_0 \leq t \leq T}(|f(0,t)|^2 \vee |g(0,t)|^2) < \infty$. Then

$$|f(x,t)|^2 \vee |g(x,t)|^2$$
$$\leq 2(|f(0,t)|^2 \vee |g(0,t)|^2) + 2(|f(x,t) - f(0,t)|^2 \vee |g(x,t) - g(0,t)|^2)$$
$$\leq 2b + 2\kappa(|x|^2) \leq 2b + 2a(1 + |x|^2) \leq 2(a + b)(1 + |x|^2).$$

That is, the linear growth condition (3.2) is fulfilled with $K = 2(a + b)$. The proof is now complete.

To close this section, let us consider a one-dimensional equation

$$dx(t) = |x(t)|^\alpha dB(t) \quad \text{on } t_0 \leq t \leq T \tag{6.10}$$

with initial value $x(t_0) = x_0$ which is bounded, where $\frac{1}{2} \leq \alpha < 1$ and $B(t)$ is a one-dimensional Brownian motion. As we pointed out before, equation (6.10) has a unique solution. Besides, the linear growth condition is clearly fulfilled. Therefore, according to Theorem 6.4, the Caratheodory approximate solutions converge to the unique solution. However, it is still open whether the Picard approximate solutions converge to the unique solution or not in this case. So far, perhaps the best conditions which guarantee the convergence of the Picard approximate solutions to the unique solution of equation (6.1) are the conditions given in Theorem 6.5, and they were obtained by Yamada (1981).

2.7 CAUCHY–MARUYAMA'S APPROXIMATE SOLUTIONS

Let us now turn to discuss the Cauchy–Maruyama approximate solutions which are defined as follows: For every integer $n \geq 1$, define $x_n(t_0) = x_0$, and then for $t_0 + (k-1)/n < t \leq (t_0 + k/n) \wedge T$, $k = 1, 2, \cdots$,

$$x_n(t) = x_n(t_0 + (k-1)/n) + \int_{t_0+(k-1)/n}^{t} f(x_n(t_0 + (k-1)/n), s)ds$$

$$+ \int_{t_0+(k-1)/n}^{t} g(x_n(t_0 + (k-1)/n), s)dB(s). \tag{7.1}$$

Note that if define

$$\hat{x}_n(t) = x_0 I_{\{t_0\}}(t) + \sum_{\kappa \geq 1} x_n(t_0 + (k-1)/n) I_{(t_0+(k-1)/n,\ t_0+k/n]}(t) \tag{7.2}$$

for $t_0 \leq t \leq T$, then it follows from (7.1) that

$$x_n(t) = x_0 + \int_{t_0}^{t} f(\hat{x}_n(s), s)ds + \int_{t_0}^{t} g(\hat{x}_n(s), s)dB(s). \tag{7.3}$$

Making use of this expression we can show the following lemmas in the same way as Lemmas 6.1 and 6.2.

Lemma 7.1 *Under the linear growth condition (3.2), the Cauchy–Maruyama approximate solutions $x_n(t)$ have the property that*

$$\sup_{t_0 \leq t \leq T} E|x_n(t)|^2 \leq C_1 := (1 + 3E|x_0|^2)e^{3K(T-t_0)(T-t_0+1)}.$$

Lemma 7.2 *Under the linear growth condition (3.2), the Cauchy–Maruyama approximate solutions $x_n(t)$ have the property that for $t_0 \leq s < t \leq T$ with $t - s \leq 1$,*
$$E|x_n(t) - x_n(s)|^2 \leq C_2(t-s),$$
where $C_2 = 4K(1 + C_1)$ and C_1 is defined in Lemma 7.1.

We can then prove the following theorem in the same way as Theorem 6.3.

Theorem 7.3 *Assume that the Lipschitz condition (3.1) and the linear growth condition (3.2) hold. Let $x(t)$ be the unique solution of equation (6.1), and $x_n(t)$, $n \geq 1$ be the Cauchy–Maruyama approximate solutions. Then*
$$E\left(\sup_{t_0 \leq t \leq T} |x_n(t) - x(t)|^2\right) \leq \frac{C_3}{n},$$
where $C_3 = 4C_2\bar{K}(T-t_0)(T-t_0+4)\exp[4\bar{K}(T-t_0)(T-t_0+4)]$ and C_2 is defined in Lemma 7.2.

We leave these proofs to the reader. Moreover, we also have the following more general result.

Theorem 7.4 *Under the same conditions as Theorem 6.4, Cauchy–Maruyama's approximate solutions $x_n(t)$ converge to the unique solution $x(t)$ of equation (6.1) in the sense of (6.6).*

This result was obtained by Kaneko & Nakao (1988). This theorem and Theorem 6.4 tell us that both Caratheodory's and Cauchy–Maruyama's approximate solutions converge to the unique solution of equation (6.1) under these quite general conditions described in Theorem 6.4. However, it is still open whether the Picard approximate solutions converge to the unique solution under these conditions.

It is interesting to see that the Cauchy–Maruyama approximation becomes much easier for the time-homogeneous stochastic differential equation
$$dx(t) = f(x(t))dt + g(x(t))dB(t). \tag{7.4}$$
In this case, the Cauchy–Maruyama approximate solutions take the following simple form: $x_n(t_0) = x_0$ and
$$\begin{aligned}x_n(t) = {} & x_n(t_0 + (k-1)/n) + f(x_n(t_0 + (k-1)/n))[t - t_0 - (k-1)/n] \\ & + g(x_n(t_0 + (k-1)/n))[B(t) - B(t_0 - (k-1)/n)].\end{aligned} \tag{7.5}$$
for $t_0 + (k-1)/n < t \leq (t_0 + k/n) \wedge T$, $k = 1, 2, \cdots$.

To close this section, let us make an important remark.

Remark 7.5 In the previous sections, the probability space $(\Omega, \mathcal{F}, P)$, the filtration $\{\mathcal{F}_t\}_{t \geq 0}$, the Brownian motion $B(t)$ and the coefficients $f(x,t)$, $g(x,t)$

are all given in advance, and then the solution $x(t)$ is constructed. Such a solution is called a *strong* solution. If we are only given the coefficients $f(x,t)$ and $g(x,t)$, and we are allowed to construct a suitable probability space, a filtration, a Brownian motion and find a solution to the equation, then such a solution is called a *weak* solution. Two solutions (weak or strong) are said to be *weakly unique* if they are identical in probability law, that is, they have the same finite-dimensional probability distribution. If two weak solutions founded under whatever probability space with a filtration and a Brownian motion are indistinguishable, we say that *pathwise uniqueness* holds for the equation. Clearly a strong solution is a weak one, but the converse is not true in general. See the Tanaka example explained in Rogers & Williams (1987), Sec.V.16. Also, the pathwise uniqueness implies the weak uniqueness. Moreover, all the conditions given above e.g. the Lipschitz condition guarantee the pathwise uniqueness, since the uniqueness has been proved under the arbitrarily given probability space etc. In this book, we are always concerned with strong solutions unless otherwise specified.

2.8 SDE AND PDE: FEYNMAN–KAC'S FORMULA

Stochastic differential equations have many applications. One of the most important applications is the stochastic representation for solutions to partial differential equations, and this is known as the Feynman–Kac formula. The formula builds a bridge between stochastic differential equations (SDE) and partial differential equations (PDE), and creates the probabilistic approach to the study of partial differential equations (cf. Friedlin (1985)).

(i) *The Dirichlet Problem*

Let us first consider the Dirichlet problem or the boundary value problem

$$\begin{cases} Lu(x) = \varphi(x) & \text{in } D, \\ u(x) = \phi(x) & \text{on } \partial D, \end{cases} \quad (8.1)$$

where L is a linear partial differential operator

$$L = \frac{1}{2}\sum_{i,j=1}^{d} a_{ij}(x)\frac{\partial^2}{\partial x_i \partial x_j} + \sum_{i=1}^{d} f_i(x)\frac{\partial}{\partial x_i} + c(x) \quad (8.2)$$

with real-valued coefficients defined in a d-dimensional domain $D \subset R^d$. It is standard to arrange a_{ij} symmetrically, i.e. $a_{ij} = a_{ji}$. Assume that D is open and bounded, and its boundary ∂D is C^2. We shall denote by $\bar{D}$ the closure of D. Assume that L is uniformly *elliptic* in D, that is, for some $\mu > 0$,

$$y^T a(x) y \geq \mu |y|^2 \quad \text{if } x \in D,\ y \in R^d, \quad (8.3)$$

where $a(x) = (a_{ij}(x))_{d\times d}$. Assume also that

$$a_{ij},\ f_i \text{ are uniformly Lipschitz continuous in } \bar{D}, \quad (8.4)$$

$$c \leq 0 \text{ and } c \text{ is uniformly Hölder continuous in } \bar{D}. \quad (8.5)$$

Under these hypotheses, it is well-known, by the theory of partial differential equations, that the Dirichlet problem (8.1) has a unique solution u for any given functions φ, ϕ satisfying:

$$\varphi \text{ is uniformly Hölder continuous in } \bar{D}, \tag{8.6}$$

$$\phi \text{ is continuous on } \partial D. \tag{8.7}$$

We shall now represent u in terms of a solution of a stochastic differential equation.

Note from (8.3) that for every $x \in D$, $a(x)$ is a $d \times d$ symmetric positive definite matrix. It is well-known that there exists a unique $d \times d$ positive definite matrix $g(x) = (g_{ij})_{d \times d}$ such that $g(x)g^T(x) = a(x)$, and $g(x)$ is called the square root of $a(x)$. Moreover, condition (8.4) guarantees that $g(x)$ is uniformly Lipschitz continuous in $\bar{D}$. Extend $g(x)$ and $f(x) = (f_1(x), \cdots, f_d(x))^T$ into the whole space R^d so that they remain uniformly Lipschitz continuous, i.e.

$$|f(x) - f(y)| \vee |g(x) - g(y)| \leq \bar{K}|x - y| \qquad \text{if } x, y \in R^d \tag{8.8}$$

for some $\bar{K} > 0$. Clearly, (8.8) implies that f and g satisfy the linear growth condition as well. Now, let $B(t) = (B_1(t), \cdots, B_d(t))^T$, $t \geq 0$ be a d-dimensional Brownian motion defined on the complete probability space $(\Omega, \mathcal{F}, P)$ with the filtration $\{\mathcal{F}_t\}_{t \geq 0}$ satisfying the usual conditions. Consider the d-dimensional stochastic differential equation

$$d\xi(t) = f(\xi(t), t)dt + g(\xi(t), t)dB(t) \qquad \text{on } t \geq 0 \tag{8.9}$$

with initial value $\xi(0) = x \in D$. By Theorem 3.6, equation (8.9) has a unique global solution, which is denoted by $\xi_x(t)$.

Theorem 8.1 *Assume that D is a bounded open subset of R^d and its boundary ∂D is C^2. Let (8.3)–(8.7) hold. Then the unique solution $u(x)$ of the Dirichlet problem (8.1) is given by*

$$u(x) = E\left[\phi(\xi_x(\tau)) \exp\left(\int_0^\tau c(\xi_x(s))ds\right)\right]$$

$$- E\left[\int_0^\tau \varphi(\xi_x(t)) \exp\left(\int_0^t c(\xi_x(s))ds\right) dt\right], \tag{8.10}$$

where τ is the first exit time of $\xi_x(t)$ from D, i.e. $\tau = \inf\{t \geq 0 : \xi_x(t) \notin D\}$.

Proof. Let $\varepsilon > 0$, and denote by U_ε the closed ε-neighbourhood of ∂D. Let $D_\varepsilon = D - U_\varepsilon$, and let τ_ε be the first exit time of $\xi_x(t)$ from D_ε. By Itô's formula, for any $T > 0$,

$$E\left[u(\xi_x(\tau_\varepsilon \wedge T)) \exp\left(\int_0^{\tau_\varepsilon \wedge T} c(\xi_x(s))ds\right)\right] - u(x)$$

$$= E\left[\int_0^{\tau_\varepsilon \wedge T} Lu(\xi_x(t)) \exp\left(\int_0^t c(\xi_x(s))ds\right) dt\right]$$

$$= E\left[\int_0^{\tau_\varepsilon \wedge T} \varphi(\xi_x(t)) \exp\left(\int_0^t c(\xi_x(s))ds\right) dt\right]. \tag{8.11}$$

Taking $\varepsilon \to 0$ and using the bounded convergence theorem, we obtain that

$$u(x) = E\left[u(\xi_x(\tau \wedge T)) \exp\left(\int_0^{\tau \wedge T} c(\xi_x(s)) ds\right)\right]$$
$$- E\left[\int_0^{\tau \wedge T} \varphi(\xi_x(t)) \exp\left(\int_0^t c(\xi_x(s)) ds\right) dt\right]. \qquad (8.12)$$

If we can prove that $\tau < \infty$ a.s., then, by letting $T \to \infty$ and using the bounded convergence theorem, we get the assertion (8.10). To show $\tau < \infty$ a.s., consider the function

$$V(x) = -e^{\lambda x_1} \qquad \text{for } x \in R^d.$$

Noting from (8.3) that $a_{11}(x) \geq \mu > 0$ in D, we can choose $\lambda > 0$ sufficiently large for

$$f_1(x) V_{x_1}(x) + \frac{1}{2} a_{11}(x) V_{x_1 x_1}(x) = \lambda e^{\lambda x_1}\left[f_1(x) - \frac{\lambda}{2} a_{11}(x)\right] \leq -1 \qquad \text{in } D.$$

By Itô's formula,

$$EV(\xi_x(\tau \wedge T)) - V(x)$$
$$= E \int_0^{\tau \wedge T} \left[f_1(\xi_x(s)) V_{x_1}(\xi_x(s)) + \frac{1}{2} a_{11}(\xi_x(s)) V_{x_1 x_1}(\xi_x(s))\right] ds$$
$$\leq -E(\tau \wedge T).$$

Since $|V(x)| \leq C$ in D for some $C > 0$, we then have $E(\tau \wedge T) \leq 2C$. Taking $T \to \infty$ and using the monotone convergence theorem we get $E\tau \leq 2C$, which implies that $\tau < \infty$ a.s. The proof is complete.

As an example, let L be the Laplace operator $\Delta = \sum_{i=1}^d \frac{\partial^2}{\partial x_i^2}$. Then the boundary value problem (8.1) reduces to

$$\begin{cases} \Delta u(x) = \varphi(x) & \text{in } D, \\ u(x) = \phi(x) & \text{on } \partial D, \end{cases} \qquad (8.13)$$

and the corresponding stochastic differential equation (8.9) takes a simple form $d\xi(t) = dB(t)$ which has the solution $\xi_x(t) = x + B(t)$. By Theorem 8.1, if (8.6) and (8.7) hold then the unique solution of equation (8.13) is given by

$$u(x) = E\left[\phi(x + B(\tau)) \exp\left(\int_0^\tau c(x + B(s)) ds\right)\right]$$
$$- E\left[\int_0^\tau \varphi(x + B(t)) \exp\left(\int_0^t c(x + B(s)) ds\right) dt\right], \qquad (8.14)$$

where $\tau = \inf\{t \geq 0 : x + B(t) \notin D\}$.

(ii) The Initial-Boundary Value Problem

Consider next the initial-boundary value problem

$$\begin{cases} \frac{\partial}{\partial t} u(x,t) + Lu(x,t) = \varphi(x) & \text{in } D \times [0,T], \\ u(x,T) = \phi(x) & \text{on } D, \\ u(x,t) = b(x,t) & \text{on } \partial D \times [0,T], \end{cases} \quad (8.15)$$

where $T > 0$, D is the same as before, and

$$L = \frac{1}{2} \sum_{i,j=1}^{d} a_{ij}(x,t) \frac{\partial^2}{\partial x_i \partial x_j} + \sum_{i=1}^{d} f_i(x,t) \frac{\partial}{\partial x_i} + c(x,t) \quad (8.16)$$

with real-valued coefficients defined in $\bar{D} \times [0,T]$. Set $a(x,t) = (a_{ij}(x,t))_{d \times d}$. We impose the following hypotheses:

$$\begin{aligned} & y^T a(x,t) y \geq \mu |y|^2 \quad \text{if } (x,t) \in D \times [0,T],\ y \in R^d, \\ & a_{ij},\ f_i \text{ are uniformly Lipschitz continuous in } (x,t) \in \bar{D} \times [0,T], \\ & c,\varphi \text{ are uniformly Hölder continuous in } (x,t) \in \bar{D} \times [0,T], \quad (8.17) \\ & \phi \text{ is continuous on } \bar{D},\ b \text{ is continuous on } \partial D \times [0,T], \\ & \phi(x) = b(x,T) \quad \text{if } x \in \partial D. \end{aligned}$$

It is well-known that the initial-boundary value problem (8.15) has a unique solution if (8.17) is fulfilled. To represent u in terms of a solution of a stochastic differential equation, set $f(x,t) = (f_1, \cdots, f_d)^T$ and let $g(x,t) = (g_{ij}(x,t))_{d \times d}$ be the square root of $a(x,t)$ in $\bar{D} \times [0,T]$, i.e. $g(x,t) g^T(x,t) = a(x,t)$. Extend $f,\ g$ to $R^d \times [0,T]$ keeping the Lipschitz continuity

$$|f(x,t) - f(y,s)| \vee |g(x,t) - g(y,s)| \leq K(|x-y| + |t-s|) \quad (K > 0).$$

For every $(x,t) \in D \times [0,T]$, consider the stochastic differential equation

$$d\xi(s) = f(\xi(s), s) ds + g(\xi(s), s) dB(s) \quad \text{on } [t,T] \quad (8.18)$$

with initial value $\xi(t) = x$. By Theorem 3.1, equation (8.18) has a unique solution, which we denote by $\xi_{x,t}(s)$ on $s \in [t,T]$.

Theorem 8.2 *Assume that D is a bounded open subset of R^d and its boundary ∂D is C^2. Let (8.17) hold. Then the unique solution $u(x,t)$ of the initial-boundary value problem (8.15) is given by*

$$\begin{aligned} u(x,t) = & E\left[I_{\{\tau < T\}} b(\xi_{x,t}(\tau), \tau) \exp\left(\int_t^\tau c(\xi_{x,t}(s), s) ds \right) \right] \\ & + E\left[I_{\{\tau = T\}} \phi(\xi_{x,t}(T)) \exp\left(\int_t^T c(\xi_{x,t}(s), s) ds \right) \right] \\ & - E\left[\int_t^\tau \varphi(\xi_{x,t}(s), s) \exp\left(\int_t^s c(\xi_{x,t}(r), r) dr \right) ds \right], \end{aligned} \quad (8.19)$$

where $\tau = T \wedge \inf\{s \in [t,T] : \xi_{x,t}(s) \notin D\}$.

The proof of this theorem is similar to that of Theorem 8.1, but here one applies Itô's formula to

$$u(\xi_{x,t}(s), s) \exp\left(\int_t^s c(\xi_{x,t}(r), r)dr\right). \tag{8.20}$$

(iii) The Cauchy Problem

When $D = R^d$ in the initial-boundary value problem (8.15), we arrive at the following Cauchy problem

$$\begin{cases} \frac{\partial}{\partial t} u(x,t) + Lu(x,t) = \varphi(x) & \text{in } R^d \times [0,T), \\ u(x,T) = \phi(x) & \text{in } R^d, \end{cases} \tag{8.21}$$

where L is given by (8.16). We shall assume:

(H1) The functions a_{ij}, f_i are bounded in $R^d \times [0,T]$ and uniformly Lipschitz continuous in (x,t) in any compact subset of $R^d \times [0,T]$. The functions a_{ij} are Hölder continuous in x, uniformly with respect to (x,t) in $R^d \times [0,T]$. Moreover, for some $\mu > 0$,

$$y^T a(x,t)y \geq \mu |y|^2 \quad \text{if } (x,t) \in R^d \times [0,T],\ y \in R^d.$$

(H2) The function c is bounded in $R^d \times [0,T]$ and uniformly Hölder continuous in (x,t) in any compact subset of $R^d \times [0,T]$.

(H3) The function f is continuous in $R^d \times [0,T]$, Hölder continuous in x uniformly with respect to (x,t) in $R^d \times [0,T]$. The function ϕ is continuous in R^d. Moreover, for some $\alpha > 0$, $\beta > 0$,

$$|f(x,t)| \vee |\phi(x)| \leq \beta(1 + |x|^\alpha) \quad \text{if } x \in R^d,\ t \in [0,T].$$

Under these hypotheses, there exists a unique solution u to the Cauchy problem (8.21). Besides, by Theorem 3.4, the stochastic differential equation (8.18) also has a unique solution denoted by $\xi_{x,t}(s)$.

Theorem 8.3 *Let (H1)–(H3) hold. Then the unique solution $u(x,t)$ of the Cauchy problem (8.21) is given by*

$$u(x,t) = E\left[\phi(\xi_{x,t}(T)) \exp\left(\int_t^T c(\xi_{x,t}(s), s)ds\right)\right]$$
$$- E\left[\int_t^T \varphi(\xi_{x,t}(s), s) \exp\left(\int_t^s c(\xi_{x,t}(r), r)dr\right)ds\right]. \tag{8.22}$$

The proof follows by applying Itô's formula to the function defined by (8.20).

We now consider some special cases of equation (8.21). First, when $\varphi = 0$ and $c = 0$, equation (8.21) becomes the Kolmogorov backward equation

$$\begin{cases} \frac{\partial}{\partial t} u(x,t) + \mathcal{L}u(x,t) = 0 & \text{in } R^d \times [0,T), \\ u(x,T) = \phi(x) & \text{in } R^d, \end{cases} \quad (8.23)$$

where

$$\mathcal{L} = \frac{1}{2} \sum_{i,j=1}^{d} a_{ij}(x,t) \frac{\partial^2}{\partial x_i \partial x_j} + \sum_{i=1}^{d} f_i(x,t) \frac{\partial}{\partial x_i}.$$

In this case, formula (8.22) reduces to the simple form

$$u(x,t) = E\phi(\xi_{x,t}(T)). \quad (8.24)$$

Next, if let $\mathcal{L} = \Delta$ and $\varphi = 0$, equation (8.21) becomes the heat equation

$$\begin{cases} \frac{\partial}{\partial t} u(x,t) + \Delta u(x,t) = 0 & \text{in } R^d \times [0,T), \\ u(x,T) = \phi(x) & \text{in } R^d. \end{cases} \quad (8.25)$$

In this case, the corresponding stochastic differential equation (8.18) reduces to

$$d\xi(s) = dB(s) \quad \text{on } [t,T]$$

with initial value $\xi(t) = x$. Clearly, this stochastic equation has the explicit solution $\xi_{x,t}(s) = x + B(s) - B(t)$. Therefore, by Theorem 8.3, the solution of the heat equation (8.25) is given by

$$u(x,t) = E\phi(x + B(T) - B(t)). \quad (8.26)$$

To close this chapter, let us point out that Feynman–Kac formula can also be applied to quasilinear parabolic partial differential equations. To explain, let us consider the following quasilinear equation

$$\begin{cases} \frac{\partial}{\partial t} u(x,t) + \mathcal{L}u(x,t) + c(x,u)u(x,t) = 0 & \text{in } R^d \times [0,T), \\ u(x,T) = \phi(x) & \text{in } R^d, \end{cases} \quad (8.27)$$

where c is now a continuous function defined on $R^d \times R$. In this case, the Feynman–Kac formula has the form

$$u(x,t) = E\left[\phi(\xi_{x,t}(T)) \exp\left(\int_t^T c(\xi_{x,t}(s), u(\xi_{x,t}(s), s)) ds \right) \right]. \quad (8.28)$$

Of course, this is no longer an explicit representation. Nevertheless it is still very useful. For example, assume $\phi(x) \geq 0$ and

$$\underline{c}(x) \leq c(x,u) \leq \bar{c}(x).$$

It then follows from (8.28) that

$$E\left[\phi(\xi_{x,t}(T))\exp\left(\int_t^T \underline{c}(\xi_{x,t}(s))ds\right)\right]$$
$$\leq u(x,t) \leq E\left[\phi(\xi_{x,t}(T))\exp\left(\int_t^T \bar{c}(\xi_{x,t}(s))ds\right)\right]. \tag{8.29}$$

If denote by $\bar{u}(x,t)$ and $\underline{u}(t,x)$ the corresponding solutions of equation (8.27) with $c(x,u)$ replaced by $\bar{c}(x)$ and $\underline{c}(x)$, respectively, we can then rewrite (8.29) as

$$\underline{u}(x,t) \leq u(x,t) \leq \bar{u}(x,t), \tag{8.30}$$

which is a comparison result.

2.9 THE SOLUTIONS AS MARKOV PROCESSES

In this section we shall discuss the Markov property of the solutions. For the convenience of the reader, let us recall some basic facts about Markov processes (for details please see Doob (1953)). In Chapter 1, we gave the definition of the conditional expectation $E(X|\mathcal{G})$. If $\mathcal{G}$ is the σ-algebra generated by a random variable Y, i.e. $\mathcal{G} = \sigma\{Y\}$, we write $E(X|\mathcal{G}) = E(X|Y)$. If X is the indicator function of set A, we write $E(I_A|\mathcal{G}) = P(A|\mathcal{G})$.

A d-dimensional $\mathcal{F}_t$-adapted process $\{\xi(t)\}_{t\geq 0}$ is called a *Markov process* if the following *Markov property* is satisfied: for all $0 \leq s \leq t < \infty$ and $A \in \mathcal{B}^d$,

$$P(\xi(t) \in A|\mathcal{F}_s) = P(\xi(t) \in A|\xi(s)). \tag{9.1}$$

In a usual definition of a Markov process, the σ-algebra $\mathcal{F}_s$ is set to be $\sigma\{\xi(r) : 0 \leq r \leq s\}$, but we here would like to make the definition slightly more general. The Markov property means that given a Markov process, the past and future are independent when the present is known. There are several equivalent formulations of the Markov property. For example, property (9.1) is equivalent to the following one: for any bounded Borel measurable function $\varphi : R^d \to R$ and $0 \leq s \leq t < \infty$,

$$E(\varphi(\xi(t))|\mathcal{F}_s) = E(\varphi(\xi(t))|\xi(s)). \tag{9.2}$$

The *transition probability* of the Markov process is a function $P(x,s;A,t)$, defined on $0 \leq s \leq t < \infty$, $x \in R^d$ and $A \in \mathcal{B}^d$, with the following properties:

a) For every $0 \leq s \leq t < \infty$ and $A \in \mathcal{B}^d$,

$$P(\xi(s), s; A, t) = P(\xi(t) \in A|\xi(s))$$

b) $P(x,s;\cdot,t)$ is a probability measure on $\mathcal{B}^d$ for every $0 \leq s \leq t < \infty$ and $x \in R^d$.

c) $P(\cdot,s;A,t)$ is Borel measurable for every $0 \leq s \leq t < \infty$ and $A \in \mathcal{B}^d$.

d) The Chapman Kolmogorov equation

$$P(x, s; A, t) = \int_{R^d} P(y, r; A, t) P(x, s; dy, r)$$

holds for any $0 \leq s \leq r \leq t < \infty$, $x \in R^d$ and $A \in \mathcal{B}^d$.
Clearly, in terms of transition probability, the Markov property (9.1) becomes

$$P(\xi(t) \in A | \mathcal{F}_s) = P(\xi(s), s; A, t). \tag{9.3}$$

We shall use the notation

$$P\{\xi(t) \in A | \xi(s) = x\} = P(x, s; A, t),$$

which is the probability that the process will be in the set A at time t given the condition that the process was in the state x at time $s \leq t$. It should be stressed that the number $P\{\xi(t) \in A | \xi(s) = x\}$ is simply defined by the equation above, even though the condition $\{\xi(s) = x\}$ may have probability 0. We shall also use the notation

$$E_{x,s}\varphi(\xi(t)) = \int_{R^d} \varphi(y) P(x, s; dy, t). \tag{9.4}$$

With this notation, the Markov property (9.2) can be written as

$$E(\varphi(\xi(t)) | \mathcal{F}_s) = E_{\xi(s),s}\varphi(\xi(t)), \tag{9.5}$$

where the right hand side is the value of the function $E_{x,s}\varphi(\xi(t))$ at $x = \xi(s)$.

A Markov process $\{\xi(t)\}_{t \geq 0}$ is said to be *homogeneous* (with respect to time) if its transition probability $P(x, s; A, t)$ is stationary, namely

$$P(x, s + u; A, t + u) = P(x, s; A, t)$$

for all $0 \leq s \leq t < \infty$, $u \geq 0$, $x \in R^d$ and $A \in \mathcal{B}^d$. In this case, the transition probability is a function of x, A and $t - s$ only, since

$$P(x, s; A, t) = P(x, 0; A, t - s).$$

We can therefore simply write $P(x, 0; A, t) = P(x; A, t)$. Clearly, $P(x; A, t)$ is the probability of transition from x to A in time t, regardless of the actual position of the interval of length t on the time axis. Moreover, the Chapman–Kolmogorov equation becomes

$$P(x; A, t + s) = \int_{R^d} P(y; A, s) P(x; dy, t).$$

Furthermore, with the notation

$$E_x \varphi(\xi(t)) = \int_{R^d} \varphi(y) P(x; dy, t),$$

the Markov property becomes

$$E(\varphi(\xi(t))|\mathcal{F}_s) = E_{\xi(s)}\varphi(\xi(t-s)).$$

A d-dimensional process $\{\xi(t)\}_{t\geq 0}$ is called a *strong Markov process* if the following *strong Markov property* is satisfied: for any bounded Borel measurable function $\varphi : R^d \to R$, any finite $\mathcal{F}_t$-stopping time τ and $t \geq 0$,

$$E(\varphi(\xi(\tau+t))|\mathcal{F}_\tau) = E(\varphi(\xi(\tau+t))|\xi(\tau)). \quad (9.6)$$

Clearly a strong Markov process is a Markov process. In terms of transition probability, the strong Markov property becomes

$$P(\xi(\tau+t) \in A|\mathcal{F}_\tau) = P(\xi(\tau), \tau; A, \tau+t).$$

Using the notation $E_{x,s}$ defined above, the strong Markov property can also be written as

$$E(\varphi(\xi(\tau+t))|\mathcal{F}_\tau) = E_{\xi(\tau),\tau}\varphi(\xi(\tau+t)).$$

Especially, in the homogeneous case, this becomes

$$E(\varphi(\xi(\tau+t))|\mathcal{F}_\tau) = E_{\xi(\tau)}\varphi(\xi(t)).$$

In general, a Markov process is not a strong one. The conditions that guarantee a Markov process possesses the strong Markov property are the right continuity of the sample paths plus the so-called *Feller property*. If for any bounded continuous function $\varphi : R^d \to R$, the mapping

$$(x,s) \to \int_{R^d} \varphi(y) P(x,s; dy, s+\lambda)$$

is continuous, for any fixed $\lambda > 0$, we then say the transition probability (or the corresponding Markov process) satisfies the *Feller property*.

We can now begin to discuss the Markov property of the solutions of stochastic differential equations.

Theorem 9.1 *Let $\xi(t)$ be a solution of the Itô equation*

$$d\xi(t) = f(\xi(t),t)dt + g(\xi(t),t)dB(t) \quad \text{on } t \geq 0, \quad (9.7)$$

whose coefficients satisfy the conditions of the existence-and-uniqueness theorem. Then $\xi(t)$ is a Markov process whose transition probability is defined by

$$P(x,s; A,t) = P\{\xi_{x,s}(t) \subset A\}, \quad (9.8)$$

where $\xi_{x,s}(t)$ is the solution of the equation

$$\xi_{x,s}(t) = x + \int_s^t f(\xi_{x,s}(r),r)dr + \int_s^t g(\xi_{x,s}(r),r)dB(r) \quad \text{on } t \geq s. \quad (9.9)$$

To prove this theorem, we need to prepare a lemma.

Lemma 9.2 Let $h(x,\omega)$ be a scalar bounded measurable random function of x, independent of $\mathcal{F}_s$. Let ζ be an $\mathcal{F}_s$-measurable random variable. Then

$$E(h(\zeta,\omega)|\mathcal{F}_s) = H(\zeta), \tag{9.10}$$

where $H(x) = Eh(x,\omega)$.

Proof. First, assume that $h(x,\omega)$ has the following simple form

$$h(x,\omega) = \sum_{i=1}^{k} u_i(x)v_i(\omega) \tag{9.11}$$

with $u_i(x)$'s bounded deterministic functions of x and $v_i(\omega)$'s bounded random variables independent of $\mathcal{F}_s$. Clearly,

$$H(x) = \sum_{i=1}^{k} u_i(x)Ev_i(\omega).$$

Moreover, for any set $G \in \mathcal{F}_s$, we compute

$$E[h(\zeta,\omega)I_G] = E\left(\sum_{i=1}^{k} u_i(\zeta)v_i(\omega)I_G\right) = \sum_{i=1}^{k} E[u_i(\zeta)I_G]Ev_i(\omega)$$

$$= E\left(\sum_{i=1}^{k} u_i(\zeta)Ev_i(\omega)I_G\right) = E[H(\zeta)I_G].$$

By definition, this means that (9.10) holds if $h(x,\omega)$ has the form of (9.11). Since any bounded measurable random function $h(x,\omega)$ can be approximated by functions of form (9.11), the general result of the lemma follows immediately.

Theorem 9.1 can now be proved easily.

Proof of Theorem 9.1 Let $\mathcal{G}_s = \sigma\{B(r) - B(s) : r \geq s\}$. Clearly, $\mathcal{G}_s$ is independent of $\mathcal{F}_s$. Moreover, the value of $\xi_{x,s}(t)$ depends completely on the increments $B(r) - B(s)$ for $r \geq s$ and so is $\mathcal{G}_s$-measurable. Hence, $\xi_{x,s}(t)$ is independent of $\mathcal{F}_s$. On the other hand, note that $\xi(t) = \xi_{\xi(s),s}(t)$ on $t \geq s$, since both $\xi(t)$ and $\xi_{\xi(s),s}(t)$ satisfy the equation

$$\xi(t) = \xi(s) + \int_s^t f(\xi(r),r)dr + \int_s^t g(\xi(r),r)dB(r)$$

whose solution is unique. For any $A \in \mathcal{B}^d$, we now apply Lemma 9.2 with $h(x,\omega) = I_A(\xi_{x,s}(t))$ to compute that

$$P(\xi(t) \in A|\mathcal{F}_s) = E(I_A(\xi(t))|\mathcal{F}_s) = E(I_A(\xi_{\xi(s),s}(t))|\mathcal{F}_s)$$
$$= E(I_A(\xi_{x,s}(t)))\big|_{x=\xi(s)} = P(x,s;A,t)\big|_{x=\xi(s)} = P(\xi(s),s;A,t).$$

if $P(x, s; A, t)$ is defined by (9.8). The proof is complete.

For the strong Markov property of the solution we need to strengthen the conditions slightly.

Theorem 9.3 *Let $\xi(t)$ be a solution of the Itô equation*

$$d\xi(t) = f(\xi(t), t)dt + g(\xi(t), t)dB(t) \quad \text{on } t \geq 0.$$

Assume the coefficients are uniformly Lipschitz continuous and satisfy the linear growth condition, that is, there are two positive constants K and $\bar{K}$ such that

$$|f(x,t) - f(y,t)|^2 \bigvee |g(x,t) - g(y,t)|^2 \leq \bar{K}|x-y|^2 \qquad (9.12)$$

and

$$|f(x,t)|^2 \bigvee |g(x,t)|^2 \leq K(1+|x|^2) \qquad (9.13)$$

for all $x, y \in R^d$ and $t \geq 0$. Then $\xi(t)$ is a strong Markov process.

We again need to prepare a lemma in order to prove the theorem.

Lemma 9.4 *Let (9.12) and (9.13) hold. For every pair $(x, s) \in R^d \times R_+$, let $\xi_{x,s}(t)$ be the solution of the equation*

$$\xi_{x,s}(t) = x + \int_s^t f(\xi_{x,s}(r), r)dr + \int_s^t g(\xi_{x,s}(r), r)dB(r) \quad \text{on } t \geq s.$$

Then for any $T > 0$ and $\delta > 0$,

$$E\left(\sup_{u \leq t \leq T} |\xi_{x,s}(t) - \xi_{y,u}(t)|^2 \right) \leq C(|x-y|^2 + |u-s|) \qquad (9.14)$$

if $0 \leq s, u \leq T$ and $|x| \vee |y| \leq \delta$, where C is a positive constant depending on T, δ, K and $\bar{K}$.

Proof. Without loss of generality we may assume that $s \leq u$. Clearly, for $u \leq t \leq T$,

$$\xi_{x,s}(t) - \xi_{y,u}(t) = \xi_{x,s}(u) - y + \int_u^t [f(\xi_{x,s}(r), r) - f(\xi_{y,u}(r), r)]dr$$

$$+ \int_u^t [g(\xi_{x,s}(r), r) - g(\xi_{y,u}(r), r)]dB(r). \qquad (9.15)$$

Note from Theorem 4.3 (condition (9.13) is used here) that

$$E|\xi_{x,s}(u) - y|^2 \leq 2E|\xi_{x,s}(u) - x|^2 + 2|x-y|^2 \leq C_1|u-s| + 2|x-y|^2, \qquad (9.16)$$

where C_1 is a positive constant depending on T, δ, K. It is now easy to drive from (9.15), (9.16) and (9.12) that if $u \leq v \leq T$,

$$E\left(\sup_{u\leq t\leq v} |\xi_{x,s}(t) - \xi_{y,u}(t)|^2\right) \leq 3C_1|u-s| + 6|x-y|^2$$
$$+ 3\bar{K}(T+4) \int_u^v E\left(\sup_{u\leq t\leq r} |\xi_{x,s}(t) - \xi_{y,u}(t)|^2\right) dr.$$

This easily implies the desired assertion (9.14).

We can now show the strong Markov property of the solution.

Proof of Theorem 9.3. The Markov property follows from Theorem 9.1 and it is known that the sample paths of the solution are continuous. Therefore we need only to verify the Feller property, this is, to show the mapping

$$(x,s) \to \int_{R^d} \varphi(y) P(x,s;dy,s+\lambda) = E\varphi(\xi_{x,s}(s+\lambda))$$

is continuous, for any bounded continuous function $\varphi : R^d \to R$ and any fixed $\lambda > 0$. Note that

$$E\varphi(\xi_{x,s}(s+\lambda)) - E\varphi(\xi_{y,u}(u+\lambda))$$
$$= E\varphi(\xi_{x,s}(s+\lambda)) - E\varphi(\xi_{x,s}(u+\lambda))$$
$$+ E\varphi(\xi_{x,s}(u+\lambda)) - E\varphi(\xi_{y,u}(u+\lambda)).$$

But, by Lemma 9.4 and the bounded convergence theorem,

$$E\varphi(\xi_{x,s}(u+\lambda)) - E\varphi(\xi_{y,u}(u+\lambda)) \to 0 \quad \text{as } (y,u) \to (x,s).$$

Also

$$E\varphi(\xi_{x,s}(s+\lambda)) - E\varphi(\xi_{x,s}(u+\lambda)) \to 0 \quad \text{as } u \to s.$$

In consequence,

$$E\varphi(\xi_{x,s}(s+\lambda)) - E\varphi(\xi_{y,u}(u+\lambda)) \to 0 \quad \text{as } (y,u) \to (x,s).$$

In other words, $E\varphi(\xi_{x,s}(s+\lambda))$ as a function of (x,s) is continuous, and that is the Feller property. The theorem has been proved.

Let us now consider the time-homogeneous stochastic differential equations. By time-homogeneous equations, we mean equations whose coefficients do not depend explicitly on time, namely equations of the form

$$d\xi(t) = f(\xi(t))dt + g(\xi(t))dB(t), \quad t \geq 0. \qquad (9.17)$$

We assume that $f : R^d \to R^d$ and $g : R^d \to R^{d\times m}$ satisfy the conditions of the existence-and-unique theorem.

Theorem 9.5 Let $\xi(t)$ be a solution of equation (9.17). Then $\xi(t)$ is a homogeneous Markov process. If f and g are uniformly Lipschitz continuous (hence the linear growth condition is satisfied), then the solution $\xi(t)$ is a homogeneous strong Markov process.

Proof. Clearly, we only need to show the homogeneous property. By Theorem 9.1, the transition probability is given by

$$P(x, s; A, s+t) = P\{\xi_{x,s}(s+t) \in A\}, \tag{9.18}$$

where $\xi_{x,s}(s+t)$ is the solution of the equation

$$\xi_{x,s}(s+t) = x + \int_s^{s+t} f(\xi_{x,s}(r))dr + \int_s^{s+t} g(\xi_{x,s}(r))dB(r) \quad \text{on } t \geq 0. \tag{9.19}$$

Write this equation as

$$\xi_{x,s}(s+t) = x + \int_0^t f(\xi_{x,s}(s+r))dr + \int_0^t g(\xi_{x,s}(s+r))d\tilde{B}(r) \quad \text{on } t \geq 0, \tag{9.20}$$

where $\tilde{B}(r) = B(s+r) - B(s)$ on $r \geq 0$ is a Brownian motion as well. On the other hand, we clearly have

$$\xi_{x,0}(t) = x + \int_0^t f(\xi_{x,0}(r))dr + \int_0^t g(\xi_{x,0}(r))dB(r) \quad \text{on } t \geq 0. \tag{9.21}$$

Comparing equation (9.20) with (9.21), we see by the weak uniqueness (Remark 7.5) that $\{\xi_{x,s}(s+t)\}_{t\geq 0}$ and $\{\xi_{x,0}(t)\}_{t\geq 0}$ are identical in probability law. In consequence,

$$P\{\xi_{x,s}(s+t) \in A\} = P\{\xi_{x,0}(t) \in A\},$$

that is

$$P(x, s; A, s+t) = P(x, 0; A, t).$$

The proof is therefore complete.

3

Linear Stochastic Differential Equations

3.1 INTRODUCTION

In the previous chapter, we discussed the solutions of stochastic differential equations. In general, nonlinear stochastic differential equations do not have explicit solutions and, in practice, we can use approximate solutions. However, it is possible to find the explicit solutions to linear equations. For example, recall the simple stochastic population growth model

$$dN(t) = r(t)N(t)dt + \sigma(t)N(t)dB(t) \quad \text{on } t \geq 0 \qquad (1.1)$$

with initial value $N(0) = N_0 > 0$. By Itô's formula,

$$\log N(t) = \log N_0 + \int_0^t \left(r(s) - \frac{\sigma^2(s)}{2}\right)ds + \int_0^t \sigma(s)dB(s).$$

This implies the explicit solution of equation (1.1)

$$N(t) = N_0 \exp\left[\int_0^t \left(r(s) - \frac{\sigma^2(s)}{2}\right)ds + \int_0^t \sigma(s)dB(s)\right]. \qquad (1.2)$$

In this chapter we wish, if possible, to get the explicit solution to the general d-dimensional linear stochastic differential equation

$$dx(t) = (F(t)x(t) + f(t))dt + \sum_{k=1}^{m}(G_k(t)x(t) + g_k(t))dB_k(t) \qquad (1.3)$$

on $[t_0, T]$, where $F(\cdot)$, $G_k(\cdot)$ are $d \times d$-matrix-valued functions, $f(\cdot)$, $g_k(\cdot)$ are R^d-valued functions and, as before, $B(t) = (B_1(t), \cdots, B_m(t))^T$ is an m-dimensional Brownian motion. The linear equation is said to be *homogeneous* if $f(t) = g_1(t) = \cdots = g_m(t) \equiv 0$. It is said to be *linear in the narrow sense* if $G_1(t) = \cdots = G_m(t) \equiv 0$. It is said to be *autonomous* if the coefficients F, f, G_k, g_k are all independent of t.

Throughout this chapter we shall assume that F, f, G_k, g_k are all Borel-measurable and bounded on $[t_0, T]$. Therefore, by the existence-and-uniqueness Theorem 2.3.1, the linear equation (1.3) has a unique continuous solution in $M^2([t_0, T]; R^d)$ for every initial value $x(t_0) = x_0$, which is $\mathcal{F}_{t_0}$-measurable and belongs to $L^2(\Omega; R^d)$. The aim of this chapter is to get, if possible, an explicit expression for this solution.

3.2 STOCHASTIC LIOUVILLE'S FORMULA

Consider the homogeneous linear stochastic differential equation

$$dx(t) = F(t)x(t)dt + \sum_{k=1}^{m} G_k(t)x(t)dB_k(t) \qquad (2.1)$$

on $[t_0, T]$. As assumed,

$$F(t) = (F_{ij}(t))_{d \times d}, \qquad G_k(t) = (G_{ij}^k(t))_{d \times d}$$

are all Borel-measurable and bounded. For every $j = 1, \cdots, d$, let e_j be the unit column-vector in the x_j-direction, i.e.

$$e_j = (\underbrace{0, \cdots, 0, 1}_{j}, 0, \cdots, 0)^T.$$

Let $\Phi_j(t) = (\Phi_{1j}(t), \cdots, \Phi_{dj}(t))^T$ be the solution of equation (2.1) with initial value $x(t_0) = e_j$. Define the $d \times d$ matrix

$$\Phi(t) = (\Phi_1(t), \cdots, \Phi_d(t)) = (\Phi_{ij}(t))_{d \times d}.$$

We call $\Phi(t)$ the *fundamental matrix* of equation (2.1). It is useful to note that $\Phi(t_0) =$ the $d \times d$ identity matrix and

$$d\Phi(t) = F(t)\Phi(t)dt + \sum_{k=1}^{m} G_k(t)\Phi(t)dB_k(t). \qquad (2.2)$$

Equation (2.2) can also be expressed as follows: For $1 \leq i, j \leq d$,

$$d\Phi_{ij}(t) = \sum_{l=1}^{d} F_{il}(t)\Phi_{lj}(t)dt + \sum_{k=1}^{m}\sum_{l=1}^{d} G_{il}^k(t)\Phi_{lj}(t)dB_k(t). \qquad (2.3)$$

The following theorem shows that any solution of equation (2.1) can be expressed in terms of $\Phi(t)$ and that is why $\Phi(t)$ is called the fundamental matrix.

Theorem 2.1 *Given the initial value $x(t_0) = x_0$, the unique solution of equation (2.1) is*
$$x(t) = \Phi(t)x_0.$$

Proof. Clearly $x(t_0) = x_0$. Moreover, by (2.2),
$$dx(t) = d\Phi(t)x_0 = F(t)\Phi(t)x_0 dt + \sum_{k=1}^{m} G_k(t)\Phi(t)x_0 dB_k(t)$$
$$= F(t)x(t)dt + \sum_{k=1}^{m} G_k(t)x(t)dB_k(t).$$

So $x(t)$ is a solution to equation (2.1). But by the existence-and-uniqueness theorem, equation (2.1) has only one solution. Hence the $x(t)$ must be the unique one.

We now denote by $W(t)$ the determinant of the fundamental matrix $\Phi(t)$, that is
$$W(t) = det.\Phi(t).$$

We call $W(t)$ the *stochastic Wronskian determinant*. Obviously, $W(t_0) = 1$. Moreover, we have the following *stochastic Liouville formula*.

Theorem 2.2 *The stochastic Wronskian determinant $W(t)$ has the explicit expression*
$$W(t) = \exp\Bigg[\int_{t_0}^{t} \Big(traceF(s) - \frac{1}{2}\sum_{k=1}^{m} trace[G_k(s)G_k^T(s)]\Big)ds$$
$$+ \sum_{k=1}^{m}\int_{t_0}^{t} traceG_k(s)dB_k(s)\Bigg]. \quad (2.4)$$

We prepare a lemma.

Lemma 2.3 *Let $a(\cdot)$, $b_k(\cdot)$ be real-valued Borel measurable bounded functions on $[t_0, T]$. Then*
$$y(t) = y_0 \exp\Bigg[\int_{t_0}^{t}\Big(a(s) - \frac{1}{2}\sum_{k=1}^{m} b_k^2(s)\Big)ds + \sum_{k=1}^{m}\int_{t_0}^{t} b_k(s)dB_k(s)\Bigg] \quad (2.5)$$

is the unique solution to the scalar linear stochastic differential equation
$$dy(t) = a(t)y(t)dt + \sum_{k=1}^{m} b_k(t)y(t)dB_k(t) \quad (2.6)$$

on $[t_0, T]$ with initial value $y(t_0) = y_0$.

Proof. Set
$$\xi(t) = \int_{t_0}^t \left(a(s) - \frac{1}{2}\sum_{k=1}^m b_k^2(s)\right) ds + \sum_{k=1}^m \int_{t_0}^t b_k(s) dB_k(s).$$

One can then write
$$y(t) = y_0 e^{\xi(t)}.$$

Clearly, $y(t_0) = y_0$. Moreover, by Itô's formula,
$$dy(t) = y(t)\left[\left(a(t) - \frac{1}{2}\sum_{k=1}^m b_k^2(t)\right) dt + \sum_{k=1}^m b_k(t) dB_k(t)\right]$$
$$+ \frac{1}{2} y(t) \sum_{k=1}^m b_k^2(t) dt$$
$$= a(t) y(t) dt + \sum_{k=1}^m b_k(t) y(t) dB_k(t).$$

In other words, $y(t)$ is a solution to equation (2.6) satisfying the initial condition. But, by Theorem 2.3.1, equation (2.6) has only one solution. So $y(t)$ must be the unique one. The lemma has been proved.

Proof of Theorem 2.2. By Itô's formula, one can show that
$$dW(t) = \sum_{i=1}^d \varphi_i + \sum_{1 \le i < j \le d} \phi_{ij}, \tag{2.7}$$

where
$$\varphi_i = \begin{vmatrix} \Phi_{11}(t), & \cdots, & \Phi_{1d}(t) \\ \vdots & & \vdots \\ d\Phi_{i1}(t), & \cdots, & d\Phi_{id}(t) \\ \vdots & & \vdots \\ \Phi_{d1}(t), & \cdots, & \Phi_{dd}(t) \end{vmatrix}$$

and
$$\phi_i = \begin{vmatrix} \Phi_{11}(t), & \cdots, & \Phi_{1d}(t) \\ \vdots & & \vdots \\ d\Phi_{i1}(t), & \cdots, & d\Phi_{id}(t) \\ \vdots & & \vdots \\ d\Phi_{j1}(t), & \cdots, & d\Phi_{jd}(t) \\ \vdots & & \vdots \\ \Phi_{d1}(t), & \cdots, & \Phi_{dd}(t) \end{vmatrix}.$$

It is not very difficult to verify by using (2.3) and the formal multiplication table defined on page 36 that

$$\varphi_i = F_{ii}(t)W(t)dt + \sum_{k=1}^{m} G_{ii}^k(t)W(t)dB_k(t) \qquad (2.8)$$

and

$$\phi_{ij} = \sum_{k=1}^{m} [G_{ii}^k(t)G_{jj}^k(t) - G_{ij}^k(t)G_{ji}^k(t)]W(t)dt. \qquad (2.9)$$

Substituting (2.8) and (2.9) into (2.7) yields that

$$dW(t) = \Big(\sum_{i=1}^{d} F_{ii}(t) + \sum_{k=1}^{m} \sum_{1 \le i < j \le d} [G_{ii}^k(t)G_{jj}^k(t) - G_{ij}^k(t)G_{ji}^k(t)]\Big)W(t)dt$$

$$+ \sum_{k=1}^{m} \sum_{i=1}^{d} G_{ii}^k(t)W(t)dB_k(t). \qquad (2.10)$$

Applying Lemma 2.3 we get that

$$W(t) = \exp\bigg[\int_{t_0}^{t}\Big(\sum_{i=1}^{d} F_{ii}(s) + \sum_{k=1}^{m}\sum_{1\le i<j\le d}[G_{ii}^k(s)G_{jj}^k(s) - G_{ij}^k(s)G_{ji}^k(s)]\Big)ds$$

$$-\frac{1}{2}\sum_{k=1}^{m}\int_{t_0}^{t}\Big(\sum_{i=1}^{d}G_{ii}^k(s)\Big)^2 ds + \sum_{k=1}^{m}\int_{t_0}^{t}\sum_{i=1}^{d}G_{ii}^k(s)dB_k(s)\bigg]. \qquad (2.11)$$

Noting that

$$\Big(\sum_{i=1}^{d}G_{ii}^k(s)\Big)^2 = \sum_{i=1}^{d}[G_{ii}^k(s)]^2 + 2\sum_{1\le i<j\le d}G_{ii}^k(s)G_{jj}^k(s),$$

we obtain immediately from (2.11) that

$$W(t) = \exp\bigg[\int_{t_0}^{t}\sum_{i=1}^{d}F_{ii}(s)ds + \sum_{k=1}^{m}\int_{t_0}^{t}\sum_{i=1}^{d}G_{ii}^k(s)dB_k(s)$$

$$-\sum_{k=1}^{m}\int_{t_0}^{t}\Big(\frac{1}{2}\sum_{i=1}^{d}[G_{ii}^k(s)]^2 + \sum_{1\le i<j\le d}G_{ij}^k(s)G_{ji}^k(s)\Big)ds\bigg],$$

which is the required (2.4). The proof is complete.

The stochastic Liouville formula (2.4) implies directly that $W(t) > 0$ a.s. for all $t \in [t_0, T]$, which in turn implies that $\Phi(t)$ is invertible. We have therefore obtained the following important result.

Theorem 2.4 *For all $t \in [t_0, T]$, the fundamental matrix $\Phi(t)$ is invertible with probability 1.*

We shall denote by $\Phi^{-1}(t)$ the inverse matrix of $\Phi(t)$.

3.3 THE VARIATION-OF-CONSTANTS FORMULA

Let us now turn to the general d-dimensional linear stochastic differential equation

$$dx(t) = (F(t)x(t) + f(t))dt + \sum_{k=1}^{m}(G_k(t)x(t) + g_k(t))dB_k(t) \quad (3.1)$$

on $[t_0, T]$ with initial value $x(t_0) = x_0$. Equation (2.1) is called the corresponding homogeneous equation of system (3.1). In this section we shall establish a useful formula, called the *variation-of-constants formula*, which represents the unique solution of equation (3.1) in terms of the fundamental matrix of the corresponding homogeneous equation (2.1).

Theorem 3.1 *The unique solution of equation (3.1) can be expressed as*

$$x(t) = \Phi(t)\left(x_0 + \int_{t_0}^{t} \Phi^{-1}(s)\left[f(s) - \sum_{k=1}^{m} G_k(s)g_k(s)\right]ds \right.$$
$$\left. + \sum_{k=1}^{m}\int_{t_0}^{t} \Phi^{-1}(s)g_k(s)dB_k(s)\right), \quad (3.2)$$

where $\Phi(t)$ is the fundamental matrix of the corresponding homogeneous equation (2.1).

Proof. Set

$$\xi(t) = x_0 + \int_{t_0}^{t} \Phi^{-1}(s)\left[f(s) - \sum_{k=1}^{m} G_k(s)g_k(s)\right]ds$$
$$+ \sum_{k=1}^{m}\int_{t_0}^{t} \Phi^{-1}(s)g_k(s)dB_k(s).$$

Then $\xi(t)$ has the differential

$$d\xi(t) = \Phi^{-1}(t)\left[f(t) - \sum_{k=1}^{m} G_k(t)g_k(t)\right]dt$$
$$+ \sum_{k=1}^{m} \Phi^{-1}(t)g_k(t)dB_k(t). \quad (3.3)$$

Let
$$\eta(t) = \Phi(t)\xi(t). \tag{3.4}$$

Clearly, $\eta(t_0) = x_0$. Moreover, by Itô's formula
$$d\eta(t) = d\Phi(t)\xi(t) + \Phi(t)d\xi(t) + d\Phi(t)d\xi(t).$$

Substituting (2.2) and (3.3) into it and using the formal multiplication table defined on page 36, we derive that

$$d\eta(t) = F(t)\eta(t)dt + \sum_{k=1}^{m} G_k(t)\eta(t)dB_k(t)$$
$$+ \left[f(t) - \sum_{k=1}^{m} G_k(t)g_k(t)\right]dt + \sum_{k=1}^{m} g_k(t)dB_k(t)$$
$$+ \left(F(t)\Phi(t)dt + \sum_{k=1}^{m} G_k(t)\Phi(t)dB_k(t)\right)$$
$$\times \left(\Phi^{-1}(t)f(t)dt + \sum_{k=1}^{m} \Phi^{-1}(t)g_k(t)dB_k(t) - \sum_{k=1}^{m} \Phi^{-1}(t)G_k(t)g_k(t)dt\right)$$
$$= (F(t)\eta(t) + f(t))dt + \sum_{k=1}^{m}(G_k(t)\eta(t) + g_k(t))dB_k(t).$$

In other words, we have shown that $\eta(t)$ is a solution to equation (3.1) satisfying the initial condition $\eta(t_0) = x_0$. On the other hand, equation (3.1) has only one solution $x(t)$. So we must have that $x(t) = \eta(t)$, which is the required formula (3.2). The proof is complete.

Since we assume that $x_0 \in L^2(\Omega; R^d)$, the first and second moments of the solution of equation (3.1) exist and are finite. The following theorem shows that one can obtain first and second moments by solving the corresponding linear ordinary differential equations.

Theorem 3.2 *For the solution of equation (3.1), we have:*
(a) $m(t) := Ex(t)$ is the unique solution of the equation
$$\dot{m}(t) = F(t)m(t) + f(t) \tag{3.5}$$
on $[t_0, T]$ with initial value $m(t_0) = Ex_0$.
(b) $P(t) := E(x(t)x^T(t))$ is the unique nonnegative-definite symmetric solution of the equation
$$\dot{P}(t) = F(t)P(t) + P(t)F^T(t) + f(t)m^T(t) + m(t)f^T(t)$$
$$+ \sum_{k=1}^{m}\left[G_k(t)P(t)G_k^T(t) + G_k(t)m(t)g_k^T(t)\right.$$
$$\left. + g_k(t)m^T(t)G_k^T(t) + g_k(t)g_k^T(t)\right] \tag{3.6}$$

on $[t_0, T]$ with initial value $P(t_0) = E(x_0 x_0^T)$. Note that (3.6) represents a system of $d(d+1)/2$ linear equations.

Proof. (a) Note that

$$x(t) = x(t_0) + \int_{t_0}^{t} (F(s)x(s) + f(s))ds + \sum_{k=1}^{m} \int_{t_0}^{t} (G_k(s)x(s) + g_k(s))dB_k(s).$$

Taking the expectation on both sides yields

$$m(t) = m(t_0) + \int_{t_0}^{t} (F(s)m(s) + f(s))ds$$

which is the integral form of equation (3.5). So the conclusion of part (a) follows.

(b) By Itô's formula,

$$d[x(t)x^T(t)] = dx(t)x^T(t) + x(t)dx^T(t)$$
$$+ \sum_{k=1}^{m} [G_k(t)x(t) + g_k(t)][G_k(t)x(t) + g_k(t)]^T dt$$
$$= \Big(F(t)x(t)x^T(t) + f(t)x^T(t) + x(t)x^T(t)F^T(t) + x(t)f^T(t)$$
$$+ \sum_{k=1}^{m} \Big[G_k(t)x(t)x^T(t)G_k^T(t) + g_k(t)x^T(t)G_k^T(t)$$
$$+ G_k(t)x(t)g_k^T(t) + g_k(t)g_k^T(t) \Big] \Big) dt$$
$$+ \sum_{k=1}^{m} \Big[(G_k(t)x(t) + g_k(t))x^T(t) + x(t)(G_k(t)x(t) + g_k(t))^T \Big] dB_k(t).$$

Now equation (3.6) follows by taking the expectation on both sides of the integral form of the above equality. Since $P(t)$ is the covariance matrix of $x(t)$, it is of course nonnegative-definite and symmetric. The proof is complete.

Theorem 3.1 tells us that we can have the explicit solution to the linear equation (3.1) provided we know the corresponding fundamental matrix $\Phi(t)$. Although we can not obtain the explicit fundamental matrix $\Phi(t)$ for every case, we can for several important cases and let us turn to these case studies.

3.4 CASE STUDIES

(i) *Scalar Linear Equations*

We first consider the general scalar linear stochastic differential equation

$$dx(t) = (a(t)x(t) + \bar{a}(t))dt + \sum_{k=1}^{m} (b_k(t)x(t) + \bar{b}_k(t))dB_k(t) \qquad (4.1)$$

on $[t_0, T]$ with initial value $x(t_0) = x_0$. Here $x_0 \in L^2(\Omega; R)$ is $\mathcal{F}_{t_0}$-measurable, and $a(t)$, $\bar{a}(t)$, $b_k(t)$, $\bar{b}_k(t)$ are Borel-measurable bounded scalar functions on $[t_0, T]$. The corresponding homogeneous linear equation is

$$dx(t) = a(t)x(t)dt + \sum_{k=1}^{m} b_k(t)x(t)dB_k(t). \qquad (4.2)$$

By Lemma 2.3, the fundamental solution of equation (4.2) is given by

$$\Phi(t) = \exp\left[\int_{t_0}^{t}\left(a(s) - \frac{1}{2}\sum_{k=1}^{m}b_k^2(s)\right)ds + \sum_{k=1}^{m}\int_{t_0}^{t}b_k(s)dB_k(s)\right].$$

Applying Theorem 3.1, we then obtain the explicit solution of equation (4.1)

$$x(t) = \Phi(t)\left(x_0 + \int_{t_0}^{t}\Phi^{-1}(s)\left[\bar{a}(s) - \sum_{k=1}^{m}b_k(s)\bar{b}_k(s)\right]ds \right.$$
$$\left. + \sum_{k=1}^{m}\int_{t_0}^{t}\Phi^{-1}(s)\bar{b}_k(s)dB_k(s)\right). \qquad (4.3)$$

(ii) Linear Equations in the Narrow Sense

We next consider the d-dimensional linear stochastic differential equation in the narrow sense

$$dx(t) = (F(t)x(t) + f(t))dt + \sum_{k=1}^{m}g_k(t)dB_k(t) \qquad (4.4)$$

on $[t_0, T]$ with initial value $x(t_0) = x_0$, where F, f, g_k and x_0 are the same as defined in Section 3.1. The corresponding homogeneous linear equation is now the ordinary differential equation

$$\dot{x}(t) = F(t)x(t). \qquad (4.5)$$

Again, let $\Phi(t)$ be the fundamental matrix of equation (4.5). Then the solution of equation (4.4) has the form

$$x(t) = \Phi(t)\left(x_0 + \int_{t_0}^{t}\Phi^{-1}(s)f(s)ds + \sum_{k=1}^{m}\int_{t_0}^{t}\Phi^{-1}(s)g_k(s)dB_k(s)\right). \qquad (4.6)$$

In particular, when $F(t)$ is independent of t, i.e. $F(t) = F$ a $d \times d$ constant matrix, the fundamental matrix $\Phi(t)$ has the simple form $\Phi(t) = e^{F(t-t_0)}$ and its inverse matrix $\Phi^{-1}(t) = e^{-F(t-t_0)}$. Therefore, in the case when $F(t) = F$, equation (4.4) has the explicit solution

$$x(t) = e^{F(t-t_0)}\left(x_0 + \int_{t_0}^t e^{-F(s-t_0)}f(s)ds + \sum_{k=1}^m \int_{t_0}^t e^{-F(s-t_0)}g_k(s)dB_k(s)\right)$$

$$= e^{F(t-t_0)}x_0 + \int_{t_0}^t e^{F(t-s)}f(s)ds + \sum_{k=1}^m \int_{t_0}^t e^{F(t-s)}g_k(s)dB_k(s). \quad (4.7)$$

(iii) *Autonomous Linear Equations*

We now consider the d-dimensional autonomous linear stochastic differential equation

$$dx(t) = (Fx(t) + f)dt + \sum_{k=1}^m (G_k x(t) + g_k)dB_k(t) \quad (4.8)$$

on $[t_0, T]$ with initial value $x(t_0) = x_0$, where $F, G_k \in R^{d\times d}$ and $f, g_k \in R^d$. The corresponding homogeneous equation is

$$dx(t) = Fx(t)dt + \sum_{k=1}^m G_k x(t)dB_k(t). \quad (4.9)$$

In general, the fundamental matrix $\Phi(t)$ can not be given explicitly. However, if the matrices $F, G_1, \cdots, G_m$ commute, that is, if

$$FG_k = G_k F, \quad G_k G_j = G_j G_k \quad \text{for all } 1 \leq k, j \leq m, \quad (4.10)$$

then the fundamental matrix of equation (4.9) has the explicit form

$$\Phi(t) = \exp\left[\left(F - \frac{1}{2}\sum_{k=1}^m G_k^2\right)(t - t_0) + \sum_{k=1}^m G_k(B_k(t) - B_k(t_0))\right]. \quad (4.11)$$

To show this, set

$$Y(t) = \left(F - \frac{1}{2}\sum_{k=1}^m G_k^2\right)(t - t_0) + \sum_{k=1}^m G_k(B_k(t) - B_k(t_0)).$$

We can then write

$$\Phi(t) = \exp(Y(t)).$$

By condition (4.10) we compute the stochastic differential

$$d\Phi(t) = \exp(Y(t))dY(t) + \frac{1}{2}\exp(Y(t))(dY(t))^2$$

$$= \Phi(t)dY(t) + \frac{1}{2}\Phi(t)\left(\sum_{k=1}^m G_k^2\right)dt$$

$$= F\Phi(t)dt + \sum_{k=1}^m G_k \Phi(t)dB_k(t).$$

That is, $\Phi(t)$ satisfies the homogeneous equation and hence is the fundamental matrix. Finally, we apply Theorem 3.1 to conclude that under condition (4.10), the autonomous linear equation (4.8) has the explicit solution

$$x(t) = \Phi(t)\left[x_0 + \left(\int_{t_0}^t \Phi^{-1}(s)ds\right)\left(f - \sum_{k=1}^m G_k g_k\right)\right.$$
$$\left. + \sum_{k=1}^m \left(\int_{t_0}^t \Phi^{-1}(s)dB_k(s)\right)g_k\right]. \tag{4.12}$$

3.5 EXAMPLES

In this section we shall investigate several important stochastic processes which are described by linear stochastic differential equations. Throughout this section, we let $B(t)$ be a 1-dimensional Brownian motion.

Example 5.1 (The Ornstein–Uhlenbeck process) We shall first discuss the historically oldest example of a stochastic differential equation. The Langevin equation

$$\dot{x}(t) = -\alpha x(t) + \sigma \dot{B}(t) \quad \text{on } t \geq 0 \tag{5.1}$$

has been used to describe the motion of a particle under the influence of friction but no other force field (cf. Uhlenbeck & Ornstein (1930)). Here $\alpha > 0$ and σ are constants, $x(t)$ is one of the three scalar velocity components of the particle and $\dot{B}(t)$ is a scalar white noise. The corresponding Itô equation

$$dx(t) = -\alpha x(t)dt + \sigma dB(t) \quad \text{on } t \geq 0 \tag{5.2}$$

is an autonomous linear equation in the narrow sense. Assume that the initial value $x(0) = x_0$ is $\mathcal{F}_0$-measurable and belongs to $L^2(\Omega; R)$. In view of (4.7), the unique solution of equation (5.2) is

$$x(t) = e^{-\alpha t}x_0 + \sigma \int_0^t e^{-\alpha(t-s)}dB(s). \tag{5.3}$$

It has the mean

$$Ex(t) = e^{-\alpha t}Ex_0$$

and the variance

$$Var(x(t)) = E|x(t) - Ex(t)|^2$$
$$= e^{-2\alpha t}E|x_0 - Ex_0|^2 + \sigma^2 e^{-2\alpha t}E\left|\int_0^t e^{\alpha s}dB(s)\right|^2$$
$$= e^{-2\alpha t}Var(x_0) + \sigma^2 e^{-2\alpha t}E\int_0^t e^{2\alpha s}ds$$
$$= e^{-2\alpha t}Var(x_0) + \frac{\sigma^2}{2\alpha}(1 - e^{-2\alpha t}).$$

Note that for arbitrary x_0,
$$\lim_{t\to\infty} e^{-\alpha t}x_0 = 0 \quad \text{a.s.}$$

and $\sigma \int_0^t e^{-\alpha(t-s)}dB(s)$ follows the normal distribution $N(0, \sigma^2(1-e^{-2\alpha t})/2\alpha)$. So the distribution of the solution $x(t)$ approaches the normal distribution $N(0, \sigma^2/2\alpha)$ as $t \to \infty$ for arbitrary x_0. If x_0 is normally distributed or constant, then the solution $x(t)$ is a Gaussian process (i.e. normally distributed process), and is called the *Ornstein–Uhlenbeck velocity process*. If start with an $N(0, \sigma^2/2\alpha)$-distributed x_0, then $x(t)$ follows the same normal distribution $N(0, \sigma^2/2\alpha)$ so the solution is a stationary Gaussian process, which is sometimes called a *coloured noise*.

Now assume that the particle starts from the initial position y_0, which is $\mathcal{F}_0$-measurable and belongs to $L^2(\Omega; R)$ as well. Then, by integration of the velocity $x(t)$, we obtain the position

$$y(t) = y_0 + \int_0^t x(s)ds \tag{5.4}$$

of the particle at time t. If y_0 and x_0 are normally distributed or constant, then $y(t)$ is a Gaussian process, the so-called *Ornstein–Uhlenbeck position process*. Of course, we can treat $x(t)$ and $y(t)$ simultaneously by combining equations (5.2) and (5.4) into the 2-dimensional linear stochastic differential equation

$$d\begin{pmatrix} x(t) \\ y(t) \end{pmatrix} = \begin{pmatrix} -\alpha & 0 \\ 1 & 0 \end{pmatrix}\begin{pmatrix} x(t) \\ y(t) \end{pmatrix}dt + \begin{pmatrix} \sigma \\ 0 \end{pmatrix}dB(t). \tag{5.5}$$

It is easy to obtain the corresponding fundamental matrix

$$\Phi(t) = \begin{pmatrix} e^{-\alpha t} & 0 \\ (1-e^{-\alpha t})/\alpha & 1 \end{pmatrix}$$

with the property $\Phi(t)\Phi^{-1}(s) = \Phi(t-s)$. Therefore, according to (4.7), the solution of equation (5.5) is

$$\begin{pmatrix} x(t) \\ y(t) \end{pmatrix} = \Phi(t)\begin{pmatrix} x_0 \\ y_0 \end{pmatrix} + \int_0^t \Phi(t-s)\begin{pmatrix} \sigma \\ 0 \end{pmatrix}dB(t).$$

This implies

$$x(t) = e^{-\alpha t}x_0 + \sigma \int_0^t e^{-\alpha(t-s)}dB(s)$$

the same as (5.3), and

$$y(t) = \frac{1}{\alpha}(1-e^{-\alpha t})x_0 + y_0 + \frac{\sigma}{\alpha}\int_0^t \left[1-e^{-\alpha(t-s)}\right]dB(s), \tag{5.6}$$

which is in fact the same as (5.4) (we leave the verification to the reader). It then follows from (5.6) that $y(t)$ has the mean

$$Ey(t) = \frac{1}{\alpha}(1 - e^{-\alpha t})Ex_0 + Ey_0$$

and the variance

$$Var(y(t)) = \frac{1}{\alpha^2}(1 - e^{-\alpha t})^2 Var(x_0) + \frac{2}{\alpha}(1 - e^{-\alpha t})Cov(x_0, y_0) + Var(y_0)$$
$$+ \frac{\sigma^2}{\alpha^2}\left[t - \frac{2}{\alpha}(1 - e^{-\alpha t}) + \frac{1}{2\alpha}(1 - e^{-2\alpha t})\right].$$

Example 5.2 (The mean-reverting Ornstein–Uhlenbeck process) If we revert the Langevin equation (5.2) by mean, we arrive at the following equation

$$dx(t) = -\alpha(x(t) - \mu)dt + \sigma dB(t) \qquad \text{on } t \geq 0 \qquad (5.7)$$

with initial value $x(0) = x_0$, where μ is a constant. Its solution is called the *mean-reverting Ornstein–Uhlenbeck process* and has the form

$$x(t) = e^{-\alpha t}\left(x_0 + \alpha\mu \int_0^t e^{\alpha s} ds + \sigma \int_0^t e^{\alpha s} dB(s)\right)$$
$$= e^{-\alpha t}x_0 + \mu(1 - e^{-\alpha t}) + \sigma \int_0^t e^{-\alpha(t-s)} dB(s). \qquad (5.8)$$

We therefore obtain that the mean

$$Ex(t) = e^{-\alpha t}Ex_0 + \mu(1 - e^{-\alpha t}) \to \mu \qquad \text{as } t \to \infty$$

and the variance

$$Var(x(t)) = e^{-2\alpha t}Var(x_0) + \frac{\sigma^2}{2\alpha}(1 - e^{-2\alpha t}) \to \frac{\sigma^2}{2\alpha} \qquad \text{as } t \to \infty.$$

It also follows from (5.8) that the distribution of the solution $x(t)$ approaches the normal distribution $N(\mu, \sigma^2/2\alpha)$ as $t \to \infty$ for arbitrary x_0. If x_0 is normally distributed or constant, then the solution $x(t)$ is a Gaussian process. If x_0 follows the normal distribution $N(\mu, \sigma^2/2\alpha)$, so does the solution $x(t)$ for all $t \geq 0$.

Example 5.3 (The Brownian motion on the unit circle) Consider the 2-dimensional linear stochastic differential equation

$$dx(t) = -\frac{1}{2}x(t)dt + Kx(t)dB(t) \qquad \text{on } t \geq 0 \qquad (5.9)$$

with initial value $x(0) = (1, 0)^T$, where

$$K = \begin{pmatrix} 0 & -1 \\ 1 & 0 \end{pmatrix}.$$

In view of (4.11), the corresponding fundamental matrix is

$$\Phi(t) = \exp\left[\left(-\frac{1}{2}I - \frac{1}{2}K^2\right)t + KB(t)\right],$$

where I is the 2×2 identity matrix. Noting that $K^2 = -I$, we obtain

$$\Phi(t) = \exp[KB(t)] = \sum_{n=0}^{\infty} \frac{K^n B^n(t)}{n!}.$$

But

$$K^{2n} = (-1)^n I \quad \text{and} \quad K^{2n+1} = (-1)^n K \quad \text{for } n = 0, 1, \cdots.$$

Thus

$$\Phi(t) = \sum_{n=0}^{\infty} \left[\frac{K^{2n} B^{2n}(t)}{(2n)!} + \frac{K^{2n+1} B^{2n+1}(t)}{(2n+1)!}\right]$$

$$= \sum_{n=0}^{\infty} \left[\frac{(-1)^n B^{2n}(t) I}{(2n)!} + \frac{(-1)^n B^{2n+1}(t) K}{(2n+1)!}\right].$$

Now, by (4.12), the unique solution of equation (5.9) is

$$x(t) = \Phi(t) \begin{pmatrix} 1 \\ 0 \end{pmatrix} = \begin{pmatrix} \sum_{n=0}^{\infty} \frac{(-1)^n B^{2n}(t)}{(2n)!} \\ \sum_{n=0}^{\infty} \frac{(-1)^n B^{2n+1}(t)}{(2n+1)!} \end{pmatrix} = \begin{pmatrix} \cos B(t) \\ \sin B(t) \end{pmatrix},$$

and this is the Brownian motion on the unit circle (see Example 2.2.3).

Example 5.4 (The Brownian bridge) Let a, b be two constants. Consider the 1-dimensional linear equation

$$dx(t) = \frac{b - x(t)}{1 - t} dt + dB(t) \qquad \text{on } t \in [0, 1) \qquad (5.10)$$

with initial value $x(0) = a$. The corresponding fundamental solution is

$$\Phi(t) = \exp\left[-\int_0^t \frac{ds}{1-s}\right] = \exp[\log(1-t)] = 1 - t.$$

Hence, by (4.3), the solution of equation (5.10) is

$$x(t) = (1-t)\left(a + b\int_0^t \frac{ds}{(1-s)^2} + \int_0^t \frac{dB(s)}{1-s}\right)$$

$$= (1-t)a + bt + (1-t)\int_0^t \frac{dB(s)}{1-s}. \qquad (5.11)$$

The solution is called the *Brownian bridge from a to b*. It is a Gaussian process with mean
$$Ex(t) = (1-t)a + bt$$
and variance
$$Var(x(t)) = t(1-t).$$

Example 5.5 (The geometric Brownian motion) The geometric Brownian motion is the solution to the 1-dimensional linear equation
$$dx(t) = \alpha x(t)dt + \sigma x(t)dB(t) \quad \text{on } t \geq 0, \quad (5.12)$$
where α, σ are constants. Given the initial value $x(0) = x_0$, the solution of the equation is
$$x(t) = x_0 \exp\left[\left(\alpha - \frac{\sigma^2}{2}\right)t + \sigma B(t)\right]. \quad (5.13)$$

If $x_0 \neq 0$ a.s., then, by the law of the iterated logarithm (i.e. Theorem 1.4.2), we obtain from (5.13) that
$$\begin{cases} \alpha < \dfrac{\sigma^2}{2} \iff \lim_{t\to\infty} x(t) = 0 \text{ a.s.} \\ \alpha = \dfrac{\sigma^2}{2} \iff \limsup_{t\to\infty} |x(t)| = \infty \text{ and } \liminf_{t\to\infty} |x(t)| = 0 \text{ a.s.} \\ \alpha > \dfrac{\sigma^2}{2} \iff \lim_{t\to\infty} |x(t)| = \infty \text{ a.s.} \end{cases} \quad (5.14)$$

We now let $p > 0$ and $x_0 \in L^p$ with $E|x_0|^p \neq 0$. It follows from (5.13) that
$$E|x(t)|^p = E\left(|x_0|^p \exp\left[p\left(\alpha - \frac{\sigma^2}{2}\right)t + p\sigma B(t)\right]\right)$$
$$= \exp\left[p\left(\alpha - \frac{(1-p)\sigma^2}{2}\right)t\right] E\left(|x_0|^p \exp\left[-\frac{p^2\sigma^2}{2}t + p\sigma B(t)\right]\right). \quad (5.15)$$

Set
$$\xi(t) = |x_0|^p \exp\left[-\frac{p^2\sigma^2}{2}t + p\sigma B(t)\right].$$
It is the unique solution to the equation
$$d\xi(t) = p\sigma\xi(t)dB(t)$$
with initial value $\xi(0) = |x_0|^p$. Hence
$$\xi(t) = |x_0|^p + p\sigma\int_0^t \xi(s)dB(s)$$

which yields that $E\xi(t) = E|x_0|^p$. Substituting this into (5.15) gives

$$E|x(t)|^p = \exp\left[p\left(\alpha - \frac{(1-p)\sigma^2}{2}\right)t\right]E|x_0|^p.$$

Consequently

$$\begin{cases} \alpha < \dfrac{(1-p)\sigma^2}{2} & \Longleftrightarrow \lim_{t\to\infty} E|x(t)|^p = 0, \\ \alpha = \dfrac{(1-p)\sigma^2}{2} & \Longleftrightarrow E|x(t)|^p = E|x_0|^p \text{ for all } t \geq 0, \\ \alpha > \dfrac{(1-p)\sigma^2}{2} & \Longleftrightarrow \lim_{t\to\infty} E|x(t)|^p = \infty. \end{cases} \qquad (5.16)$$

Example 5.6 (Equations driven by a coloured noise) Instead of a white noise, it is often to use a coloured noise to describe stochastic perturbations. For example, consider the linear equation driven by a coloured noise

$$dx(t) = ax(t)dt + by(t)dt \qquad \text{on } t \geq 0 \qquad (5.17)$$

with initial value $x(0) = x_0$, where $y(t)$ is the coloured noise, i.e. the solution to the equation

$$dy(t) = -\alpha y(t) + \sigma dB(t) \qquad \text{on } t \geq 0 \qquad (5.18)$$

with initial value $y(0) = y_0 \sim N(0, \sigma^2/2\alpha)$. We now treat $x(t)$ and $y(t)$ simultaneously by combining equations (5.17) and (5.18) into the 2-dimensional linear stochastic differential equation

$$d\begin{pmatrix} x(t) \\ y(t) \end{pmatrix} = F\begin{pmatrix} x(t) \\ y(t) \end{pmatrix} dt + \begin{pmatrix} 0 \\ \sigma \end{pmatrix} dB(t), \qquad (5.19)$$

where

$$F = \begin{pmatrix} a & b \\ 0 & -\alpha \end{pmatrix}.$$

Hence the solution is

$$\begin{pmatrix} x(t) \\ y(t) \end{pmatrix} = e^{Ft}\begin{pmatrix} x_0 \\ y_0 \end{pmatrix} + \int_0^t e^{F(t-s)}\begin{pmatrix} 0 \\ \sigma \end{pmatrix} dB(s).$$

4

Stability of Stochastic Differential Equations

4.1 INTRODUCTION

In 1892, A.M. Lyapunov introduced the concept of stability of a dynamic system. Roughly speaking, the stability means insensitivity of the state of the system to small changes in the initial state or the parameters of the system. For a stable system, the trajectories which are "close" to each other at a specific instant should therefore remain close to each other at all subsequent instants.

To make the stochastic stability theory more understandable, let us recall a few basic facts on the theory of stability of deterministic systems described by ordinary differential equations. For the details please see Hahn (1967) and Lakshmikantham et al. (1989). Consider a d-dimensional ordinary differential equation

$$\dot{x}(t) = f(x(t), t) \quad \text{on } t \geq t_0. \tag{1.1}$$

Assume that for every initial value $x(t_0) = x_0 \in R^d$, there exists a unique global solution which is denoted by $x(t; t_0, x_0)$. Assume furthermore that

$$f(0, t) = 0 \quad \text{for all } t \geq t_0.$$

So equation (1.1) has the solution $x(t) \equiv 0$ corresponding to the initial value $x(t_0) = 0$. This solution is called the *trivial solution* or *equilibrium position*. The trivial solution is said to be *stable* if, for every $\varepsilon > 0$, there exists a $\delta = \delta(\varepsilon, t_0) > 0$ such that

$$|x(t; t_0, x_0)| < \varepsilon \quad \text{for all } t \geq t_0$$

whenever $|x_0| < \delta$. Otherwise, it is said to be *unstable*. The trivial solution is said to be *asymptotically stable* if it is stable and if there exists a $\delta_0 = \delta_0(t_0) > 0$ such that
$$\lim_{t \to \infty} x(t; t_0, x_0) = 0$$
whenever $|x_0| < \delta_0$.

If equation (1.1) can be solved explicitly, it would be rather easy to determine whether the trivial solution is stable. However, equation (1.1) can only be solved explicitly in some special cases. Fortunately, Lyapunov in 1892 developed a method for determining stability without solving the equation, and this method is now known as the Lyapunov direct or second method. To explain the method, let us introduce a few necessary notations. Let $\mathcal{K}$ denote the family of all continuous nondecreasing functions $\mu : R_+ \to R_+$ such that $\mu(0) = 0$ and $\mu(r) > 0$ if $r > 0$. For $h > 0$, let $S_h = \{x \in R^d : |x| < h\}$. A continuous function $V(x, t)$ defined on $S_h \times [t_0, \infty)$ is said to be *positive-definite* (in the sense of Lyapunov) if $V(0, t) \equiv 0$ and, for some $\mu \in \mathcal{K}$,
$$V(x, t) \geq \mu(|x|) \qquad \text{for all } (x, t) \in S_h \times [t_0, \infty).$$

A function V is said to be *negative-definite* if $-V$ is positive-definite. A continuous nonnegative function $V(x, t)$ is said to be *decrescent* (i.e. to have an arbitrarily small upper bound) if for some $\mu \in \mathcal{K}$,
$$V(x, t) \leq \mu(|x|) \qquad \text{for all } (x, t) \in S_h \times [t_0, \infty).$$

A function $V(x, t)$ defined on $R^d \times [t_0, \infty)$ is said to be *radially unbounded* if
$$\lim_{|x| \to \infty} \inf_{t \geq t_0} V(x, t) = \infty.$$

Let $C^{1,1}(S_h \times [t_0, \infty); R_+)$ denote the family of all continuous functions $V(x, t)$ from $S_h \times [t_0, \infty)$ to R_+ with continuous first partial derivatives with respect to every component of x and to t. Let $x(t)$ be a solution of equation (1.1) and $V(x, t) \in C^{1,1}(S_h \times [t_0, \infty); R_+)$. Then $v(t) = V(x(t), t)$ represents a function of t with the derivative
$$\dot{v}(t) = V_t(x(t), t) + V_x(x(t), t) f(x(t), t)$$
$$= \frac{\partial V}{\partial t}(x(t), t) + \sum_{i=1}^{d} \frac{\partial V}{\partial x_i}(x(t), t) f_i(x(t), t).$$

If $\dot{v}(t) \leq 0$, then $v(t)$ will not increase so the "distance" of $x(t)$ from the equilibrium point measured by $V(x(t), t)$ does not increase. If $\dot{v}(t) < 0$, then $v(t)$ will decrease to zero so the distance will decrease to zero, that is $x(t) \to 0$. These are the basic ideas of the Lyapunov direct method and lead to the following well-known Lyapunov theorem.

Theorem 1.1 *(i) If there exists a positive-definite function $V(x,t) \in C^{1,1}(S_h \times [t_0, \infty); R_+)$ such that*

$$\dot{V}(x,t) := V_t(x(t),t) + V_x(x(t),t)f(x(t),t) \leq 0$$

for all $(x,t) \in S_h \times [t_0, \infty)$, then the trivial solution of equation (1.1) is stable.

(ii) If there exists a positive-definite decrescent function $V(x,t) \in C^{1,1}(S_h \times [t_0, \infty); R_+)$ such that $\dot{V}(x,t)$ is negative-definite, then the trivial solution of equation (1.1) is asymptotically stable.

A function $V(x,t)$ that satisfies the stability conditions of Theorem 1.1 is called a *Lyapunov function* corresponding to the ordinary differential equation.

When we try to carry over the principles of the Lyapunov stability theory for deterministic systems to stochastic ones, we face the following problems:

- What is a suitable definition of stochastic stability?
- What conditions should a stochastic Lyapunov function satisfy?
- With what should the inequality $\dot{V}(x,t) \leq 0$ be replaced in order to get stability assertions?

It turns out that there are at least three different types of stochastic stability: stability in probability, moment stability and almost sure stability. In 1965, Bucy recognized that a stochastic Lyapunov function should have the supermartingale property and gave surprisingly simple sufficient criteria for stability in probability as well as for moment stability. Almost sure stability was considered by Has'minskii (1967) for linear stochastic differential equations. Stochastic stability has been one of the most active areas in stochastic analysis and many mathematicians have devoted their interests to it. We here mention Arnold, Baxendale, Chow, Curtain, Elworthy, Friedman, Ichikawa, Kliemann, Kolmanovskii, Kushner, Ladde, Lakshmikantham, Mohammed, Pardoux, Pinsky, Pritchard, Truman, Wihstutz, Zabczyk and myself among others.

In this chapter we shall investigate various types of stability for the d-dimensional stochastic differential equation

$$dx(t) = f(x(t),t)dt + g(x(t),t)dB(t) \quad \text{on } t \geq t_0. \quad (1.2)$$

For the stability purpose of this chapter, it is enough (we shall explain why later) to consider the constant initial value $x_0 \in R^d$ only, instead of the $\mathcal{F}_{t_0}$-measurable random variable $x_0 \in L^2(\Omega; R^d)$. Throughout this chapter we shall assume that the assumptions of the existence-and-uniqueness Theorem 2.3.6 are fulfilled. Hence, for any given initial value $x(t_0) = x_0 \in R^d$, equation (1.2) has a unique global solution that is denoted by $x(t; t_0, x_0)$. We know that the solution has continuous sample paths and its every moment is finite. Assume furthermore that

$$f(0,t) = 0 \quad \text{and} \quad g(0,t) = 0 \quad \text{for all } t \geq t_0.$$

So equation (1.2) has the solution $x(t) \equiv 0$ corresponding to the initial value $x(t_0) = 0$. This solution is called the *trivial solution* or *equilibrium position*.

Besides, we shall need a few more notations. Let $0 < h \leq \infty$. Denote by $C^{2,1}(S_h \times R_+; R_+)$ the family of all nonnegative functions $V(x,t)$ defined on $S_h \times R_+$ such that they are continuously twice differentiable in x and once in t. Define the differential operator L associated with equation (1.2) by

$$L = \frac{\partial}{\partial t} + \sum_{i=1}^{d} f_i(x,t) \frac{\partial}{\partial x_i} + \frac{1}{2} \sum_{i,j=1}^{d} [g^T(x,t)g(x,t)]_{ij} \frac{\partial^2}{\partial x_i \partial x_i}.$$

If L acts on a function $V \in C^{2,1}(S_h \times R_+; R_+)$, then

$$LV(x,t) = V_t(x,t) + V_x(x,t)f(x,t) + \frac{1}{2} \text{trace}\left[g^T(x,t)V_{xx}(x,t)g(x,t)\right].$$

(See page 31 for the definition of V_t, V_x and V_{xx}). By Itô's formula, if $x(t) \in S_h$, then

$$dV(x(t),t) = LV(x(t),t)dt + V_x(x(t),t)g(x(t),t)dB(t)$$

and this explains why the differential operator L is defined as above. We shall see that the inequality $\dot{V}(x,t) \leq 0$ will be replaced by $LV(x,t) \leq 0$ in order to get the stochastic stability assertions.

4.2 STABILITY IN PROBABILITY

In this section, we shall discuss the stability in probability. Let us stress that throughout this chapter, we shall let the initial value x_0 be a constant (in R^d) but not a random variable. We shall explain why we need only discuss this case of constant initial values after the definition of stability in probability.

Definition 2.1 *(i) The trivial solution of equation (1.2) is said to be stochastically stable or stable in probability if for every pair of $\varepsilon \in (0,1)$ and $r > 0$, there exists a $\delta = \delta(\varepsilon, r, t_0) > 0$ such that*

$$P\{|x(t; t_0, x_0)| < r \text{ for all } t \geq t_0\} \geq 1 - \varepsilon$$

whenever $|x_0| < \delta$. Otherwise, it is said to be stochastically unstable.

(ii) The trivial solution is said to be stochastically asymptotically stable if it is stochastically stable and, moreover, for every $\varepsilon \in (0,1)$, there exists a $\delta_0 = \delta_0(\varepsilon, t_0) > 0$ such that

$$P\{\lim_{t \to \infty} x(t; t_0, x_0) = 0\} \geq 1 - \varepsilon$$

whenever $|x_0| < \delta_0$.

(iii) The trivial solution is said to be *stochastically asymptotically stable in the large* if it is stochastically stable and, moreover, for all $x_0 \in R^d$

$$P\{\lim_{t\to\infty} x(t;t_0,x_0) = 0\} = 1.$$

Let us now explain why we need only to discuss the case of constant initial values. Suppose one would like to let the initial value x_0 be a random variable. He then should replace e.g. "$|x_0| < \delta$" by "$|x_0| < \delta$ a.s." in the definition accordingly. This seems more general but is in fact equivalent to the above definition. For example, suppose we have (i), then for any random variable x_0 with $|x_0| < \delta$ a.s., we have

$$P\{|x(t;t_0,x_0)| < r \text{ for all } t \geq t_0\}$$
$$= \int_{S_\delta} P\{|x(t;t_0,y)| < r \text{ for all } t \geq t_0\} P\{x_0 \in dy\}$$
$$\geq \int_{S_\delta} (1-\varepsilon) P\{x_0 \in dy\} = 1 - \varepsilon.$$

It should also be pointed out that when $g(x,t) \equiv 0$, these definitions reduce to the corresponding deterministic ones. We now extend the Lyapunov Theorem 1.1 to the stochastic case.

Theorem 2.2 *If there exists a positive-definite function $V(x,t) \in C^{2,1}(S_h \times [t_0,\infty); R_+)$ such that*
$$LV(x,t) \leq 0$$
for all $(x,t) \in S_h \times [t_0,\infty)$, then the trivial solution of equation (1.2) is stochastically stable.

Proof. By the definition of a positive-definite function, we know that $V(0,t) \equiv 0$ and there is a function $\mu \in \mathcal{K}$ such that

$$V(x,t) \geq \mu(|x|) \quad \text{for all } (x,t) \in S_h \times [t_0,\infty). \tag{2.1}$$

Let $\varepsilon \in (0,1)$ and $r > 0$ be arbitrary. Without loss of generality we may assume that $r < h$. By the continuity of $V(x,t)$ and the fact $V(0,t_0) = 0$, we can find a $\delta = \delta(\varepsilon,r,t_0) > 0$ such that

$$\frac{1}{\varepsilon} \sup_{x \in S_\delta} V(x,t_0) \leq \mu(r). \tag{2.2}$$

It is easy to see that $\delta < r$. Now fix the initial value $x_0 \in S_\delta$ arbitrarily and write $x(t;t_0,x_0) = x(t)$ simply. Let τ be the first exit time of $x(t)$ from S_r, that is

$$\tau = \inf\{t \geq t_0 : x(t) \notin S_r\}.$$

By Itô's formula, for any $t \geq t_0$,

$$V(x(\tau \wedge t), \tau \wedge t) = V(x_0, t_0) + \int_{t_0}^{\tau \wedge t} LV(x(s), s)ds$$
$$+ \int_{t_0}^{\tau \wedge t} V_x(x(s), s)g(x(s), s)dB(s).$$

Taking the expectation on both sides and making use of the condition $LV \leq 0$, we obtain that

$$EV(x(\tau \wedge t), \tau \wedge t) \leq V(x_0, t_0). \tag{2.3}$$

Note that $|x(\tau \wedge t)| = |x(\tau)| = r$ if $\tau \leq t$. Hence, by (2.1),

$$EV(x(\tau \wedge t), \tau \wedge t) \geq E\Big[I_{\{\tau \leq t\}} V(x(\tau), \tau)\Big] \geq \mu(r) P\{\tau \leq t\}.$$

This, together with (2.3) and (2.2), implies

$$P\{\tau \leq t\} \leq \varepsilon.$$

Letting $t \to \infty$ we get $P\{\tau < \infty\} \leq \varepsilon$, that is

$$P\{|x(t)| < r \text{ for all } t \geq t_0\} \geq 1 - \varepsilon$$

as required.

Theorem 2.3 *If there exists a positive-definite decrescent function $V(x,t) \in C^{2,1}(S_h \times [t_0, \infty); R_+)$ such that $LV(x,t)$ is negative-definite, then the trivial solution of equation (1.2) is stochastically asymptotically stable.*

Proof. We know from Theorem 2.2 that the trivial solution is stochastically stable. So we only need to show that for any $\varepsilon \in (0,1)$, there is a $\delta_0 = \delta_0(\varepsilon, t_0) > 0$ such that

$$P\{\lim_{t \to \infty} x(t; t_0, x_0) = 0\} \geq 1 - \varepsilon \tag{2.4}$$

whenever $|x_0| < \delta_0$. Note that the assumptions on function $V(x,t)$ mean that $V(0,t) \equiv 0$ and, moreover, there are three functions $\mu_1, \mu_2, \mu_3 \in \mathcal{K}$ such that

$$\mu_1(|x|) \leq V(x,t) \leq \mu_2(|x|) \quad \text{and} \quad LV(x,t) \leq -\mu_3(|x|) \tag{2.5}$$

for all $(x,t) \in S_h \times [t_0, \infty)$. Fix $\varepsilon \in (0,1)$ arbitrarily. By Theorem 2.2, there is a $\delta_0 = \delta_0(\varepsilon, t_0) > 0$ such that

$$P\{|x(t; t_0, x_0)| < h/2\} \geq 1 - \frac{\varepsilon}{4} \tag{2.6}$$

whenever $x_0 \in S_{\delta_0}$. Fix any $x_0 \in S_{\delta_0}$ and write $x(t; t_0, x_0) = x(t)$ simply. Let $0 < \beta < |x_0|$ be arbitrary, and choose $0 < \alpha < \beta$ sufficiently small for

$$\frac{\mu_2(\alpha)}{\mu_1(\beta)} \leq \frac{\varepsilon}{4}. \tag{2.7}$$

Define the stopping times

$$\tau_\alpha = \inf\{t \geq t_0 : |x(t)| \leq \alpha\}$$

and

$$\tau_h = \inf\{t \geq t_0 : |x(t)| \geq h/2\}.$$

By Itô's formula and (2.5), we can derive that for any $t \geq t_0$,

$$0 \leq EV(x(\tau_\alpha \wedge \tau_h \wedge t), \tau_\alpha \wedge \tau_h \wedge t)$$
$$= V(x_0, t_0) + E\int_{t_0}^{\tau_\alpha \wedge \tau_h \wedge t} LV(x(s), s)ds$$
$$\leq V(x_0, t_0) - \mu_3(\alpha)E(\tau_\alpha \wedge \tau_h \wedge t - t_0).$$

Consequently

$$(t - t_0)P\{\tau_\alpha \wedge \tau_h \geq t\} \leq E(\tau_\alpha \wedge \tau_h \wedge t - t_0) \leq \frac{V(x_0, t_0)}{\mu_3(\alpha)}.$$

This implies immediately that

$$P\{\tau_\alpha \wedge \tau_h < \infty\} = 1.$$

But, by (2.6), $P\{\tau_h < \infty\} \leq \varepsilon/4$. Hence

$$1 = P\{\tau_\alpha \wedge \tau_h < \infty\} \leq P\{\tau_\alpha < \infty\} + P\{\tau_h < \infty\} \leq P\{\tau_\alpha < \infty\} + \frac{\varepsilon}{4},$$

which yields

$$P\{\tau_\alpha < \infty\} \geq 1 - \frac{\varepsilon}{4}. \tag{2.8}$$

Choose θ sufficiently large for

$$P\{\tau_\alpha < \theta\} \geq 1 - \frac{\varepsilon}{2}.$$

Then

$$P\{\tau_\alpha < \tau_h \wedge \theta\} \geq P(\{\tau_\alpha < \theta\} \cap \{\tau_h = \infty\})$$
$$\geq P\{\tau_\alpha < \theta\} - P\{\tau_h < \infty\} \geq 1 - \frac{3\varepsilon}{4}. \tag{2.9}$$

Now, define two stopping times

$$\sigma = \begin{cases} \tau_\alpha & \text{if } \tau_\alpha < \tau_h \wedge \theta, \\ \infty & \text{otherwise} \end{cases}$$

and

$$\tau_\beta = \inf\{t > \sigma : |x(t)| \geq \beta\}.$$

We can then show by Itô's formula that for any $t \geq \theta$,

$$EV(x(\tau_\beta \wedge t), \tau_\beta \wedge t) \leq EV(x(\sigma \wedge t), \sigma \wedge t).$$

Noting that $V(x(\tau_\beta \wedge t), \tau_\beta \wedge t) = V(x(\sigma \wedge t), \sigma \wedge t) = V(x(t), t)$ on $\omega \in \{\tau_\alpha \geq \tau_h \wedge \theta\}$, we get

$$E\Big[I_{\{\tau_\alpha < \tau_h \wedge \theta\}} V(x(\tau_\beta \wedge t), \tau_\beta \wedge t)\Big] \leq E\Big[I_{\{\tau_\alpha < \tau_h \wedge \theta\}} EV(x(\tau_\alpha), \tau_\alpha)\Big].$$

Using (2.6) and the fact $\{\tau_\beta \leq t\} \subset \{\tau_\alpha < \tau_h \wedge \theta\}$ we further obtain

$$\mu_1(\beta) P\{\tau_\beta \leq t\} \leq \mu_2(\alpha).$$

This, together with (2.7), yields

$$P\{\tau_\beta \leq t\} \leq \frac{\varepsilon}{4}.$$

Letting $t \to \infty$ we have

$$P\{\tau_\beta < \infty\} \leq \frac{\varepsilon}{4}.$$

It then follows, using (2.9) as well, that

$$P\{\sigma < \infty \text{ and } \tau_\beta = \infty\} \geq P\{\tau_\alpha < \tau_h \wedge \theta\} - P\{\tau_\beta < \infty\} \geq 1 - \varepsilon.$$

But this means that

$$P\{\omega : \limsup_{t \to \infty} |x(t)| \leq \beta\} \geq 1 - \varepsilon.$$

Since β is arbitrary, we must have

$$P\{\omega : \limsup_{t \to \infty} x(t) = 0\} \geq 1 - \varepsilon$$

as required. The proof is complete.

Theorem 2.4 *If there exists a positive-definite decrescent radially unbounded function $V(x,t) \in C^{2,1}(R^d \times [t_0, \infty); R_+)$ such that $LV(x,t)$ is negative-definite, then the trivial solution of equation (1.2) is stochastically asymptotically stable in the large.*

Proof. By Theorem 2.2, the trivial solution of equation is stochastically stable. So we only need to show that

$$P\{\lim_{t \to \infty} x(t; t_0, x_0) = 0\} = 1 \tag{2.10}$$

for all $x_0 \in R^d$. Fix any x_0 and write $x(t;t_0,x_0) = x(t)$ again. Let $\varepsilon \in (0,1)$ be arbitrary. Since $V(x,t)$ is radially unbounded, we can find an $h > |x_0|$ sufficiently large for

$$\inf_{|x|\geq h, t\geq t_0} V(x,t) \geq \frac{4V(x_0,t_0)}{\varepsilon}. \tag{2.11}$$

Define the stopping time

$$\tau_h = \inf\{t \geq t_0 : |x(t)| \geq h\}.$$

By Itô's formula, we can show that for any $t \geq t_0$,

$$EV(x(\tau_h \wedge t), \tau_h \wedge t) \leq V(x_0, t_0). \tag{2.12}$$

But, by (2.11), we see that

$$EV(x(\tau_h \wedge t), \tau_h \wedge t) \geq \frac{4V(x_0,t_0)}{\varepsilon} P\{\tau_h \leq t\}$$

It then follows from (2.12) that

$$P\{\tau_h \leq t\} \leq \frac{\varepsilon}{4}.$$

Letting $t \to \infty$ gives $P\{\tau_h < \infty\} \leq \varepsilon/4$. That is

$$P\{|x(t)| \leq h \text{ for all } t \geq t_0\} \geq 1 - \frac{\varepsilon}{4}. \tag{2.13}$$

From here, we can show in the same way as the proof of Theorem 2.3 that

$$P\{\lim_{t\to\infty} x(t) = 0\} \geq 1 - \varepsilon.$$

Since ε is arbitrary, the required (2.10) must hold and the proof is complete.

The functions $V(x,t)$ used in Theorems 2.2–2.4 are called *stochastic Lyapunov functions*, and the use of these theorems depends on the construction of the functions. As in the deterministic case, there are a number of techniques that can be used to find suitable functions. For example, the quadratic function

$$V(x,t) = x^T Q x,$$

where Q is a symmetric positive-definite matrix, will do if

$$LV(x,t) = 2x^T Q f(x,t) + \text{trace}[g^T(x,t) Q g(x,t)] \leq 0$$

or is negative-definite in some neighbourhood of $x = 0$ for $t \geq t_0$. Besides, one can seek a positive-definite solution of the equation $LV(x,t) = 0$ or of the inequality $LV(x,t) \leq 0$. We now discuss a few examples to illustrate the theory.

Example 2.5 Consider a one-dimensional stochastic differential equation

$$dx(t) = f(x(t), t)dt + g(x(t), t)dB(t) \quad \text{on } t \geq t_0 \quad (2.14)$$

with initial value $x(t_0) = x_0 \in R$. Assume that $f : R \times R_+ \to R$ and $g : R \times R_+ \to R^m$ have the expansions

$$f(x,t) = a(t)x + o(|x|), \quad g(x,t) = (b_1(t)x, \cdots, b_m(t)x)^T + o(|x|) \quad (2.15)$$

in a neighbourhood of $x = 0$ uniformly with respect to $t \geq t_0$, where $a(t)$, $b_i(t)$ are all bounded Borel-measurable real-valued functions. We impose a condition that there is a pair of positive constants θ and K such that

$$-K \leq \int_{t_0}^t \left(a(s) - \frac{1}{2} \sum_{i=1}^m b_i^2(s) + \theta \right) ds \leq K \quad \text{for all } t \geq t_0. \quad (2.16)$$

Let

$$0 < \varepsilon < \frac{\theta}{\sup_{t \geq t_0} \sum_{i=1}^m b_i^2(t)}$$

and define the stochastic Lyapunov function

$$V(x,t) = |x|^\varepsilon \exp\left[-\varepsilon \int_{t_0}^t \left(a(s) - \frac{1}{2} \sum_{i=1}^m b_i^2(s) + \theta \right) ds \right].$$

By condition (2.16),

$$|x|^\varepsilon e^{-\varepsilon K} \leq V(x,t) \leq |x|^\varepsilon e^{\varepsilon K}.$$

Hence $V(x,t)$ is positive-definite and decrescent. On the other hand, by (2.15),

$$LV(x,t) = \varepsilon |x|^\varepsilon \exp\left[-\varepsilon \int_{t_0}^t \left(a(s) - \frac{1}{2} \sum_{i=1}^m b_i^2(s) + \theta \right) ds \right]$$

$$\times \left(\frac{\varepsilon}{2} \sum_{i=1}^m b_i^2(t) - \theta \right) + o(|x|^\varepsilon)$$

$$\leq -\frac{1}{2} \varepsilon \theta e^{-\varepsilon K} |x|^\varepsilon + o(|x|^\varepsilon).$$

We hence see that $LV(x,t)$ is negative-definite in a sufficiently small neighbourhood of $x = 0$ for $t \geq t_0$. By Theorem 2.4 we therefore conclude that under (2.15) and (2.16), the trivial solution of equation (2.14) is stochastically asymptotically stable.

Example 2.6 Assume that the coefficients f and g of equation (1.2) have the expansions

$$f(x,t) = F(t)x + o(|x|), \quad g(x,t) = (G_1(t)x, \cdots, G_m(t)x) + o(|x|) \quad (2.17)$$

in a neighbourhood of $x = 0$ uniformly with respect to $t \geq t_0$, where $F(t)$, $G_i(t)$ are all bounded Borel-measurable $d \times d$-matrix-valued functions. Assume that there is a symmetric positive-definite matrix Q such that the symmetric matrix

$$QF(t) + F^T(t)Q + \sum_{i=1}^{m} G_i^T(t)QG_i(t)$$

is negative-definite uniformly in $t \geq t_0$, that is

$$\lambda_{\max}\left(QF(t) + F^T(t)Q + \sum_{i=1}^{m} G_i^T(t)QG_i(t)\right) \leq -\lambda < 0 \qquad (2.18)$$

for all $t \geq t_0$, where (and throughout this book) $\lambda_{\max}(A)$ denotes the largest eigenvalue of matrix A. Now, define the stochastic Lyapunov function $V(x,t) = x^T Q x$. It is obviously positive-definite and decrescent. Moreover,

$$LV(x,t) = x^T\left(QF(t) + F^T(t)Q + \sum_{i=1}^{m} G_i^T(t)QG_i(t)\right)x + o(|x|^2)$$

$$\leq -\lambda|x|^2 + o(|x|^2).$$

Hence $LV(x,t)$ is negative-definite in a sufficiently small neighbourhood of $x = 0$ for $t \geq t_0$. By Theorem 2.4 we therefore conclude that under (2.17) and (2.18), the trivial solution of equation (1.2) is stochastically asymptotically stable.

In the case of linear stochastic differential equations, one may make use of the explicit solutions to determine whether the equations are stochastically stable or not. The following example demonstrates the idea.

Example 2.7 Consider a one-dimensional linear stochastic differential equation

$$dx(t) = a(t)x(t)dt + \sum_{i=1}^{m} b_i(t)x(t)dB_i(t) \qquad \text{on } t \geq t_0 \qquad (2.19)$$

with initial value $x(t_0) = x_0$, where $a(t)$, $b_i(t)$ are continuous real-valued functions on $[t_0, \infty)$. By Lemma 3.2.3, the unique solution of equation (2.19) is

$$x(t) = x_0 \exp\left[\int_{t_0}^{t}\left(a(s) - \frac{1}{2}\sum_{i=1}^{m} b_i^2(s)\right)ds + \sum_{i=1}^{m}\int_{t_0}^{t} b_i(s)dB_i(s)\right]. \qquad (2.20)$$

Set $\sigma(t) = \sum_{i=1}^{m} \int_{t_0}^{t} b_i^2(s)ds$ for $t_0 \leq t \leq \infty$. We divide the discussion of stability into two cases.

Case (i) : $\sigma(\infty) < \infty$. In this case, $\sum_{i=1}^{m} \int_{t_0}^{t} b_i(s)dB_i(s)$ approaches the normal distribution $N(0, \sigma(\infty))$. It therefore follows from (2.20) that the trivial solution of equation (2.20) is stochastically stable if and only if

$$\limsup_{t \to \infty} \int_{t_0}^{t} a(s)ds < \infty,$$

while the trivial solution is stochastically asymptotically stable in the large if and only if
$$\lim_{t\to\infty} \int_{t_0}^{t} a(s)ds = -\infty.$$

Case (ii) : $\sigma(\infty) = \infty$. Let $\tau(s)$, $s \geq 0$, be the inverse function of $\sigma(t)$, that is
$$\tau(s) = \inf\{t \geq t_0 : \sigma(t) = s\}.$$
Clearly, $\sigma(\tau(s)) = s$ if $s \geq 0$ and $\tau(\sigma(t)) = t$ if $t \geq t_0$. Define
$$\bar{B}(s) = \sum_{i=1}^{m} \int_{t_0}^{\tau(s)} b_i(t) dB_i(t) \qquad \text{on } s \geq 0.$$

Then $\bar{B}(s)$ is a continuous martingale with $\bar{B}(0) = 0$ and the quadratic variation
$$\langle \bar{B}, \bar{B} \rangle_s = \sum_{i=1}^{m} \int_{t_0}^{\tau(s)} b_i^2(t) dt = \sigma(\tau(s)) = s.$$

By Lévy's theorem (i.e. Theorem 1.4.4), $\bar{B}(s)$ is a Brownian motion. So, by the law of the iterated logarithm,
$$\limsup_{s\to\infty} \frac{\bar{B}(s)}{\sqrt{2s \log\log s}} = 1 \qquad a.s.$$

Consequently
$$\limsup_{t\to\infty} \frac{\sum_{i=1}^{m} \int_{t_0}^{t} b_i(s) dB_i(s)}{\sqrt{2\sigma(t)\log\log \sigma(t)}} = \limsup_{t\to\infty} \frac{\bar{B}(\sigma(t))}{\sqrt{2\sigma(t)\log\log \sigma(t)}} = 1 \qquad a.s.$$

Applying this to (2.20) we can conclude that the trivial solution of equation (2.19) is stochastically asymptotically stable in the large if
$$\limsup_{t\to\infty} \frac{\int_{t_0}^{t} a(s)ds - \frac{1}{2}\sigma(t)}{\sqrt{2\sigma(t)\log\log \sigma(t)}} < -1 \qquad a.s. \qquad (2.21)$$

As a special case, let
$$a(t) = a, \quad b_i(t) = b_i \qquad (2.22)$$
be all constants. In this case, (2.21) holds if and only if
$$a < \frac{1}{2} \sum_{i=1}^{m} b_i^2. \qquad (2.23)$$

Hence, under (2.22) and (2.23), the solution of equation (2.19) will tend to the equilibrium position $x = 0$. On the other hand, we can compute more precisely how fast the solution tends to zero. In fact, under (2.22), the unique solution of equation (2.19) is

$$x(t; t_0, x_0) = x_0 \exp\left[\left(a - \frac{1}{2}\sum_{i=1}^m b_i^2\right)(t - t_0) + \sum_{i=1}^m b_i(B_i(t) - B_i(t_0))\right].$$

So

$$\log|x(t; t_0, x_0)| = \log|x_0| + \left(a - \frac{1}{2}\sum_{i=1}^m b_i^2\right)(t - t_0) + \sum_{i=1}^m b_i(B_i(t) - B_i(t_0)).$$

Noting from the law of the iterated logarithm that

$$\lim_{t\to\infty} \frac{B_i(t) - B_i(t_0)}{t} = 0 \quad a.s.$$

we then derive that, if (2.23) holds,

$$\lim_{t\to\infty} \frac{1}{t}\log|x(t; t_0, x_0)| = a - \frac{1}{2}\sum_{i=1}^m b_i^2 < 0 \quad a.s. \quad (2.24)$$

that is the sample Lyapunov exponent is negative. Hence, for any $0 < \varepsilon < \frac{1}{2}\sum_{i=1}^m b_i^2 - a$, one can find a positive random variable $\xi = \xi(t_0, x_0, \varepsilon)$ such that

$$|x(t; t_0, x_0)| \leq \xi \exp\left[-\left(\frac{1}{2}\sum_{i=1}^m b_i^2 - a - \varepsilon\right)(t - t_0)\right] \quad \text{for all } t \geq t_0.$$

In other words, almost all sample paths of the solution will tend to the equilibrium position $x = 0$ exponentially fast. Such a property will be called the almost sure exponential stability. Let us now turn to study this type of stability in detail.

4.3 ALMOST SURE EXPONENTIAL STABILITY

We first give the formal definition of the almost sure exponential stability.

Definition 3.1 *The trivial solution of equation (1.2) is said to be almost surely exponentially stable if*

$$\limsup_{t\to\infty} \frac{1}{t}\log|x(t; t_0, x_0)| < 0 \quad a.s. \quad (3.1)$$

for all $x_0 \in R^d$.

As defined in Section 2.5, the left-hand side of (3.1) is called the sample Lyapunov exponents of the solution. We therefore see that the trivial solution is almost surely exponentially stable if and only if the sample Lyapunov exponents are negative. As explained in the end of previous section, the almost sure exponential stability means that almost all sample paths of the solution will tend to the equilibrium position $x = 0$ exponentially fast. Moreover, let us explain once again why we only need to discuss the case of constant initial values. For a general initial value x_0 (i.e. x_0 is $\mathcal{F}_{t_0}$-measurable and belongs to $L^2(\Omega; R^d)$), it follows from (3.1) that

$$P\{\limsup_{t\to\infty} \frac{1}{t} \log |x(t; t_0, x_0)| < 0\}$$
$$= \int_{R^d} P\{\limsup_{t\to\infty} \frac{1}{t} \log |x(t; t_0, y)| < 0\} P\{x_0 \in dy\}$$
$$= \int_{R^d} P\{x_0 \in dy\} = 1,$$

that is

$$\limsup_{t\to\infty} \frac{1}{t} \log |x(t; t_0, x_0)| < 0 \quad a.s.$$

To establish the theorems on the almost sure exponential stability, we need prepare a useful lemma. Recall that we assume, throughout this chapter, that the assumptions of the existence-and-uniqueness Theorem 2.3.6 are fulfilled and, moreover, $f(0, t) \equiv 0$, $g(0, t) \equiv 0$. Under these standing hypotheses, we have the following useful lemma.

Lemma 3.2 For all $x_0 \neq 0$ in R^d

$$P\{x(t; t_0, x_0) \neq 0 \text{ on } t \geq t_0\} = 1. \tag{3.2}$$

That is, almost all the sample path of any solution starting from a non-zero state will never reach the origin.

Proof. If (3.2) were false, there would exist some $x_0 \neq 0$ such that $P\{\tau < \infty\} > 0$, where τ is the first time of zero of the corresponding solution, i.e.

$$\tau = \inf\{t \geq t_0 : x(t) = 0\}$$

in which we write $x(t; t_0, x_0) = x(t)$ simply. So we can find a pair of constants $T > t_0$ and $\theta > 1$ sufficiently large for $P(B) > 0$, where

$$B = \{\tau \leq T \text{ and } |x(t)| \leq \theta - 1 \text{ for all } t_0 \leq t \leq \tau\}.$$

But, by the standing hypotheses, there exists a positive constant K_θ such that

$$|f(x, t)| \vee |g(x, t)| \leq K_\theta |x| \qquad \text{for all } |x| \leq \theta, \ t_0 \leq t \leq T.$$

Let $V(x,t) = |x|^{-1}$. Then, for $0 < |x| \leq \theta$ and $t_0 \leq t \leq T$,

$$LV(x,t) = -|x|^{-3}x^T f(x,t) + \frac{1}{2}\left(-|x|^{-3}|g(x,t)|^2 + 3|x|^{-5}|x^T g(x,t)|^2\right)$$
$$\leq |x|^{-2}|f(x,t)| + |x|^{-3}|g(x,t)|^2$$
$$\leq K_\theta |x|^{-1} + K_\theta^2 |x|^{-1} = K_\theta(1 + K_\theta)V(x,t).$$

Now, for any $\varepsilon \in (0, |x_0|)$, define the stopping time

$$\tau_\varepsilon = \inf\{t \geq t_0 : |x(t)| \notin (\varepsilon, \theta)\}.$$

By Itô's formula,

$$E\left[e^{-K_\theta(1+K_\theta)(\tau_\varepsilon \wedge T - t_0)}V(x(\tau_\varepsilon \wedge T), \tau_\varepsilon \wedge T)\right] = V(x_0, t_0)$$
$$+ E\int_{t_0}^{\tau_\varepsilon \wedge T} e^{-K_\theta(1+K_\theta)(s-t_0)}\left[-(K_\theta(1+K_\theta))V(x(s),s) + LV(x(s),s)\right]ds$$
$$\leq |x_0|^{-1}.$$

Note that for $\omega \in B$, $\tau_\varepsilon \leq T$ and $|x(\tau_\varepsilon)| = \varepsilon$. The above inequality therefore implies that

$$E\left[e^{-K_\theta(1+K_\theta)(T-t_0)}\varepsilon^{-1}I_B\right] \leq |x_0|^{-1}.$$

Hence

$$P(B) \leq \varepsilon |x_0|^{-1} e^{K_\theta(1+K_\theta)(T-t_0)}.$$

Letting $\varepsilon \to 0$ yields that $P(B) = 0$, but this contradicts the definition of B. The proof is complete.

Theorem 3.3 *Assume that there exists a function $V \in C^{2,1}(R^d \times [t_0, \infty); R_+)$, and constants $p > 0$, $c_1 > 0$, $c_2 \in R$, $c_3 \geq 0$, such that for all $x \neq 0$ and $t \geq t_0$,*

(i) $c_1|x|^p \leq V(x,t)$,

(ii) $LV(x,t) \leq c_2 V(x,t)$,

(iii) $|V_x(x,t)g(x,t)|^2 \geq c_3 V^2(x,t)$.

Then

$$\limsup_{t\to\infty} \frac{1}{t} \log|x(t; t_0, x_0)| \leq -\frac{c_3 - 2c_2}{2p} \quad a.s. \qquad (3.3)$$

for all $x_0 \in R^d$. In particular, if $c_3 > 2c_2$, the trivial solution of equation (1.2) is almost surely exponentially stable.

Proof. Clearly, (3.3) holds for $x_0 = 0$ since $x(t; t_0, 0) \equiv 0$. We therefore only need to show (3.3) for $x_0 \neq 0$. Fix any $x_0 \neq 0$ and write $x(t; t_0, x_0) = x(t)$. By Lemma 3.2, $x(t) \neq 0$ for all $t \geq t_0$ almost surely. Thus, one can apply Itô's formula and condition (ii) to show that, for $t \geq t_0$,

$$\log V(x(t), t) \leq \log V(x_0, t_0) + c_2(t - t_0) + M(t)$$
$$- \frac{1}{2}\int_{t_0}^t \frac{|V_x(x(s),s)g(x(s),s)|^2}{V^2(x(s),s)}ds, \qquad (3.4)$$

where
$$M(t) = \int_{t_0}^{t} \frac{V_x(x(s),s)g(x(s),s)}{V(x(s),s)} dB(s)$$

is a continuous martingale with initial value $M(t_0) = 0$. Assign $\varepsilon \in (0,1)$ arbitrarily and let $n = 1, 2, \cdots$. By the exponential martingale inequality,

$$P\left\{ \sup_{t_0 \le t \le t_0+n} \left[M(t) - \frac{\varepsilon}{2} \int_{t_0}^{t} \frac{|V_x(x(s),s)g(x(s),s)|^2}{V^2(x(s),s)} ds \right] > \frac{2}{\varepsilon} \log n \right\} \le \frac{1}{n^2}.$$

Applying the Borel–Cantelli lemma we see that for almost all $\omega \in \Omega$, there is an integer $n_0 = n_0(\omega)$ such that if $n \ge n_0$,

$$M(t) \le \frac{2}{\varepsilon} \log n + \frac{\varepsilon}{2} \int_{t_0}^{t} \frac{|V_x(x(s),s)g(x(s),s)|^2}{V^2(x(s),s)} ds$$

holds for all $t_0 \le t \le t_0+n$. Substituting this into (3.4) and then using condition (iii) we obtain that

$$\log V(x(t),t) \le \log V(x_0, t_0) - \frac{1}{2}[(1-\varepsilon)c_3 - 2c_2](t-t_0) + \frac{2}{\varepsilon} \log n$$

for all $t_0 \le t \le t_0+n$, $n \ge n_0$ almost surely. Consequently, for almost all $\omega \in \Omega$, if $t_0+n-1 \le t \le t_0+n$ and $n \ge n_0$,

$$\frac{1}{t} \log V(x(t),t) \le -\frac{t-t_0}{2t}[(1-\varepsilon)c_3 - 2c_2] + \frac{\log V(x_0,t_0) + \frac{2}{\varepsilon}\log n}{t_0+n-1}.$$

This implies

$$\limsup_{t \to \infty} \frac{1}{t} \log V(x(t),t) \le -\frac{1}{2}[(1-\varepsilon)c_3 - 2c_2] \quad a.s.$$

Finally, using condition (i) we obtain

$$\limsup_{t \to \infty} \frac{1}{t} \log |x(t)| \le -\frac{(1-\varepsilon)c_3 - 2c_2}{2p} \quad a.s.$$

and the required assertion (3.3) follows since $\varepsilon > 0$ is arbitrary. The proof is complete.

Corollary 3.4 *Assume that there exists a function $V \in C^{2,1}(R^d \times [t_0, \infty); R_+)$, and positive constants p, α, λ, such that for all $x \ne 0$, $t \ge t_0$,*

$$\alpha |x|^p \le V(x,t) \quad \text{and} \quad LV(x,t) \le -\lambda V(x,t).$$

Then

$$\limsup_{t \to \infty} \frac{1}{t} \log |x(t; t_0, x_0)| \le -\frac{\lambda}{p} \quad a.s.$$

for all $x_0 \in R^d$. In other words, the trivial solution of equation (1.2) is almost surely exponentially stable.

This corollary follows from Theorem 3.3 immediately by letting $c_1 = \alpha$, $c_2 = -\lambda$ and $c_3 = 0$. These results have given the upper bound for the sample Lyapunov exponents. Let us now turn to the study of the lower bound.

Theorem 3.5 *Assume that there exists a function $V \in C^{2,1}(R^d \times [t_0, \infty); R_+)$, and constants $p > 0$, $c_1 > 0$, $c_2 \in R$, $c_3 > 0$, such that for all $x \neq 0$ and $t \geq t_0$,*

(i) $c_1|x|^p \geq V(x,t) > 0$,

(ii) $LV(x,t) \geq c_2 V(x,t)$,

(iii) $|V_x(x,t)g(x,t)|^2 \leq c_3 V^2(x,t)$.

Then

$$\liminf_{t \to \infty} \frac{1}{t} \log |x(t; t_0, x_0)| \geq \frac{2c_2 - c_3}{2p} \quad a.s. \quad (3.5)$$

for all $x_0 \neq 0$ in R^d. In particular, if $2c_2 > c_3$, then almost all the sample paths of $|x(t; t_0, x_0)|$ will tend to infinity, and we say in this case that the trivial solution of equation (1.2) is almost surely exponentially unstable.

Proof. Fix any $x_0 \neq 0$ and write $x(t; t_0, x_0) = x(t)$. By Itô's formula, conditions (ii) and (iii), we can easily show that for $t \geq t_0$,

$$\log V(x(t), t) \geq \log V(x_0, t_0) + \frac{1}{2}(2c_2 - c_3)(t - t_0) + M(t), \quad (3.6)$$

where

$$M(t) = \int_{t_0}^t \frac{V_x(x(s), s)g(x(s), s)}{V(x(s), s)} dB(s)$$

is a continuous martingale with the quadratic variation

$$\langle M(t), M(t) \rangle = \int_{t_0}^t \frac{|V_x(x(s), s)g(x(s), s)|^2}{V^2(x(s), s)} ds \leq c_3(t - t_0).$$

By the strong law of large numbers (i.e. Theorem 1.3.4), $\lim_{t \to \infty} M(t)/t = 0$ a.s. It therefore follows from (3.6) that

$$\liminf_{t \to \infty} \frac{1}{t} \log V(x(t), t) \geq \frac{1}{2}(2c_2 - c_3) \quad a.s.$$

which implies the required assertion (3.5) by using condition (i).

We have already known that for the scalar linear stochastic differential equation

$$dx(t) = ax(t) + \sum_{i=1}^m b_i x(t) dB_i(t) \quad \text{on } t \geq t_0, \quad (3.7)$$

the sample Lyapunov exponent is

$$\lim_{t\to\infty} \frac{1}{t} \log |x(t; t_0, x_0)| = a - \frac{1}{2}\sum_{i=1}^{m} b_i^2 \quad a.s. \qquad (3.8)$$

We now apply Theorems 3.3 and 3.5 to obtain the same conclusion. Let $V(x,t) = x^2$. Then

$$LV(x,t) = \left(2a + \sum_{i=1}^{m} b_i^2\right)|x|^2$$

and, writing $g(x,t) = (b_1 x, \cdots, b_m x)$,

$$|V_x(x,t)g(x,t)|^2 = 4\sum_{i=1}^{m} b_i^2 |x|^4.$$

Hence, by Theorem 3.3 with $p = 2$, $c_1 = 1$, $c_2 = 2a + \sum_{i=1}^{m} b_i^2$ and $c_3 = 4\sum_{i=1}^{m} b_i^2$, we have

$$\limsup_{t\to\infty} \frac{1}{t} \log |x(t; t_0, x_0)| \leq a - \frac{1}{2}\sum_{i=1}^{m} b_i^2 \quad a.s. \qquad (3.9)$$

But, by Theorem 3.5,

$$\liminf_{t\to\infty} \frac{1}{t} \log |x(t; t_0, x_0)| \geq a - \frac{1}{2}\sum_{i=1}^{m} b_i^2 \quad a.s. \qquad (3.10)$$

Combining (3.9) and (3.10) gives (3.8). This show that results obtained in Theorems 3.3 and 3.5 are very sharp. Let us now discuss a few more examples.

Example 3.6 Consider the two-dimensional stochastic differential equation

$$dx(t) = f(x(t))dt + Gx(t)dB(t) \quad \text{on } t \geq t_0 \qquad (3.11)$$

with initial value $x(t_0) = x_0 \subset R^2$, where $B(t)$ is a one-dimensional Brownian motion,

$$f(x) = \begin{pmatrix} x_2 \cos x_1 \\ 2x_1 \sin x_2 \end{pmatrix}, \quad G = \begin{pmatrix} 3 & -0.3 \\ -0.3 & 3 \end{pmatrix}.$$

Let $V(x,t) = |x|^2$. It is easy to verify that

$$4.29|x|^2 \leq LV(x,t) = 2x_1 x_2 \cos x_1 + 4x_1 x_2 \sin x_2 + |Gx|^2 \leq 13.89|x|^2$$

and

$$29.16|x|^2 \leq |V_x(x,t)Gx|^2 = |2x^T Gx|^2 \leq 43.56|x|^4.$$

Applying Theorems 3.3 and 3.5 we obtain the following lower and upper bound for the sample Lyapunov exponents of the solutions of equation (3.11)

$$-8.745 \leq \liminf_{t\to\infty} \frac{1}{t} \log|x(t;t_0,x_0)| \leq \limsup_{t\to\infty} \frac{1}{t} \log|x(t;t_0,x_0)| \leq -0.345$$

almost surely. Hence the trivial solution of equation (3.11) is almost surely exponentially stable.

Example 3.7 It is known that a linear oscillator $\ddot{y}(t) + a\dot{y}(t) + by(t) = 0$ is exponentially stable if $a > 0$ and $b > 0$. Assume that the oscillator is now driven by an external disturbance of white noise described by $(c\dot{y}(t) + hy(t))\dot{B}(t)$. In other words, we arrive at the scalar linear stochastic oscillator

$$\ddot{y}(t) + a\dot{y}(t) + by(t) = (c\dot{y}(t) + hy(t))\dot{B}(t) \tag{3.12}$$

on $t \geq 0$ with initial value $(y(0), \dot{y}(0)) = (y_1, y_2) \in R^2$. Here $\dot{B}(t)$ is a scalar white noise (i.e. $B(t)$ is a Brownian motion), and c, h are constants which represent the intensity of the stochastic disturbance. Introduce a vector $x = (x_1, x_2)^T = (y, \dot{y})^T$. Then equation (3.12) can be written as the two-dimensional Itô equation

$$\begin{cases} dx_1(t) = x_2(t)dt, \\ dx_2(t) = (-bx_1(t) - ax_2(t))dt + (cx_2(t) + hx_1(t))dB(t). \end{cases} \tag{3.13}$$

For the Lyapunov function, we try a quadratic function

$$V(x,t) = \alpha x_1^2 + \beta x_1 x_2 + x_2^2.$$

Compute

$$LV(x,t) = -(\beta b - h^2)x_1^2 - (2a - \beta - c^2)x_2^2 + (2\alpha - \beta a - 2b + 2ch)x_1 x_2.$$

In order to convert $LV(x,t)$ to be negative-definite (i.e. $LV(x,t) \leq -\varepsilon|x|^2$ for some $\varepsilon > 0$), we set

$$2\alpha - \beta a - 2b + 2ch = 0, \text{ that is } \alpha = \frac{1}{2}(\beta a + 2b - 2ch).$$

Then V and LV become

$$V(x,t) = \frac{1}{2}(\beta a + 2b - 2ch)x_1^2 + \beta x_1 x_2 + x_2^2$$

and

$$LV(x,t) = -(\beta b - h^2)x_1^2 - (2a - \beta - c^2)x_2^2.$$

For $LV(x,t)$ to be negative-definite, we must have $\beta b - h^2 > 0$ and $2a - \beta - c^2 > 0$, that is

$$\frac{h^2}{b} < \beta < 2a - c^2. \tag{3.14}$$

For V to be positive-definite (i.e. $V(x,t) \geq \varepsilon |x|^2$ for some $\varepsilon > 0$), we must have

$$2(\beta a + 2b - 2ch) > \beta^2,$$

this is equivalent to

$$a - \sqrt{a^2 + 4(b - ch)} < \beta < a + \sqrt{a^2 + 4(b - ch)}. \tag{3.15}$$

Combining (3.14) and (3.15) we see that if

$$\max\left\{\frac{h^2}{b},\ a - \sqrt{a^2 + 4(b - ch)}\right\} < \beta < \min\left\{2a - c^2,\ a + \sqrt{a^2 + 4(b - ch)}\right\},$$

then V is positive-definite and LV negative-definite. We therefore conclude, by Corollary 3.4, that if

$$\max\left\{\frac{h^2}{b},\ a - \sqrt{a^2 + 4(b - ch)}\right\} < \min\left\{2a - c^2,\ a + \sqrt{a^2 + 4(b - ch)}\right\} \tag{3.16}$$

then

$$\limsup_{t \to \infty} \frac{1}{t} \log(|y(t)| + |\dot{y}(t)|) < 0 \quad \text{a.s.}$$

that is the trivial solution $(y(t), \dot{y}(t)) = 0$ of the stochastic oscillator (3.12) is almost surely exponentially stable. A more restrictive but possibly more convenient condition than (3.16) is

$$ch \leq b, \qquad h^2 + bc^2 < 2ab. \tag{3.17}$$

This condition gives a quite clear estimate for the intensity of the external stochastic disturbance in the sense that the disturbance can be tolerated by the stable deterministic oscillator $\ddot{y}(t) + a\dot{y}(t) + by(t) = 0$ without loss of the stability property.

Example 3.8 Consider the linear homogeneous Itô equation

$$dx(t) = Fx(t)dt + \sum_{i=1}^{m} G_k x(t) dB_i(t) \qquad \text{on } t \geq t_0 \tag{3.18}$$

with initial value $x(t_0) = x_0 \in R^d$. Assume that all the $d \times d$ matrices $F, G_1, \cdots, G_m$ commute, that is,

$$FG_i = G_i F, \quad G_i G_j = G_j G_i \qquad \text{for all } 1 \leq i, j \leq m. \tag{3.19}$$

In Section 3.4 we have shown that equation (3.18) has the explicit solution

$$x(t; t_0, x_0) = \exp\left[\left(F - \frac{1}{2}\sum_{i=1}^m G_i^2\right)(t - t_0) + \sum_{i=1}^m G_i(B_i(t) - B_i(t_0))\right] x_0. \quad (3.20)$$

We now assume that all the eigenvalues of $F - \frac{1}{2}\sum_{i=1}^m G_i^2$ have negative real parts. This is equivalent to that there is a pair of positive constants C and λ such that

$$\left\|\exp\left[\left(F - \frac{1}{2}\sum_{i=1}^m G_i^2\right)(t - t_0)\right]\right\| \leq C e^{-\lambda(t-t_0)}. \quad (3.21)$$

It then follows from (3.20) that

$$|x(t; t_0, x_0)| \leq C|x_0| \exp\left[-\lambda(t - t_0) + \sum_{i=1}^m \|G_i\| \, |B_i(t) - B_i(t_0)|\right].$$

Using the property $\lim_{t\to\infty} |B_i(t) - B_i(t_0)|/t = 0$ a.s. of the Brownian motion, we obtain immediately that

$$\limsup_{t\to\infty} \frac{1}{t} \log |x(t; t_0, x_0)| \leq -\lambda \quad a.s. \quad (3.22)$$

In other words, we have shown that under conditions (3.19) and (3.21) the trivial solution of equation (3.18) is almost surely exponentially stable.

4.4 MOMENT EXPONENTIAL STABILITY

In this section we shall discuss the pth moment exponential stability for equation (1.2) and we shall always let $p > 0$. Let us first give the definition of the pth moment exponential stability.

Definition 4.1 *The trivial solution of equation (1.2) is said to be pth moment exponentially stable if there is a pair of positive constants λ and C such that*

$$E|x(t; t_0, x_0)|^p \leq C|x_0|^p e^{-\lambda(t-t_0)} \quad \text{on } t \geq t_0 \quad (4.1)$$

for all $x_0 \in R^d$. When $p = 2$, it is usually said to be exponentially stable in mean square.

Clearly, the pth moment exponential stability means that the pth moment of the solution will tend to 0 exponentially fast. It also follows from (4.1) that

$$\limsup_{t\to\infty} \frac{1}{t} \log\left(E|x(t; t_0, x_0)|^p\right) < 0. \quad (4.2)$$

As defined in Section 2.5, the left-hand side of (4.2) is called the pth moment Lyapunov exponent of the solution. So, in this case, the pth moment Lyapunov exponent is negative. Moreover, if one wishes to consider the initial value of an $\mathcal{F}_{t_0}$-measurable random variable $x_0 \in L^p(\Omega; R^d)$, then, by (4.1),

$$E|x(t;t_0,x_0)|^p = \int_{R^d} E|x(t;t_0,y)|^p P\{x_0 \in dy\}$$

$$\leq \int_{R^d} C|y|^p e^{-\lambda(t-t_0)} P\{x_0 \in dy\} = CE|x_0|^p e^{-\lambda(t-t_0)}.$$

Besides, noting $(E|x(t)|^{\hat{p}})^{1/\hat{p}} \leq (E|x(t)|^p)^{1/p}$ for $0 < \hat{p} < p$ we see that the pth moment exponential stability implies the $\hat{p}$th moment exponential stability.

Generally speaking, the pth moment exponential stability and the almost sure exponential stability do not imply each other and additional conditions are required in order to deduce one from the other. The following theorem gives the conditions under which the pth moment exponential stability implies the almost sure exponential stability.

Theorem 4.2 *Assume that there is a positive constant K such that*

$$x^T f(x,t) \vee |g(x,t)|^2 \leq K|x|^2 \quad \text{for all } (x,t) \in R^d \times [t_0, \infty). \quad (4.3)$$

Then the pth moment exponential stability of the trivial solution of equation (1.2) implies the almost sure exponential stability.

Proof. Fix any $x_0 \neq 0$ in R^d and write $x(t;t_0,x_0) = x(t)$ simply. By the definition of the pth moment exponential stability, there is a pair of positive constants λ and C such that

$$E|x(t)|^p \leq C|x_0|^p e^{-\lambda(t-t_0)} \quad \text{on } t \geq t_0. \quad (4.4)$$

Let $n = 1, 2, \cdots$. By Itô's formula and condition (4.3), one can show that for $t_0 + n - 1 \leq t \leq t_0 + n$,

$$|x(t)|^p = |x(t_0+n-1)|^p + \int_{t_0+n-1}^t p|x(s)|^{p-2} x^T(s) f(x(s),s) ds$$

$$+ \frac{1}{2} \int_{t_0+n-1}^t \left[p|x(s)|^{p-2}|g(x(s),s)|^2 + p(p-2)|x|^{p-4}|x^T(s)g(x(s),s)|^2 \right] ds$$

$$+ \int_{t_0+n-1}^t p|x(s)|^{p-2} x^T(s) g(x(s),s) dB(s)$$

$$\leq |x(t_0+n-1)|^p + c_1 \int_{t_0+n-1}^t |x(s)|^p ds$$

$$+ \int_{t_0+n-1}^t p|x(s)|^{p-2} x^T(s) g(x(s),s) dB(s),$$

where $c_1 = pK + p(1 + |p-2|)K/2$. Hence

$$E\left(\sup_{t_0+n-1 \leq t \leq t_0+n} |x(t)|^p \right) \leq E|x(t_0+n-1)|^p + c_1 \int_{t_0+n-1}^{t_0+n} E|x(s)|^p ds$$

$$+ E\left(\sup_{t_0+n-1 \leq t \leq t_0+n} \int_{t_0+n-1}^t p|x(s)|^{p-2} x^T(s) g(x(s),s) dB(s) \right). \quad (4.5)$$

On the other hand, by the Burkholder–Davis–Gundy inequality (i.e. Theorem 1.7.3), we have that

$$E\left(\sup_{t_0+n-1\leq t\leq t_0+n} \int_{t_0+n-1}^{t} p|x(s)|^{p-2}x^T(s)g(x(s),s)dB(s)\right)$$

$$\leq 4\sqrt{2}E\left(\int_{t_0+n-1}^{t_0+n} p^2|x(s)|^{2(p-2)}|x^T(s)g(x(s),s)|^2 ds\right)^{\frac{1}{2}}$$

$$\leq 4\sqrt{2}E\left(\sup_{t_0+n-1\leq s\leq t_0+n} |x(s)|^p \int_{t_0+n-1}^{t_0+n} p^2 K|x(s)|^p ds\right)^{\frac{1}{2}}$$

$$\leq \frac{1}{2}E\left(\sup_{t_0+n-1\leq s\leq t_0+n} |x(s)|^p\right) + 16p^2 K \int_{t_0+n-1}^{t_0+n} E|x(s)|^p ds,$$

where we have also used the elementary inequality $\sqrt{ab} \leq (a+b)/2$. Substituting this into (4.5) yields that

$$E\left(\sup_{t_0+n-1\leq t\leq t_0+n} |x(t)|^p\right) \leq 2E|x(t_0+n-1)|^p + c_2 \int_{t_0+n-1}^{t_0+n} E|x(s)|^p ds,$$

where $c_2 = 2c_1 + 32p^2 K$. Applying (4.4) we obtain that

$$E\left(\sup_{t_0+n-1\leq t\leq t_0+n} |x(t)|^p\right) \leq c_3 e^{-\lambda(n-1)}, \tag{4.6}$$

where $c_3 = C|x_0|^p(2+c_2)$. Now, let $\varepsilon \in (0,\lambda)$ be arbitrary. It follows from (4.6) that

$$P\left\{\sup_{t_0+n-1\leq t\leq t_0+n} |x(t)|^p > e^{-(\lambda-\varepsilon)(n-1)}\right\}$$

$$\leq e^{(\lambda-\varepsilon)(n-1)} E\left(\sup_{t_0+n-1\leq t\leq t_0+n} |x(t)|^p\right) \leq c_3 e^{-\varepsilon(n-1)}.$$

In view of the Borel–Cantelli lemma we see that for almost all $\omega \in \Omega$,

$$\sup_{t_0+n-1\leq t\leq t_0+n} |x(t)|^p \leq e^{-(\lambda-\varepsilon)(n-1)} \tag{4.7}$$

holds for all but finitely many n. Hence, there exists an $n_0 = n_0(\omega)$, for all $\omega \in \Omega$ excluding a P-null set, for which (4.7) holds whenever $n \geq n_0$. Consequently, for almost all $\omega \in \Omega$,

$$\frac{1}{t}\log|x(t)| \leq \frac{1}{pt}\log(|x(t)|^p) \leq -\frac{(\lambda-\varepsilon)(n-1)}{p(t_0+n-1)}$$

if $t_0 + n - 1 \leq t \leq t_0 + n$, $n \geq n_0$. Hence

$$\limsup_{t\to\infty} \frac{1}{t}\log|x(t)| \leq -\frac{(\lambda-\varepsilon)}{p} \quad a.s.$$

Since $\varepsilon > 0$ is arbitrary, we must have
$$\limsup_{t\to\infty} \frac{1}{t} \log|x(t)| \le -\frac{\lambda}{p} \qquad a.s.$$
By definition, the trivial solution of equation (1.2) is almost surely exponentially stable. The proof is complete.

Although condition (4.3) is not guaranteed by the assumptions of the existence-and-uniqueness Theorem 2.3.6 which are assumed throughout this chapter, it is satisfied in many important cases. For example, if the coefficients $f(x,t)$ and $g(x,t)$ are uniformly Lipschitz continuous, then (4.3) is fulfilled bearing in mind that we always assume $f(0,t) \equiv 0$ and $g(0,t) \equiv 0$ in this chapter. Moreover, for the d-dimensional linear stochastic differential equation
$$dx(t) = F(t)x(t)dt + \sum_{i=1}^{m} G_i(t)x(t)dB_i(t), \tag{4.8}$$
condition (4.3) is fulfilled if F, G_i are all bounded $d \times d$-matrix-valued functions. Hence, we obtain a useful corollary.

Corollary 4.3 *Let F, G_i be all bounded $d \times d$-matrix-valued functions. Then the pth moment exponential stability of the trivial solution of the linear equation (4.8) implies the almost sure exponential stability.*

We shall now establish a sufficient criterion for the pth moment exponential stability via a Lyapunov function.

Theorem 4.4 *Assume that there is a function $V(x,t) \in C^{2,1}(R^d \times [t_0, \infty); R_+)$, and positive constants c_1–c_3, such that*
$$c_1|x|^p \le V(x,t) \le c_2|x|^p \qquad and \qquad LV(x,t) \le -c_3 V(x,t) \tag{4.9}$$
for all $(x,t) \in R^d \times [t_0, \infty)$. Then
$$E|x(t;t_0,x_0)|^p \le \frac{c_2}{c_1}|x_0|^p e^{-c_3(t-t_0)} \qquad on\ t \ge t_0 \tag{4.10}$$
for all $x_0 \in R^d$. In other words, the trivial solution of equation (1.2) is pth moment exponentially stable and the pth moment Lyapunov exponent should not be greater than $-c_3$.

Proof. Fix any $x_0 \in R^d$ and write $x(t;t_0,x_0) = x(t)$. For each $n \ge |x_0|$, define the stopping time
$$\tau_n = \inf\{t \ge t_0 : |x(t)| \ge n\}.$$
Obviously, $\tau_n \to \infty$ as $n \to \infty$ almost surely. By Itô's formula, we can derive that for $t \ge t_0$,
$$E\left[e^{c_3(t\wedge\tau_n - t_0)} V(x(t\wedge\tau_n), t\wedge\tau_n)\right] = V(x_0, t_0)$$
$$+ E\int_{t_0}^{t\wedge\tau_n} e^{c_3(s-t_0)}\left[c_3 V(x(s),s) + LV(x(s),s)\right]ds.$$

Using condition (4.9) we then obtain that

$$c_1 e^{c_3(t\wedge\tau_n - t_0)} E|x(t\wedge\tau_n)|^p \leq E\left[e^{c_3(t\wedge\tau_n - t_0)} V(x(t\wedge\tau_n), t\wedge\tau_n)\right]$$
$$\leq V(x_0, t_0) \leq c_2 |x_0|^p.$$

Letting $n \to \infty$ yields that

$$c_1 e^{c_3(t-t_0)} E|x(t)|^p \leq c_2 |x_0|^p$$

which implies the desired assertion (4.10).

Similarly we can prove the following theorem that gives a sufficient criterion for the qth moment exponential instability.

Theorem 4.5 *Let $q > 0$. Assume that there is a function $V(x,t) \in C^{2,1}(R^d \times [t_0, \infty); R_+)$, and positive constants c_1–c_3, such that*

$$c_1 |x|^q \leq V(x,t) \leq c_2 |x|^q \quad \text{and} \quad LV(x,t) \geq c_3 V(x,t)$$

for all $(x,t) \in R^d \times [t_0, \infty)$. Then

$$E|x(t; t_0, x_0)|^q \geq \frac{c_1}{c_2} |x_0|^q e^{c_3(t-t_0)} \quad \text{on } t \geq t_0$$

for all $x_0 \in R^d$, and we say in this case that the trivial solution of equation (1.2) is qth moment exponentially unstable.

Since $(E|x(t)|^{\hat{q}})^{1/\hat{q}} \geq (E|x(t)|^q)^{1/q}$ for $\hat{q} > q$, the qth moment exponential instability implies the $\hat{q}$th moment exponential instability. We now use Theorem 4.4 to establish a useful corollary.

Corollary 4.6 *Assume that there exists a symmetric positive-definite $d \times d$ matrix Q, and constants α_1–α_3, such that for all $(x,t) \in R^d \times [t_0, \infty)$,*

$$x^T Q f(x,t) + \frac{1}{2}\text{trace}[g^T(x,t) Q g(x,t)] \leq \alpha_1 x^T Q x \tag{4.11}$$

and

$$\alpha_2 x^T Q x \leq |x^T Q g(x,t)| \leq \alpha_3 x^T Q x. \tag{4.12}$$

(i) *If $\alpha_1 < 0$, then the trivial solution of equation (1.2) is pth moment exponentially stable provided $p < 2 + 2|\alpha_1|/\alpha_3^2$.*

(ii) *If $0 \leq \alpha_1 < \alpha_2^2$, then the trivial solution of equation (1.2) is pth moment exponentially stable provided $p < 2 - 2\alpha_1/\alpha_2^2$.*

Proof. Let $V(x,t) = (x^T Q x)^{\frac{p}{2}}$. Then

$$\lambda_{\min}^{\frac{p}{2}}(Q) |x|^p \leq V(x,t) \leq \lambda_{\max}^{\frac{p}{2}}(Q) |x|^p,$$

where $\lambda_{\min}(Q)$ and $\lambda_{\max}(Q)$ denote the smallest and largest eigenvalue of Q, respectively. It is also easy to verify that

$$LV(x,t) = p(x^T Q x)^{\frac{p}{2}-1}\left(x^T Q f(x,t) + \frac{1}{2} trace[g^T(x,t) Q g(x,t)]\right)$$
$$+ p\left(\frac{p}{2}-1\right)(x^T Q x)^{\frac{p}{2}-2}|x^T Q g(x,t)|^2. \qquad (4.13)$$

(i) Assume that $\alpha_1 < 0$ and $p < 2 + 2|\alpha_1|/\alpha_3^2$. Without loss of generality, we can let $p \geq 2$. Using (4.11) and (4.12), we then derive from (4.13) that

$$LV(x,t) \leq -p\left[|\alpha_1| - \left(\frac{p}{2}-1\right)\alpha_3^2\right]V(x,t).$$

An application of Theorem 4.4 implies that the trivial solution of equation (1.2) is pth moment exponentially stable.

(ii) Assume that $0 \leq \alpha_1 < \alpha_2^2$ and $p < 2 - 2\alpha_1/\alpha_2^2$. In this case we have

$$LV(x,t) \leq -p\left[\left(\frac{p}{2}-1\right)\alpha_2^2 - \alpha_1\right]V(x,t).$$

So the conclusion follows from Theorem 4.4 again. The proof is complete.

Similarly, we can use Theorem 4.5 to show the following result on the moment exponential instability.

Corollary 4.7 *Assume that there exists a symmetric positive-definite $d \times d$ matrix Q, and positive constants β_1, β_2, such that for all $(x,t) \in R^d \times [t_0, \infty)$,*

$$x^T Q f(x,t) + \frac{1}{2} trace[g^T(x,t) Q g(x,t)] \geq \beta_1 x^T Q x \qquad (4.14)$$

and

$$|x^T Q g(x,t)| \leq \beta_2 x^T Q x. \qquad (4.15)$$

Then the trivial solution of equation (1.2) is qth moment exponentially unstable provided $q > 0 \vee (2 - 2\beta_1/\beta_2^2)$.

Let us now discuss a few example for illustration.

Example 4.8 Consider the scalar linear Itô equation

$$dx(t) = ax(t) + \sum_{i=1}^{m} b_i x(t) dB_i(t) \qquad \text{on } t \geq t_0. \qquad (4.16)$$

Here a, b_i are all constants, and we assume that

$$0 < a < \frac{1}{2}\sum_{i=1}^{m} b_i^2. \qquad (4.17)$$

With $f(x,t) = ax$ and $g(x,t) = (b_1 x, \cdots, b_m x)$, we have

$$xf(x,t) + \frac{1}{2}\text{trace}[g^T(x,t)g(x,t)] = \left(a + \frac{1}{2}\sum_{i=1}^{m} b_i^2\right)x^2$$

and

$$|xg(x,t)| = \sqrt{\sum_{i=1}^{m} b_i^2}\,|x|^2.$$

Hence, by Corollary 4.6, the trivial solution of equation (4.16) is pth moment exponentially stable if

$$p < 1 - \frac{a}{\frac{1}{2}\sum_{i=1}^{m} b_i^2},$$

while, by Corollary 4.7, it is qth moment exponentially unstable if

$$q > 1 - \frac{a}{\frac{1}{2}\sum_{i=1}^{m} b_i^2}.$$

Example 4.9 This example is from the satellite dynamics. Sagirow (1970) derived the equation

$$\ddot{y}(t) + \beta(1 + \alpha\dot{B}(t))\dot{y}(t) + (1 + \alpha\dot{B}(t))y(t) - \gamma\sin(2y(t)) = 0 \quad (4.18)$$

in the study of the influence of a rapidly fluctuating density of the atmosphere of the earth on the motion of a satellite in a circular orbit. Here $\dot{B}(t)$ is a scalar white noise, α is a constant representing the intensity of the disturbance, and β, γ are two positive constants. Introducing $x = (x_1, x_2)^T = (y, \dot{y})^T$, we can write equation (4.18) as the two-dimensional Itô equation

$$\begin{cases} dx_1(t) = x_2(t)dt, \\ dx_2(t) = [-x_1(t) + \gamma\sin(2x_1(t)) - \beta x_2(t)]dt - \alpha[x_1(t) + \beta x_2(t)]dB(t). \end{cases}$$

For the Lyapunov function, we try an expression consisting of a quadratic form and integral of the nonlinear component:

$$V(x,t) = ax_1^2 + bx_1 x_2 + x_2^2 + c\int_0^{x_1} \sin(2y)dy$$
$$= ax_1^2 + bx_1 x_2 + x_2^2 + c\sin^2 x_1.$$

This yields

$$LV(x,t) = -(b - \alpha^2)x_1^2 + b\gamma x_1 \sin(2x_1) - (2\beta - b - \alpha^2\beta^2)x_2^2$$
$$+ (2a - b\beta - 2 + 2\alpha^2\beta)x_1 x_2 + (c + 2\gamma)x_2 \sin(2x_1).$$

Setting $2a - b\beta - 2 + 2\alpha^2\beta = 0$ and $c + 2\gamma = 0$ we obtain

$$V(x,t) = \frac{1}{2}(b\beta + 2 - 2\alpha^2\beta)x_1^2 + bx_1x_2 + x_2^2 - 2\gamma\sin^2 x_1$$

and

$$LV(x,t) = -(b - \alpha^2)x_1^2 + b\gamma x_1 \sin(2x_1) - (2\beta - b - \alpha^2\beta^2)x_2^2.$$

Note that

$$V(x,t) \geq \frac{1}{2}(b\beta + 2 - 2\alpha^2\beta - 4\gamma)x_1^2 + bx_1x_2 + x_2^2.$$

So $V(x,t) \geq \varepsilon|x|^2$ for some $\varepsilon > 0$ if

$$2(b\beta + 2 - 2\alpha^2\beta - 4\gamma) \geq b^2$$

or equivalently

$$\beta - \sqrt{\beta^2 + 4 - 8\gamma - 4\alpha^2\beta} < b < \beta + \sqrt{\beta^2 + 4 - 8\gamma - 4\alpha^2\beta}. \quad (4.19)$$

Note also that

$$LV(x,t) \leq -(b - \alpha^2 - 2b\gamma)x_1^2 - (2\beta - b - \alpha^2\beta^2)x_2^2.$$

So $LV(x,t) \leq -\bar{\varepsilon}|x|^2$ for some $\bar{\varepsilon} > 0$ provided both $b - \alpha^2 - 2b\gamma > 0$ and $2\beta - b - \alpha^2\beta^2 > 0$, that is

$$2\gamma < 1 \quad \text{and} \quad \alpha^2/(1 - 2\gamma) < b < 2\beta - \alpha^2\beta^2. \quad (4.20)$$

We therefore conclude, by Theorem 4.4, that if $\gamma < 1/2$ and

$$\max\{\alpha^2/(1 - 2\gamma),\ \beta - \sqrt{\beta^2 + 4 - 8\gamma - 4\alpha^2\beta}\}$$
$$< \min\{2\beta - \alpha^2\beta^2,\ \beta + \sqrt{\beta^2 + 4 - 8\gamma - 4\alpha^2\beta}\} \quad (4.21)$$

then the trivial solution of equation (4.18) is exponentially stable in mean square.

Example 4.10 In the case of linear stochastic differential equations, the explicit solutions would of course be very useful in determining the pth moment exponential stability. We now explain this idea through this example. Consider the scalar linear Itô equation

$$dx(t) = a(t)x(t)dt + \sum_{i=1}^{m} b_i(s)x(s)dB_i(s) \quad (4.22)$$

on $t \geq t_0$ with initial value $x(t_0) = x_0 \in R^d$, where $a(t)$, $b_i(t)$ are all continuous functions on $[t_0, \infty)$. It has been shown that equation (4.22) has the explicit solution

$$x(t) = x_0 \exp\left[\int_{t_0}^{t}\left(a(s) - \frac{1}{2}\sum_{i=1}^{m} b_i^2(s)\right)ds + \sum_{i=1}^{m}\int_{t_0}^{t} b_i(s)dB_i(s)\right].$$

Therefore

$$E|x(t)|^p = |x_0|^p E \exp\left[p \int_{t_0}^t \left(a(s) - \frac{1}{2}\sum_{i=1}^m b_i^2(s)\right) ds + p \sum_{i=1}^m \int_{t_0}^t b_i(s) dB_i(s)\right].$$

But one can show (as an exercise for the reader) that

$$E \exp\left[-\frac{p^2}{2}\sum_{i=1}^m \int_{t_0}^t b_i^2(s) ds + p \sum_{i=1}^m \int_{t_0}^t b_i(s) dB_i(s)\right] = 1.$$

Thus

$$E|x(t)|^p = |x_0|^p \exp\left[p \int_{t_0}^t \left(a(s) - \frac{1-p}{2}\sum_{i=1}^m b_i^2(s)\right) ds\right]. \quad (4.23)$$

We therefore see that the trivial solution of equation (4.22) is pth moment exponentially stable if and only if

$$\limsup_{t\to\infty} \frac{1}{t} \int_{t_0}^t \left(a(s) - \frac{1-p}{2}\sum_{i=1}^m b_i^2(s)\right) ds < 0; \quad (4.24)$$

while it is qth moment exponentially unstable if and only if

$$\liminf_{t\to\infty} \frac{1}{t} \int_{t_0}^t \left(a(s) - \frac{1-q}{2}\sum_{i=1}^m b_i^2(s)\right) ds > 0. \quad (4.25)$$

If $a(t) = a$, $b_i(t) = b_i$ are all constants, equation (4.22) reduces to equation (4.16). In this case, (4.24) holds if and only if

$$a - \frac{1-p}{2}\sum_{i=1}^m b_i^2 < 0, \quad \text{i.e.} \quad p < 1 - \frac{a}{\frac{1}{2}\sum_{i=1}^m b_i^2}; \quad (4.26)$$

while (4.25) holds if and only if

$$a - \frac{1-q}{2}\sum_{i=1}^m b_i^2 > 0, \quad \text{i.e.} \quad q > 1 - \frac{a}{\frac{1}{2}\sum_{i=1}^m b_i^2}. \quad (4.27)$$

Clearly, these conclusions are the same as those of Example 4.8.

4.5 STOCHASTIC STABILIZATION AND DESTABILIZATION

It is not surprising that noise can destabilize a stable system. For example, suppose that a given 2-dimensional exponentially stable system

$$\dot{y}(t) = -y(t) \quad \text{on } t \geq t_0, \quad y(t_0) = x_0 \in R^2 \quad (5.1)$$

is perturbed by noise and the stochastically perturbed system is described by the Itô equation

$$dx(t) = -x(t)dt + Gx(t)dB(t) \quad \text{on } t \geq t_0, \ x(t_0) = x_0 \in R^2. \quad (5.2)$$

Here $B(t)$ is a one-dimensional Brownian motion and

$$G = \begin{pmatrix} 0 & -2 \\ 2 & 0 \end{pmatrix}$$

It has been shown that equation (5.2) has the explicit solution

$$x(t) = \exp\left[\left(-I - \frac{1}{2}G^2\right)(t - t_0) + G(B(t) - B(t_0))\right]x_0$$
$$= \exp\left[I(t - t_0) + G(B(t) - B(t_0))\right],$$

where I is the 2×2 identity matrix. Consequently

$$\lim_{t \to \infty} \frac{1}{t} \log |x(t)| = 1 \quad a.s.$$

That is, the stochastically perturbed system (5.2) becomes almost surely exponentially unstable.

On the other hand, it has also been observed that noise can have a stabilizing effect as well. For example, consider a scalar unstable system

$$\dot{y}(t) = y(t) \quad \text{on } t \geq t_0, \ y(t_0) = x_0 \in R \quad (5.3)$$

Perturb this system by noise and suppose that the perturbed system has the form

$$dx(t) = x(t)dt + 2x(t)dB(t) \quad \text{on } t \geq t_0, \ x(t_0) = x_0 \in R, \quad (5.4)$$

where $B(t)$ is again a one-dimensional Brownian motion. Equation (5.4) has the explicit solution

$$x(t) = x_0 \exp\left[-(t - t_0) + 2(B(t) - B(t_0))\right],$$

which yields immediately that

$$\lim_{t \to \infty} \frac{1}{t} \log |x(t)| = -1 \quad a.s.$$

That is, the perturbed system (5.4) becomes stable. In other words, the noise has stabilized the unstable system (5.3).

In this section we shall establish a general theory of stochastic stabilization and destabilization for a given nonlinear system. Suppose that the given system is described by a nonlinear ordinary differential equation

$$\dot{y}(t) = f(y(t), t) \quad \text{on } t \geq t_0, \ y(t_0) = x_0 \in R^d. \quad (5.5)$$

Here $f : R^d \times R_+ \to R^d$ is a locally Lipschitz continuous function and particularly, for some $K > 0$,

$$|f(x,t)| \leq K|x| \quad \text{for all } (x,t) \in R^d \times R_+. \quad (5.6)$$

We now use the m-dimensional Brownian motion $B(t) = (B_1(t), \cdots, B_m(t))^T$ as the source of noise to perturb the given system. For simplicity, suppose the stochastic perturbation is of a linear form, that is the stochastically perturbed system is described by the semilinear Itô equation

$$dx(t) = f(x(t),t)dt + \sum_{i=1}^{m} G_i x(t) dB_i(t) \quad \text{on } t \geq t_0, \ x(t_0) = x_0 \in R^d, \quad (5.7)$$

where G_i, $1 \leq i \leq m$, are all $d \times d$ matrices. Clearly, equation (5.7) has a unique solution denoted by $x(t; t_0, x_0)$ again and, moreover, it admits a trivial solution $x(t) \equiv 0$. Let us begin to discuss how the stochastic perturbation affect the property of stability or instability of the given system (5.5), and we shall see that different choices of G_i make the thing different.

Theorem 5.1 Let (5.6) hold. Assume that there are two constants $\lambda > 0$ and $\rho \geq 0$ such that

$$\sum_{i=1}^{m} |G_i x|^2 \leq \lambda |x|^2 \quad \text{and} \quad \sum_{i=1}^{m} |x^T G_i x|^2 \geq \rho |x|^4 \quad (5.8)$$

for all $x \in R^d$. Then

$$\limsup_{t \to \infty} \frac{1}{t} \log |x(t; t_0, x_0)| \leq -\left(\rho - K - \frac{\lambda}{2}\right) \quad a.s. \quad (5.9)$$

for all $x_0 \in R^d$. In particular, if $\rho > K + \frac{1}{2}\lambda$, then the trivial solution of equation (5.7) is almost surely exponentially stable.

Proof. Let $V(x,t) = |x|^2$. Then

$$LV(x,t) = 2x^T f(x,t) + \sum_{i=1}^{m} |G_i x|^2 \leq (2K + \lambda)|x|^2.$$

Moreover, with $g(x,t) = (G_1 x, \cdots, G_m x)$,

$$|V_x(x,t)g(x,t)|^2 = 4 \sum_{i=1}^{m} |x^T G_i x|^2 \geq 4\rho |x|^4.$$

An application of Theorem 3.3 yields the desired assertion (5.8).

Let us now consider some special cases of equation (5.7). First of all, let $G_i = \sigma_i I$ for $1 \leq i \leq m$, where I is the $d \times d$ identity matrix and σ_i a constant.

These σ_i's represent the intensity of the stochastic perturbation. In this case, equation (5.7) becomes

$$dx(t) = f(x(t),t)dt + \sum_{i=1}^{m} \sigma_i x(t) dB_i(t). \tag{5.10}$$

Moreover,

$$\sum_{i=1}^{m} |G_i x|^2 = \sum_{i=1}^{m} \sigma_i^2 |x|^2 \quad \text{and} \quad \sum_{i=1}^{m} |x^T G_i x|^2 = \sum_{i=1}^{m} \sigma_i^2 |x|^4.$$

By Theorem 5.1, the solution of equation (5.10) has the property

$$\limsup_{t \to \infty} \frac{1}{t} \log |x(t;t_0,x_0)| \le -\left(\frac{1}{2}\sum_{i=1}^{m} \sigma_i^2 - K\right) \quad a.s.$$

Therefore, the trivial solution of equation (5.10) is almost surely exponentially stable provided $\frac{1}{2}\sum_{i=1}^{m} \sigma_i^2 > K$. An even simpler case is that when $\sigma_i = 0$ for $2 \le i \le m$, i.e. the equation

$$dx(t) = f(x(t),t)dt + \sigma_1 x(t) dB_1(t).$$

The trivial solution of this equation is almost surely exponentially stable provided $\frac{1}{2}\sigma_1^2 > K$. These show that if we add a strong enough stochastic perturbation to the given system (5.5), then the system is stabilized. We summarize these as a theorem.

Theorem 5.2 *Any nonlinear system $\dot{y}(t) = f(y(t),t)$ can be stabilized by Brownian motions provided (5.6) is satisfied. Moreover, one can even use only a scalar Brownian motion to stabilize the system.*

Theorem 5.1 ensures that there are many choices for the matrices B_i in order to stabilize a given system and of course the above choices are just the simplest ones. For illustration, we give one more example here. For each i, choose a positive-definite matrix D_i such that

$$x^T D_i x \ge \frac{\sqrt{3}}{2} ||D_i|| \, |x|^2.$$

Obviously, there are many such matrices. Let σ be a constant and $G_i = \sigma D_i$. Then

$$\sum_{i=1}^{m} |G_i x|^2 \le \sigma^2 \sum_{i=1}^{m} ||D_i||^2 |x|^2$$

and

$$\sum_{i=1}^{m} |x^T G_i x|^2 \ge \frac{3\sigma^2}{4} \sum_{i=1}^{m} ||D_i||^2 |x|^4.$$

By Theorem 5.1, the solution of equation (5.7) satisfies

$$\limsup_{t\to\infty} \frac{1}{t}\log|x(t;t_0,x_0)| \le -\left(\frac{\sigma^2}{4}\sum_{i=1}^{m}\|D_i\|^2 - K\right) \quad a.s.$$

Therefore the trivial solution of equation (5.7) is almost surely exponentially stable if

$$\sigma^2 > \frac{4K}{\sum_{i=1}^{m}\|D_i\|^2}.$$

Let us now turn to consider the opposite problem—stochastic destabilization. It is not difficult to apply Theorem 3.5 to show the following result and the details are left to the reader.

Theorem 5.3 Let (5.6) hold. Assume that there are two positive constants λ and ρ such that

$$\sum_{i=1}^{m}|G_i x|^2 \ge \lambda |x|^2 \quad \text{and} \quad \sum_{i=1}^{m}|x^T G_i x|^2 \le \rho |x|^4$$

for all $x \in R^d$. Then

$$\liminf_{t\to\infty} \frac{1}{t}\log|x(t;t_0,x_0)| \ge \left(\frac{\lambda}{2} - K - \rho\right) \quad a.s.$$

for all $x_0 \ne 0$. In particular, if $\lambda > 2(K+\rho)$, then the trivial solution of equation (5.7) is almost surely exponentially unstable.

We now employ this theorem to show how one can use stochastic perturbation to destabilize the given system. First of all, let the dimension of the state space $d \ge 3$ and choose the dimension of the Brownian motion to be the same, i.e. $m = d$. Let σ be a constant. For each $i = 1, 2, \cdots, d-1$, define the $d \times d$ matrix $G_i = (g^i_{uv})$ by $g^i_{uv} = \sigma$ if $u = i$ and $v = i+1$ or otherwise $g^i_{uv} = 0$. Moreover, define $G_d = (g^d_{uv})$ by $g^d_{uv} = \sigma$ if $u = d$ and $v = 1$ or otherwise $g^d_{uv} = 0$. Then equation (5.7) becomes

$$dx(t) = f(x(t),t)dt + \sigma \begin{bmatrix} x_2(t)dB_1(t) \\ \vdots \\ x_d(t)dB_{d-1}(t) \\ x_1(t)dB_d(t) \end{bmatrix}. \quad (5.11)$$

Compute that

$$\sum_{i=1}^{m}|G_i x|^2 = \sum_{i=1}^{m}(\sigma x_{i+1})^2 = \sigma^2|x|^2$$

and

$$\sum_{i=1}^{m}|x^T G_i x|^2 = \sigma^2 \sum_{i=1}^{m} x_i^2 x_{i+1}^2,$$

where we use $x_{d+1} = x_1$. Noting

$$\sum_{i=1}^{m} x_i^2 x_{i+1}^2 \leq \frac{1}{2} \sum_{i=1}^{m} (x_i^4 + x_{i+1}^4) = \sum_{i=1}^{m} x_i^4,$$

we have

$$3 \sum_{i=1}^{m} x_i^2 x_{i+1}^2 \leq 2 \sum_{i=1}^{m} x_i^2 x_{i+1}^2 + \sum_{i=1}^{m} x_i^4 \leq |x|^4.$$

Therefore

$$\sum_{i=1}^{m} |x^T G_i x|^2 \leq \frac{\sigma^2}{3} |x|^4.$$

By Theorem 5.3, the solution of equation (5.11) has the property that

$$\liminf_{t \to \infty} \frac{1}{t} \log |x(t; t_0, x_0)| \geq \left(\frac{\sigma^2}{2} - K - \frac{\sigma^2}{3} \right) = \frac{\sigma^2}{6} - K \quad a.s.$$

for any $x_0 \neq 0$. If $\sigma^2 > 6K$, then the trivial solution of equation (5.11) will be almost surely exponentially unstable.

Secondly, let the dimension of the state space d be an even number, say $d = 2k (k \geq 1)$. let σ be a constant. Define

$$G_1 = \begin{bmatrix} 0 & \sigma & & & 0 \\ -\sigma & 0 & & & \\ & & \ddots & & \\ & & & 0 & \sigma \\ 0 & & & -\sigma & 0 \end{bmatrix}$$

but set $G_i = 0$ for $2 \leq i \leq m$. So equation (5.7) becomes

$$dx(t) = f(x(t), t)dt + \sigma \begin{bmatrix} x_2(t) \\ -x_1(t) \\ \vdots \\ x_{2k}(t) \\ -x_{2k-1}(t) \end{bmatrix} dB_1(t). \quad (5.12)$$

In this case we have

$$\sum_{i=1}^{m} |G_i x|^2 = \sigma^2 |x|^2 \quad \text{and} \quad \sum_{i=1}^{m} |x^T G_i x|^2 = 0$$

Hence, by Theorem 5.3, the solution of equation (5.12) has the property that

$$\liminf_{t \to \infty} \frac{1}{t} \log |x(t; t_0, x_0)| \geq \frac{\sigma^2}{2} - K \quad a.s.$$

for any $x_0 \neq 0$. If $\sigma^2 > 2K$, then the trivial solution of equation (5.12) will be almost surely exponentially unstable. Summarizing these results we obtain the following conclusion.

Theorem 5.4 *Any d-dimensional nonlinear system $\dot{y}(t) = f(y(t), t)$ can be destabilized by Brownian motions provided the dimension $d \geq 2$ and (5.6) is satisfied.*

Naturally one may ask what happens to one-dimensional systems. To answer this let us look at the scalar linear Itô equation

$$dx(t) = -ax(t) + \sum_{i=1}^{m} b_i x(t) dB_i(t) \qquad \text{on } t \geq t_0 \qquad (5.13)$$

with initial data $x(t_0) = x_0$. This equation is regarded as the stochastically perturbed system of the exponentially stable system

$$\dot{y}(t) = -ay(t) \qquad (a > 0).$$

It has been shown that the sample Lyapunov exponent of the solution is

$$\lim_{t \to \infty} \frac{1}{t} \log |x(t; t_0, x_0)| = -a - \frac{1}{2} \sum_{i=1}^{m} b_i^2 < 0 \qquad a.s.$$

That is, the perturbed system (5.13) remains stable. We therefore see that the exponentially stable system $\dot{y}(t) = ay(t)$ ($a < 0$) cannot be destabilized by Brownian motions if we restrict the stochastic perturbation in the linear form of $\sum_{i=1}^{m} b_i x(t) dB_i(t)$.

4.6 FURTHER TOPICS

If the coefficients f and g in equation (1.2) are such that $f(0, t) \neq 0$ and $g(0, t) \neq 0$ but f has the decomposition $f(x, t) = f_1(x, t) + f_2(x, t)$ with $f_1(0, t) \equiv 0$, then we can regard the equation

$$dx(t) = [f_1(x(t), t) + f_2(x(t), t)]dt + g(x(t), t)dw(t) \qquad (6.1)$$

as the stochastically perturbed system of the ordinary differential equation

$$\dot{y}(t) = f_1(y(t), t). \qquad (6.2)$$

In this case, the equilibrium position is a solution of the unperturbed system (6.2) but no longer of the perturbed system (6.1). However, we can in principle apply our definitions of stability. For example, consider a d-dimensional linear stochastic differential equation in the narrow sense

$$dx(t) = [Ax(t) + F(t)]dt + G(t)dB(t) \qquad \text{on } t \geq t_0 \qquad (6.3)$$

with initial value $x(t_0) = x_0 \in R^d$, where

$$A \in R^{d\times d}, \qquad F: R_+ \to R^d, \qquad G: R_+ \to R^{d\times m}.$$

We impose two hypotheses: (i) The eigenvalues of A have negative real parts. This is equivalent to that there is a pair of positive constants β_1 and λ_1 such that

$$\|e^{At}\|^2 \leq \beta_1 e^{-\lambda_1 t} \qquad \text{for } t \geq 0. \tag{6.4}$$

(ii) There is also a pair of positive constants β_2 and λ_2 such that

$$|F(t)|^2 \vee |G(t)|^2 \leq \beta_2 e^{-\lambda_2 t} \qquad \text{for } t \geq 0. \tag{6.5}$$

It was shown in Chapter 3 that the solution of equation (6.3) is

$$x(t) = e^{A(t-t_0)} x_0 + \int_{t_0}^t e^{A(t-s)} F(s) ds + \int_{t_0}^t e^{A(t-s)} G(s) dB(s). \tag{6.6}$$

So

$$E|x(t)|^2$$
$$\leq 3|e^{A(t-t_0)} x_0|^2 + 3(t-t_0) \int_{t_0}^t |e^{A(t-s)} F(s)|^2 ds + 3\int_{t_0}^t |e^{A(t-s)} G(s)|^2 ds$$
$$\leq 3\beta_1 |x_0|^2 e^{-\lambda_1(t-t_0)} + 3\beta_1\beta_2 (t-t_0+1) \int_{t_0}^t e^{-\lambda_1(t-s)-\lambda_2 s} ds$$
$$\leq 3\beta_1 |x_0|^2 e^{-\lambda_1(t-t_0)} + 3\beta_1\beta_2 (t-t_0+1) \int_{t_0}^t e^{-(\lambda_1\wedge\lambda_2)(t-s)-(\lambda_1\wedge\lambda_2)s} ds$$
$$\leq 3\beta_1 |x_0|^2 e^{-\lambda_1(t-t_0)} + 3\beta_1\beta_2 (t-t_0+1)(t-t_0) e^{-(\lambda_1\wedge\lambda_2)t}. \tag{6.7}$$

This implies

$$\limsup_{t\to\infty} \frac{1}{t} \log(E|x(t)|^2) \leq -(\lambda_1 \wedge \lambda_2). \tag{6.8}$$

Now let $0 < \varepsilon < (\lambda_1 \wedge \lambda_2)/2$ be arbitrary. Set

$$c_1 = 3\beta_1 |x_0|^2 + 3\beta_1\beta_2 \sup_{t\geq t_0}\left[(t-t_0+1)(t-t_0) e^{-\varepsilon t}\right].$$

It then follows from (6.7) that

$$E|x(t)|^2 \leq c_1 e^{-(\lambda_1\wedge\lambda_2 - \varepsilon)(t-t_0)} \qquad \text{on } t \geq t_0.$$

Let $n = 1, 2, \cdots$. Note that for $t_0 + n - 1 \leq t \leq t_0 + n$,

$$x(t) = x(t_0+n-1) + \int_{t_0+n-1}^t [Ax(s) + F(s)] ds + \int_{t_0+n-1}^t G(s) dB(s).$$

Using Hölder's inequality, Doob's martingale inequality etc. we can derive that

$$E\left(\sup_{t_0+n-1\leq t\leq t_0+n} |x(t)|^2\right) \leq 3E|x(t_0+n-1)|^2$$
$$+ 3E\int_{t_0+n-1}^{t_0+n} |Ax(s)+F(s)|^2 ds + 12\int_{t_0+n-1}^{t_0+n} |G(s)|^2 ds$$
$$\leq 3c_1 e^{-(\lambda_1\wedge\lambda_2-\varepsilon)(n-1)} + 6\int_{t_0+n-1}^{t_0+n} \left(c_1\|A\|^2 e^{-(\lambda_1\wedge\lambda_2-\varepsilon)(s-t_0)} + 3\beta_2 e^{-\lambda_2 s}\right)ds$$
$$\leq c_2 e^{-(\lambda_1\wedge\lambda_2-\varepsilon)(n-1)},$$

where c_2 is a constant. From this we can show in the same way as the proof of Theorem 4.2 that

$$\limsup_{t\to\infty} \frac{1}{t}\log|x(t)| \leq -\frac{\lambda_1\wedge\lambda_2-2\varepsilon}{2} \quad a.s.$$

Since ε is arbitrary, we must have

$$\limsup_{t\to\infty} \frac{1}{t}\log|x(t)| \leq -\frac{\lambda_1\wedge\lambda_2}{2} \quad a.s. \quad (6.9)$$

In other words, we have shown that, under hypotheses (i) and (ii), the solution of equation (6.3) will tend to zero exponentially in mean square and almost surely as well. For the further details in this direction please see the author's earlier books Mao (1991a, 1994a).

Let us now turn to the another topic. In the case of stochastic asymptotic stability in the large, we know that all the solutions will tend to zero almost surely but we do not know how fast. To improve this situation we introduce the almost sure exponential stability, and in this case we do know that the solutions will tend to zero almost surely exponentially fast. However, we may sometimes find that the solutions will tend to zero but not so fast as exponentially or faster, and we wish to determine more precisely how fast they tend to zero. To explain, let us consider a scalar linear stochastic differential equation

$$dx(t) = -\frac{p}{1+t}x(t)dt + (1+t)^{-p}dB(t) \quad \text{on } t\geq t_0 \quad (6.10)$$

with initial value $x(t_0) = x_0 \in R$, where $p > \frac{1}{2}$ and $B(t)$ is a scalar Brownian motion. The solution of equation (6.10) is

$$x(t) = x_0\exp\left(-\int_{t_0}^t \frac{p}{1+r}dr\right) + \int_{t_0}^t \exp\left(-\int_s^t \frac{p}{1+r}dr\right)(1+s)^{-p}dB(s)$$
$$= x_0\left(\frac{1+t}{1+t_0}\right)^{-p} + \int_{t_0}^t \left(\frac{1+t}{1+s}\right)^{-p}(1+s)^{-p}dB(s)$$
$$= [x_0(1+t_0)^p + B(t)-B(t_0)](1+t)^{-p}. \quad (6.11)$$

Therefore the sample Lyapunov exponent

$$\lim_{t\to\infty} \frac{1}{t} \log |x(t)| = 0 \quad a.s.$$

which indicates that almost all the sample paths of the solution will not tend to zero exponentially. On the other hand, by the law of the iterated logarithm, we note that for almost all $\omega \in \Omega$ there is a sufficiently large $T = T(\omega)$ such that

$$|B(t) - B(t_0)| \leq 2\sqrt{2(t-t_0)\log\log(t-t_0)} \quad \text{if } t \geq T.$$

It therefore follows from (6.11) that, almost surely,

$$|x(t)| \leq \Big[|x_0|(1+t_0)^p + 2\sqrt{2(t-t_0)\log\log(t-t_0)}\Big](1+t)^{-p}$$

whenever $t \geq T$. Thus, for any $0 < \varepsilon < p - \frac{1}{2}$, there is a finite random variable ξ such that

$$|x(t)| \leq \xi t^{-(p-\frac{1}{2}-\varepsilon)} \quad \text{for all } t \geq t_0 \tag{6.12}$$

almost surely. This means that the solution will tend to zero almost surely polynomially. It is much nicer to express (6.12) as

$$\limsup_{t\to\infty} \frac{\log|x(t)|}{\log t} \leq -\Big(p - \frac{1}{2}\Big) \quad a.s. \tag{6.13}$$

since (6.12) implies

$$\limsup_{t\to\infty} \frac{\log|x(t)|}{\log t} \leq -\Big(p - \frac{1}{2} - \varepsilon\Big) \quad a.s.$$

and ε is arbitrary. Motivated by this example the author introduced in 1991 the concept of almost sure polynomial stability. A detailed study of such stability can be found in Mao (1991a).

We shall now take one further step to introduce a more general type of stability. Note that the almost sure exponential stability means $|x(t)| \leq \xi e^{-\lambda t}$ a.s. while the almost sure polynomial stability means $|x(t)| \leq \xi t^{-\lambda}$ a.s. Replacing the function $e^{-\lambda t}$ or $t^{-\lambda}$ with a more general function $\lambda(t)$ leads to the following new definition.

Definition 6.1 *Let $\lambda : R_+ \to (0, \infty]$ be a continuous nonincreasing function such that $\lambda(t) \to 0$ as $t \to \infty$. The trivial solution of equation (1.2) is said to be almost surely asymptotically stable with rate function $\lambda(t)$ if*

$$|x(t; t_0, x_0)| \leq \xi \lambda(t) \quad \text{for all } t \geq t_0 \tag{6.14}$$

almost surely, where ξ is a finite random variable which depends on x_0 and t_0.

Due to the page limit we establish only one simple criterion on such stability here.

Theorem 6.2 *Let $p > 0$ and $V(x,t) \in C^{2,1}(R^d \times [t_0, \infty); R_+)$. Let $\gamma : R_+ \to R_+$ be a continuous nondecreasing function such that $\gamma(t) \to \infty$ as $t \to \infty$. Let $\eta : R_+ \to R_+$ be a continuous function such that $\int_0^\infty \eta(t)dt < \infty$. If*

$$\gamma(t)|x|^p \leq V(x,t) \quad \text{and} \quad LV(x,t) \leq \eta(t) \tag{6.15}$$

for all $(x,t) \in R^d \times [t_0, \infty)$, then the trivial solution of equation (1.2) is almost surely asymptotically stable with rate function $\lambda(t) = (\gamma(t))^{-1/p}$.

Proof. Fix any initial value x_0 and write $x(t; t_0, x_0) = x(t)$. By Itô's formula,

$$V(x,t) = V(x_0, t_0) + \int_{t_0}^t LV(x(s), s)ds + M(t),$$

where

$$M(t) = \int_{t_0}^t V_x(x(s), s)g(x(s), s)dB(s)$$

is a continuous local martingale on $[t_0, \infty)$ with $M(t_0) = 0$. Using condition (6.15) we obtain that

$$0 \leq \gamma(t)|x(t)|^p \leq V(x_0, t_0) + \int_{t_0}^t \eta(s)ds + M(t).$$

In view of Theorem 1.3.9, $\lim_{t \to \infty} M(t)$ exists and is finite almost surely, and hence there is a finite random variable ξ such that

$$\gamma(t)|x(t)|^p \leq \xi \quad \text{i.e.} \quad |x(t)| \leq \left(\frac{\xi}{\gamma(t)}\right)^{\frac{1}{p}} \quad a.s.$$

The proof is complete.

For illustration we first apply this theorem to equation (6.10). Let $0 < \varepsilon < p - \frac{1}{2}$ be arbitrary and

$$V(x,t) = (t+1)^{2p-1-2\varepsilon}x^2.$$

Compute

$$\begin{aligned}
LV(x,t) &= (2p - 1 - 2\varepsilon)(t+1)^{2p-2-2\varepsilon}x^2 \\
&\quad - 2p(t+1)^{2p-2-2\varepsilon}x^2 + (t+1)^{-(1+2\varepsilon)} \\
&\leq (t+1)^{-(1+2\varepsilon)}
\end{aligned}$$

and note

$$\int_0^\infty (t+1)^{-(1+2\varepsilon)}dt = \frac{1}{2\varepsilon} < \infty.$$

By Theorem 6.2, with $p = 2$, $\gamma(t) = (t+1)^{2p-1-2\varepsilon}$ and $\eta(t) = (t+1)^{-(1+2\varepsilon)}$, we see that the trivial solution of equation (6.10) is almost surely asymptotically stable with rate function $\lambda(t) = (t+1)^{-(p-1/2-\varepsilon)}$. In other words, the solution of equation (6.10) has the property that

$$|x(t;t_0,x_0)| \leq \xi(t+1)^{-(p-\frac{1}{2}-\varepsilon)}$$

for all $t \geq t_0$ almost surely, where ξ is a finite random variable. This implies, for ε is arbitrary, that

$$\limsup_{t \to \infty} \frac{\log|x(t;t_0,x_0)|}{\log t} \leq -\left(p - \frac{1}{2}\right) \quad a.s.$$

which is the same as (6.13).

To close this chapter let us discuss one more example. Consider a stochastic differential equation in R^d of the form

$$dx(t) = f(x(t),t)dt + \sigma(t)dB(t) \quad \text{on } t \geq t_0 \quad (6.16)$$

with initial value $x(t_0) = x_0$, where f is the same as before but $\sigma : R_+ \to R^{d \times m}$. Assume that, for some $p > 0$,

$$2x^T f(x,t) \leq -\frac{p|x|^2}{(t+1)\log(t+1)} \quad \text{and} \quad \int_0^\infty \log(t+1)|\sigma(t)|^2 dt < \infty.$$

Let $V(x,t) = \log^p(t+1)|x|^2$. Then

$$LV(x,t) = \frac{p\log^{p-1}(t+1)}{t+1}|x|^2 + 2\log^p(t+1)x^T f(x,t) + \log^p(t+1)|\sigma(t)|^2$$

$$\leq \log^p(t+1)|\sigma(t)|^2.$$

By Theorem 6.2, with $p = 2$, $\gamma(t) = \log^p(t+1)$ and $\eta(t) = \log^p(t+1)|\sigma(t)|^2$, we see that the trivial solution of equation (6.16) is almost surely asymptotically stable with rate function $\lambda(t) = \log^{-p/2}(t+1)$.

5

Stochastic Functional Differential Equations

5.1 INTRODUCTION

In many applications, one assumes that the system under consideration is governed by a principle of causality; that is, the future state of the system is independent of the past states and is determined solely by the present. However, under closer scrutiny, it becomes apparent that the principle of causality is often only a first approximation to the true situation and that a more realistic model would include some of the past states of the system. Stochastic functional differential equations give a mathematical formulation for such system.

The simplest type of past dependence in a differential equation is that in which the past dependence is through the state variable but not the derivative of the state variable. Lord Cherwell (see Wright (1961)) has encountered the differential difference equation

$$\dot{x}(t) = -\alpha x(t-1)[1+x(t)] \tag{1.1}$$

in his study of the distribution of primes. Dunkel (1968) suggested the more general equation

$$\dot{x}(t) = -\alpha \left[\int_{-1}^{0} x(t+\theta) d\eta(\theta) \right] [1+x(t)] \tag{1.2}$$

for the growth of a single species. In his study of predator-prey models, Volterra

(1928) had earlier investigated the equation

$$\begin{cases} \dot{x}(t) = \left(\varepsilon_1 - \gamma_1 y(t) - \int_{-r}^{0} F_1(\theta) y(t+\theta) d\theta\right) x(t) \\ \dot{y}(t) = \left(\varepsilon_2 + \gamma_2 x(t) + \int_{-r}^{0} F_2(\theta) y(t+\theta) d\theta\right) y(t) \end{cases} \quad (1.3)$$

where x and y are the number of preies and predators, respectively. Under suitable assumptions, the equation

$$\dot{x}(t) = \sum_{i=1}^{k} A_i x(t - \tau_i) \quad (1.4)$$

is a suitable model for describing the mixing of a dye from a central tank as dyed water circulates through a number of pipes. The equation

$$\dot{x}(t) = -\int_{t-\tau}^{t} a(t-\theta) g(x(\theta)) d\theta \quad (1.5)$$

was encountered by Ergen (1954) in the theory of a circulating fuel nuclear reactor. Taking into account the transmission time in the triode oscillator, Rubanik (1969) has studied the van der Pol equation

$$\ddot{x}(t) + \alpha \dot{x}(t) - f(x(t-\tau))\dot{x}(t-\tau) + x(t) = 0 \quad (1.6)$$

with the delayed argument τ. All these equations are special cases of the general functional differential equation

$$\dot{x}(t) = f(x_t, t), \quad (1.7)$$

where $x_t = \{x(t+\theta) : -\tau \leq \theta \leq 0\}$ is the past history of the state. Taking into account the environmental noise we are led to the stochastic functional differential equation

$$dx(t) = f(x_t, t)dt + g(x_t, t)dB(t). \quad (1.8)$$

When we try to carry over the theory of stochastic differential equations to stochastic functional differential equations, the following natural questions arise:

· What is the initial-value problem for equation (1.8)?
· What are the conditions to guarantee the existence and uniqueness of the solution?
· What properties does the solution have?
· Is there any explicit solution or otherwise how can one obtain the approximate solution?

5.2 EXISTENCE-AND-UNIQUENESS THEOREMS

In this chapter we shall answer these questions one by one. Moreover, we shall introduce a new technique—the Razumikhin argument to investigate the stability problem. We shall also introduce and investigate the problem of stochastic self-stabilization.

5.2 EXISTENCE-AND-UNIQUENESS THEOREMS

As before, we are working on the given complete probability space $(\Omega, \mathcal{F}, P)$ with the filtration $\{\mathcal{F}_t\}_{t \geq 0}$ satisfying the usual conditions, and $B(t)$ is the given m-dimensional Brownian motion defined on the space. Let $\tau > 0$ and denote by $C([-\tau, 0]; R^d)$ the family of continuous functions φ from $[-\tau, 0]$ to R^d with the norm $\|\varphi\| = \sup_{-\tau \leq \theta \leq 0} |\varphi(\theta)|$. Let $0 \leq t_0 < T < \infty$. Let

$$f : C([-\tau, 0]; R^d) \times [t_0, T] \to R^d \quad \text{and} \quad g : C([-\tau, 0]; R^d) \times [t_0, T] \to R^{d \times m}$$

be both Borel measurable. Consider the d-dimensional stochastic functional differential equation

$$dx(t) = f(x_t, t)dt + g(x_t, t)dB(t) \quad \text{on } t_0 \leq t \leq T, \tag{2.1}$$

where $x_t = \{x(t+\theta) : -\tau \leq \theta \leq 0\}$ is regarded as a $C([-\tau, 0]; R^d)$-valued stochastic process.

The first question is the following: What is the initial-value problem for this equation? More specifically, what is the minimum amount of initial data that must be specified in order for equation (2.1) to define a stochastic process $x(t)$ on $t_0 \leq t \leq T$? A moment of reflection indicates that a stochastic process must be specified on the entire interval $[t_0 - \tau, t_0]$. We therefore impose the initial data:

$$x_{t_0} = \xi = \{\xi(\theta) : -\tau \leq \theta \leq 0\} \text{ is an } \mathcal{F}_{t_0}\text{-measurable} \\ C([-\tau, 0]; R^d)\text{-valued random variable such that } E\|\xi\|^2 < \infty. \tag{2.2}$$

The initial-value problem for equation (2.1) is now to find the solution of equation (2.1) satisfying the initial data (2.2). But, what is the solution?

Definition 2.1 An R^d-valued stochastic process $x(t)$ on $t_0 - \tau \leq t \leq T$ is called a solution to equation (2.1) with initial data (2.2) if it has the following properties:

(i) it is continuous and $\{x_t\}_{t_0 \leq t \leq T}$ is $\mathcal{F}_t$-adapted;
(ii) $\{f(x_t, t)\} \in \mathcal{L}^1([t_0, T]; R^d)$ and $\{g(x_t, t)\} \in \mathcal{L}^2([t_0, T]; R^{d \times m})$;
(iii) $x_{t_0} = \xi$ and, for every $t_0 \leq t \leq T$,

$$x(t) = \xi(0) + \int_{t_0}^t f(x_s, s)ds + \int_{t_0}^t g(x_s, s)dB(s) \quad a.s.$$

A solution $x(t)$ is said to be unique if any other solution $\bar{x}(t)$ is indistinguishable from it, that is

$$P\{x(t) = \bar{x}(t) \text{ for all } t_0 - \tau \leq t \leq T\} = 1.$$

Let us now begin to establish the theory of the existence and uniqueness of the solution. We first show that the Lipschitz condition and the linear growth condition again guarantee the existence and uniqueness.

Theorem 2.2 *Assume that there exist two positive constants $\bar{K}$ and K such that*

(i) *(uniform Lipschitz condition) for all $\varphi, \phi \in C([-\tau, 0]; R^d)$ and $t \in [t_0, T]$*

$$|f(\varphi, t) - f(\phi, t)|^2 \bigvee |g(\varphi, t) - g(\phi, t)|^2 \leq \bar{K} ||\varphi - \phi||^2; \tag{2.3}$$

(ii) *(linear growth condition) for all $(\varphi, t) \in C([-\tau, 0]; R^d) \times [t_0, T]$*

$$|f(\varphi, t)|^2 \bigvee |g(\varphi, t)|^2 \leq K(1 + ||\varphi||^2). \tag{2.4}$$

Then there exists a unique solution $x(t)$ to equation (2.1) with initial data (2.2). Moreover, the solution belongs to $\mathcal{M}^2([t_0 - \tau, T]; R^d)$.

We prepare a lemma in order to prove this theorem.

Lemma 2.3 *Let the linear growth condition (2.4) hold. If $x(t)$ is a solution to equation (2.1) with initial data (2.2), then*

$$E\left(\sup_{t_0-\tau \leq t \leq T} |x(t)|^2\right) \leq (1 + 4E||\xi||^2)e^{3K(T-t_0)(T-t_0+4)}. \tag{2.5}$$

In particular, $x(t)$ belongs to $\mathcal{M}^2([t_0 - \tau, T]; R^d)$.

Proof. For every integer $n \geq 1$, define the stopping time

$$\tau_n = T \wedge \inf\{t \subset [t_0, T] : ||x_t|| \geq n\}.$$

Clearly, $\tau_n \uparrow T$ a.s. Set $x^n(t) = x(t \wedge \tau_n)$ for $t \in [t_0 - \tau, T]$. Then, for $t_0 \leq t \leq T$,

$$x^n(t) = \xi(0) + \int_{t_0}^{t} f(x_s^n, s) I_{[[t_0, \tau_n]]}(s) ds + \int_{t_0}^{t} g(x_s^n, s) I_{[[t_0, \tau_n]]}(s) dB(s).$$

By Hölder's inequality, Doob's martingale inequality and the linear growth condition, we then show that

$$E\left(\sup_{t_0 \leq s \leq t} |x^n(s)|^2\right) \leq 3E|\xi(0)|^2 + 3K(T - t_0 + 4)\int_{t_0}^{t}(1 + E||x_s^n||^2)ds.$$

Existence-and-Uniqueness Theorems

Noting that $\sup_{t_0-\tau \le s \le t} |x^n(s)|^2 \le ||\xi||^2 + \sup_{t_0 \le s \le t} |x^n(s)|^2$, we obtain

$$1 + E\Big(\sup_{t_0-\tau \le s \le t} |x^n(s)|^2\Big)$$

$$\le 1 + 4E||\xi||^2 + 3K(T - t_0 + 4) \int_{t_0}^{t} \Big[1 + E\Big(\sup_{t_0-\tau \le r \le s} |x^n(r)|^2\Big)\Big] ds.$$

Now the Gronwall inequality yields that

$$1 + E\Big(\sup_{t_0-\tau \le t \le T} |x^n(t)|^2\Big) \le (1 + 4E||\xi||^2) e^{3K(T-t_0)(T-t_0+4)}.$$

Consequently

$$E\Big(\sup_{t_0-\tau \le t \le \tau_n} |x(t)|^2\Big) \le (1 + 4E|\xi|^2) e^{3K(T-t_0)(T-t_0+4)}.$$

Finally the required inequality (2.5) follows by letting $n \to \infty$.

Proof of Theorem 2.2. Uniqueness. Let $x(t)$ and $\bar{x}(t)$ be the two solutions. By Lemma 2.3, both of them belong to $\mathcal{M}^2([t_0 - \tau, T]; R^d)$. Noting

$$x(t) - \bar{x}(t) = \int_{t_0}^{t} [f(x_s, s) - f(\bar{x}_s, s)] ds + \int_{t_0}^{t} [g(x_s, s) - g(\bar{x}_s, s)] dB(s),$$

we can easily show that

$$E\Big(\sup_{t_0 \le s \le t} |x(s) - \bar{x}(s)|^2\Big) \le 2\bar{K}(T + 4) \int_{t_0}^{t} E||x_s - \bar{x}_s||^2 ds$$

$$\le 2\bar{K}(T + 4) \int_{t_0}^{t} E\Big(\sup_{t_0 \le r \le s} |x(r) - \bar{x}(r)|^2\Big) ds.$$

The Gronwall inequality then yields that

$$E\Big(\sup_{t_0 \le t \le T} |x(t) - \bar{x}(t)|^2\Big) = 0.$$

This implies that $x(t) = \bar{x}(t)$ for $t_0 \le t \le T$, hence for all $t_0 - \tau \le t \le T$, almost surely. The uniqueness has been proved.

Existence. Define $x_{t_0}^0 = \xi$ and $x^0(t) = \xi(0)$ for $t_0 \le t \le T$. For each $n = 1, 2, \cdots$, set $x_{t_0}^n = \xi$ and define, by the Picard iterations,

$$x^n(t) = \xi(0) + \int_{t_0}^{t} f(x_s^{n-1}, s) ds + \int_{t_0}^{t} g(x_s^{n-1}, s) dB(s) \qquad (2.6)$$

for $t \in [t_0, T]$. It is easy to show that $x^n(\cdot) \in \mathcal{M}^2([t_0 - \tau, T]; R^d)$ (the details are left to the reader). We claim that for all $n \geq 0$,

$$E\left(\sup_{t_0 \leq s \leq t} |x^{n+1}(s) - x^n(s)|^2\right) \leq \frac{C[M(t - t_0)]^n}{n!} \quad \text{on } t_0 \leq t \leq T, \quad (2.7)$$

where $M = 2\bar{K}(T - t_0 + 4)$ and C will be defined below. First we compute

$$E\left(\sup_{t_0 \leq t \leq T} |x^1(t) - x^0(t)|^2\right)$$

$$\leq 2K(T - t_0) \int_{t_0}^{T} (1 + E||x_s^0||^2) ds + 8K \int_{t_0}^{T} (1 + E||x_s^0||^2) ds$$

$$\leq 2K(T - t_0 + 4)(T - t_0)(1 + E||\xi||^2) := C.$$

So (2.7) holds for $n = 0$. Next, assume (2.7) holds for some $n \geq 0$. Then

$$E\left(\sup_{t_0 \leq s \leq t} |x^{n+2}(s) - x^{n+1}(s)|^2\right)$$

$$\leq 2\bar{K}(t - t_0 + 4) E \int_{t_0}^{t} ||x_s^{n+1} - x_s^n||^2 ds$$

$$\leq M \int_{t_0}^{t} E\left(\sup_{t_0 \leq r \leq s} |x^{n+1}(r) - x^n(r)|^2\right) ds$$

$$\leq M \int_{t_0}^{t} \frac{C[M(s - t_0)]^n}{n!} ds = \frac{C[M(t - t_0)]^{n+1}}{(n+1)!}.$$

That is, (2.7) holds for $n + 1$. Hence, by induction, (2.7) holds for all $n \geq 0$. From (2.7), we can then show in the same way as in the proof of Theorem 2.3.1 that $x^n(\cdot)$ converges to $x(t)$ in $\mathcal{M}^2([t_0 - \tau, T]; R^d)$ in the sense of L^2 as well as probability 1, and the $x(t)$ is a solution to equation (2.1) satisfying the initial condition (2.2). The existence has also been proved.

In the proof above we have shown that the Picard iterations $x^n(t)$ converge to the unique solution $x(t)$ of equation (2.1). The following theorem gives an estimate on the difference between $x^n(t)$ and $x(t)$, and it clearly shows that one can use the Picard iteration procedure to obtain the approximate solutions to equation (2.1).

Theorem 2.4 *Let the assumptions of Theorem 2.2 hold. Let $x(t)$ be the unique solution of equation (2.1) with initial data (2.2) and $x^n(t)$ be the Picard iterations defined by (2.6). Then, for all $n \geq 1$,*

$$E\left(\sup_{t_0 \leq t \leq T} |x^n(t) - x(t)|^2\right) \leq \frac{2C[M(T - t_0)]^n}{n!} e^{2M(T - t_0)} \quad (2.8)$$

where $C = 2K(T - t_0 + 4)(T - t_0)(1 + E||\xi||^2)$ and $M = 2\bar{K}(T - t_0 + 4)$.

Proof. It is easy to derive that

$$E\left(\sup_{t_0 \le s \le t} |x^n(s) - x(s)|^2\right) \le M \int_{t_0}^t E\|x_s^{n-1} - x_s\|^2 ds$$

$$\le 2M \int_{t_0}^t E\left(\sup_{t_0 \le r \le s} |x^n(r) - x^{n-1}(r)|^2\right) ds$$

$$+ 2M \int_{t_0}^t E\left(\sup_{t_0 \le r \le s} |x^n(r) - x(r)|^2\right) ds.$$

Substituting (2.7) into this yields that

$$E\left(\sup_{t_0 \le s \le t} |x^n(s) - x(s)|^2\right)$$

$$\le 2M \int_{t_0}^T \frac{C[M(s-t_0)]^{n-1}}{(n-1)!} ds + 2M \int_{t_0}^t E\left(\sup_{t_0 \le r \le s} |x^n(r) - x(r)|^2\right) ds$$

$$\le \frac{2C[M(T-t_0)]^n}{n!} + 2M \int_{t_0}^t E\left(\sup_{t_0 \le r \le s} |x^n(r) - x(r)|^2\right) ds.$$

The required inequality (2.8) now follows by applying the Gronwall inequality. The proof is complete.

As pointed out in the study of stochastic differential equations, the uniform Lipschitz condition is somewhat restrictive. Fortunately, the following generalization ensures that one can replace it by the local Lipschitz condition.

Theorem 2.5 *Assume that the linear growth condition (2.4) is satisfied but the uniform Lipschitz condition (2.3) is replaced by the following local Lipschitz condition: For every integer $n \ge 1$, there exists a positive constant K_n such that, for all $t \in [t_0, T]$ and those $\varphi, \phi \in C([-\tau, 0]; R^d)$ with $\|\varphi\| \vee \|\phi\| \le n$,*

$$|f(\varphi, t) - f(\phi, t)|^2 \vee |g(\varphi, t) - g(\phi, t)|^2 \le K_n \|\varphi - \phi\|^2. \tag{2.9}$$

Then there exists a unique solution $x(t)$ to the initial-value problem (2.1)-(2.2), and the solution belongs to $\mathcal{M}^2([t_0 - \tau, T]; R^d)$.

This theorem can be proved by a truncation procedure as outlined in the proof of Theorem 2.3.4 but the details are left to the reader.

In what follows we often discuss the stochastic functional differential equation on $[t_0, \infty)$, namely

$$dx(t) = f(x_t, t)dt + g(x_t, t)dB(t) \qquad \text{on } t \in [t_0, \infty) \tag{2.10}$$

with initial data (2.2), where f and g are of course now the mappings from $C([-\tau, 0]; R^d) \times [t_0, \infty)$ to R^d and $R^{d \times m}$, respectively. If the assumptions of the existence-and-uniqueness theorem hold on every finite subinterval $[t_0, T]$ of

$[t_0, \infty)$, then equation (2.10) has a unique solution $x(t)$ on the entire interval $[t_0 - \tau, \infty)$. Such a solution is called a *global* solution. The following theorem is immediate.

Theorem 2.6 *Assume that for every real number $T > t_0$ and integer $n \geq 1$, there exists a positive constant $K_{T,n}$ such that, for all $t \in [t_0, T]$ and all $\varphi, \phi \in C([-\tau, 0]; R^d)$ with $\|\varphi\| \vee \|\phi\| \leq n$,*

$$|f(\varphi, t) - f(\phi, t)|^2 \bigvee |g(\varphi, t) - g(\phi, t)|^2 \leq K_{T,n} \|\varphi - \phi\|^2.$$

Assume also that for every $T > t_0$, there exists a positive constant K_T such that for all $(\varphi, t) \in C([-\tau, 0]; R^d) \times [t_0, T]$,

$$|f(\varphi, t)|^2 \bigvee |g(\varphi, t)|^2 \leq K_T(1 + \|\varphi\|^2).$$

Then there exists a unique global solution $x(t)$ to equation (2.10) and the solution belongs to $\mathcal{M}^2([t_0 - \tau, \infty); R^d)$.

In this book we shall occasionally encounter a more general type of stochastic functional differential equations in which the future state is determined by the entire of the past states rather than some of them. For example, we shall meet the stochastic integrodifferential equation

$$dx(t) = F(x(t), t)dt + \left(\int_{t_0}^{t} |x(s)|ds \right) G(x(t), t)dB(t) \tag{2.11}$$

as well as the functional equation

$$dx(t) = F(x(t), t)dt + \left(\sup_{t_0 \leq s \leq t} |r(s)x(s)| \right) G(x(t), t)dB(t). \tag{2.12}$$

To formulate such equations in a general way, let us introduce a few more notations. For each $t \geq t_0$, denote by $C([t_0 - \tau, t]; R^d)$ the family of continuous functions φ from $[t_0 - \tau, t]$ to R^d with the norm $\|\varphi\| = \sup_{t_0 - \tau \leq \theta \leq t} |\varphi(\theta)|$. Also, let $f(\cdot, t)$ and $g(\cdot, t)$ be mappings from $C([t_0 - \tau, t]; R^d)$ to R^d and $R^{d \times m}$, respectively. Moreover, define $x_{\tau, t} = \{x(\theta) : t_0 - \tau \leq \theta \leq t\}$. Consider the stochastic functional differential equation

$$dx(t) = f(x_{\tau,t}, t)dt + g(x_{\tau,t}, t)dB(t) \quad \text{on } t \in [t_0, \infty) \tag{2.13}$$

with initial data (2.2). Obviously, equations (2.11) and (2.12) are the special cases of equation (2.13). We now state an existence-and-uniqueness theorem for this equation which can be proved in the same way as before.

Theorem 2.7 *Assume that for every real number $T > t_0$ and integer $n \geq 1$, there exists a positive constant $K_{T,n}$ such that, for all $t \in [t_0, T]$ and all $\varphi, \phi \in C([t_0 - \tau, t]; R^d)$ with $\|\varphi\| \vee \|\phi\| \leq n$,*

$$|f(\varphi, t) - f(\phi, t)|^2 \bigvee |g(\varphi, t) - g(\phi, t)|^2 \leq K_{T,n} \|\varphi - \phi\|^2.$$

Assume also that for every $T > t_0$, there exists a positive constant K_T such that for all $t \in [t_0, T]$ and $\varphi \in C([t_0 - \tau, t]; R^d)$,

$$|f(\varphi, t)|^2 \vee |g(\varphi, t)|^2 \leq K_T(1 + ||\varphi||^2).$$

Then there exists a unique global solution $x(t)$ to equation (2.13) and the solution belongs to $\mathcal{M}^2([t_0 - \tau, \infty); R^d)$.

5.3 STOCHASTIC DIFFERENTIAL DELAY EQUATIONS

A special but important class of stochastic functional differential equations is the stochastic differential delay equations. Let us begin with the discussion of the following delay equation

$$dx(t) = F(x(t), x(t-\tau), t)dt + G(x(t), x(t-\tau), t)dB(t) \qquad (3.1)$$

on $t \in [t_0, T]$ with initial data (2.2), where $F : R^d \times R^d \times [t_0, T] \to R^d$ and $G : R^d \times R^d \times [t_0, T] \to R^{d \times m}$. If we define

$$f(\varphi, t) = F(\varphi(0), \varphi(-\tau), t) \quad \text{and} \quad g(\varphi, t) = G(\varphi(0), \varphi(-\tau), t)$$

for $(\varphi, t) \in C([-\tau, 0]; R^d) \times [t_0, T]$, then equation (3.1) can be written as equation (2.1) so one can apply the existence-and-uniqueness theorems established in the previous section to the delay equation (3.1). For example, let F and G satisfy the local Lipschitz condition and the linear growth condition. That is, for every integer $n \geq 1$, there exists a positive constant K_n such that for all $t \in [t_0, T]$ and all $x, y, \bar{x}, \bar{y} \in R^d$ with $|x| \vee |y| \vee |\bar{x}| \vee |\bar{y}| \leq n$,

$$|F(x,y,t) - F(\bar{x},\bar{y},t)|^2 \vee |G(x,y,t) - G(\bar{x},\bar{y},t)|^2 \leq K_n(|x-\bar{x}|^2 + |y-\bar{y}|^2); \quad (3.2)$$

and there is moreover a $K > 0$ such that for all $(x, y, t) \in R^d \times R^d \times [t_0, T]$,

$$|F(x,y,t)|^2 \vee |G(x,y,t)|^2 \leq K(1 + |x|^2 + |y|^2). \qquad (3.3)$$

Then there is a unique solution to the delay equation (3.1). However, we can take one further step to weaken these conditions slightly. Note that on $[t_0, t_0+\tau]$, equation (3.1) becomes

$$dx(t) = F(x(t), \xi(t - t_0 - \tau), t)dt + G(x(t), \xi(t - t_0 - \tau), t)dB(t)$$

with initial value $x(t_0) = \xi(0)$. But this is a stochastic differential equation (without delay), and it will have a unique solution if the linear growth condition (3.3) holds and $F(x, y, t)$, $G(x, y, t)$ are locally Lipschitz continuous in x only. Once the solution $x(t)$ on $[t_0, t_0 + \tau]$ is known, we can proceed this argument on $[t_0 + \tau, t_0 + 2\tau]$, $[t_0 + 2\tau, t_0 + 3\tau]$ etc. and hence obtain the solution on the entire interval $[t_0 - \tau, T]$. This argument shows that it is unnecessary to require that the functions $F(x, y, t)$ and $G(x, y, t)$ be locally Lipschitz continuous in y. We describe this result in the following theorem.

Theorem 3.1 *Assume that the linear growth condition (3.3) is fulfilled. Assume also that both $F(x, y, t)$ and $G(x, y, t)$ are locally Lipschitz continuous in x, that is, for every integer $n \geq 1$, there exists a positive constant K_n such that for all $t \in [t_0, T]$, $y \in R^d$ and $x, \bar{x} \in R^d$ with $|x| \vee |\bar{x}| \leq n$,*

$$|F(x, y, t) - F(\bar{x}, y, t)|^2 \bigvee |G(x, y, t) - G(\bar{x}, y, t)|^2 \leq K_n |x - \bar{x}|^2. \tag{3.4}$$

Then there exists a unique solution to the delay equation (3.1).

This result becomes evident in the case when both F and G are independent of the present state $x(t)$, namely for the equation

$$dx(t) = F(x(t - \tau), t)dt + G(x(t - \tau), t)dB(t).$$

In this case we have explicitly that

$$x(t) = x(t_0) + \int_{t_0}^{t} F(x(s - \tau), s)ds + \int_{t_0}^{t} G(x(s - \tau), s)dB(s)$$

$$= \xi(t_0) + \int_{t_0}^{t} F(\xi(s - t_0 - \tau), s)ds + \int_{t_0}^{t} G(\xi(s - t_0 - \tau), s)dB(s)$$

for $t_0 \leq t \leq t_0 + \tau$. Then, for $t_0 + \tau \leq t \leq t_0 + 2\tau$,

$$x(t) = x(t_0 + \tau) + \int_{t_0+\tau}^{t} F(x(s - \tau), s)ds + \int_{t_0+\tau}^{t} G(x(s - \tau), s)dB(s).$$

Repeating this procedure over the intervals $[t_0 + 2\tau, t_0 + 3\tau]$ etc. we can obtain the explicit solution. Clearly, all we here require is the condition that guarantees the integrals are well defined, and the linear growth condition will do. In other words, we do not require the local Lipschitz condition in this case.

Let us now proceed to discuss the equations in which the delay is time dependent. Let $\delta : [t_0, T] \to [0, \tau]$ be a Borel measurable function. Consider the stochastic differential delay equation

$$dx(t) = F(x(t), x(t - \delta(t)), t)dt + G(x(t), x(t - \delta(t)), t)dB(t) \tag{3.5}$$

on $t \in [t_0, T]$ with initial data (2.2). This is again a special case of equation (2.1) if we define

$$f(\varphi, t) = F(\varphi(0), \varphi(-\delta(t)), t) \quad \text{and} \quad g(\varphi, t) = G(\varphi(0), \varphi(-\delta(t)), t)$$

for $(\varphi, t) \in C([-\tau, 0]; R^d) \times [t_0, T]$. Hence, conditions (3.2) and (3.3) will guarantee the existence and uniqueness of the solution to this delay equation. On the other hand, if the delay is really "true" in the sense that $\sup_{t_0 \leq t \leq T} \delta(t) > 0$, then the above argument which led to Theorem 3.1 still works, and therefore conditions (3.3) and (3.4) will guarantee the existence and uniqueness of the solution to equation (3.5).

This argument can be further extended without any difficulty to the more general stochastic system with several delays, namely

$$dx(t) = F(x(t), x(t - \delta_1(t)), \cdots, x(t - \delta_k(t)), t)dt$$
$$+ G(x(t), x(t - \delta_1(t)), \cdots, x(t - \delta_k(t)), t)dB(t) \quad (3.6)$$

on $t \in [t_0, T]$ with initial data (2.2). Here

$$F : R^{d \times (k+1)} \times [t_0, T] \to R^d, \qquad G : R^{d \times (k+1)} \times [t_0, T] \to R^{d \times m},$$

and $\delta_i : [t_0, T] \to [0, \tau]$ are all Borel-measurable. The following result is immediate.

Theorem 3.2 *Assume that for every integer $n \geq 1$, there exists a positive constant K_n such that for all $t \in [t_0, T]$ and all $x, y_i, \bar{x}, \bar{y}_i \in R^d$ with $|x| \vee |y_i| \vee |\bar{x}| \vee |\bar{y}_i| \leq n$,*

$$|F(x, y_1, \cdots, y_k, t) - F(\bar{x}, \bar{y}_i, \cdots, \bar{y}_k, t)|^2$$
$$\vee |G(x, y_1, \cdots, y_k, t) - G(\bar{x}, \bar{y}_i, \cdots, \bar{y}_k, t)|^2$$
$$\leq K_n \left(|x - \bar{x}|^2 + \sum_{i=1}^{k} |y_i - \bar{y}_i|^2 \right). \quad (3.7)$$

Assume also that there is a $K > 0$ such that for all $(x, y_1, \cdots, y_k, t) \in R^{d \times (k+1)} \times [t_0, T]$,

$$|F(x, y_1, \cdots, y_k, t)|^2 \vee |G(x, y_1, \cdots, y_k, t)|^2 \leq K\left(1 + |x|^2 + \sum_{i=1}^{k} |y_i|^2\right). \quad (3.8)$$

Then there is a unique solution to equation (3.6). If,

$$\sup_{t_0 \leq t \leq T} \delta_i(t) > 0 \quad \text{for every } i = 1, \cdots, k,$$

then condition (3.7) can be replaced by the weaker one: For every integer $n \geq 1$, there exists a positive constant K_n such that for all $(y_1, \cdots, y_k, t) \in R^{d \times k} \times [t_0, T]$ and all $x, \bar{x} \in R^d$ with $|x| \vee |\bar{x}| \leq n$,

$$|F(x, y_1, \cdots, y_k, t) - f(\bar{x}, y_i, \cdots, y_k, t)|^2$$
$$\vee |G(x, y_1, \cdots, y_k, t) - G(\bar{x}, y_i, \cdots, y_k, t)|^2$$
$$\leq K_n |x - \bar{x}|^2. \quad (3.9)$$

5.4 EXPONENTIAL ESTIMATES

In this section we shall give the exponential estimates for the solution of equation (2.10), namely

$$dx(t) = f(x_t, t)dt + g(x_t, t)dB(t) \quad \text{on } t \in [t_0, \infty) \quad (4.1)$$

with initial data $x_{t_0} = \xi$ satisfying (2.2). We assume that this equation has a unique global solution $x(t)$. We also impose the linear growth condition: There is a $K > 0$ such that

$$|f(\varphi, t)|^2 \vee |g(\varphi, t)|^2 \leq K(1 + ||\varphi||^2) \quad (4.2)$$

for all $(\varphi, t) \in C([-\tau, 0]; R^d) \times [t_0, \infty)$. Let us first establish an L^p-estimate.

Theorem 4.1 Let $p \geq 2$, $E||\xi||^p < \infty$ and let (4.2) hold. Then

$$E\left(\sup_{t_0-\tau \leq s \leq t} |x(s)|^p\right) \leq \frac{3}{2} 2^{\frac{p}{2}} (1 + E||\xi||^p) e^{C(t-t_0)} \quad (4.3)$$

for all $t \geq t_0$, where $C = p[2\sqrt{K} + (33p-1)K]$.

Proof. By the Itô formula and the linear growth condition, one can derive that, for $t \geq t_0$,

$$[1 + |x(t)|^2]^{\frac{p}{2}} = [1 + |\xi(0)|^2]^{\frac{p}{2}} + p \int_{t_0}^{t} [1 + |x(s)|^2]^{\frac{p-2}{2}} x^T(s) f(x_s, s) ds$$

$$+ \frac{p}{2} \int_{t_0}^{t} [1 + |x(s)|^2]^{\frac{p-2}{2}} |g(x_s, s)|^2 ds$$

$$+ \frac{p(p-2)}{2} \int_{t_0}^{t} [1 + |x(s)|^2]^{\frac{p-4}{2}} |x^T(s) g(x_s, s)|^2 ds$$

$$+ p \int_{t_0}^{t} [1 + |x(s)|^2]^{\frac{p-2}{2}} x^T(s) g(x_s, s) dB(s)$$

$$\leq 2^{\frac{p-2}{2}} (1 + |\xi(0)|^p) + p \int_{t_0}^{t} [1 + |x(s)|^2]^{\frac{p-2}{2}}$$

$$\times \left(\frac{\sqrt{K}}{2} |x(s)|^2 + \frac{1}{2\sqrt{K}} |f(x_s, s)|^2 + \frac{p-1}{2} |g(x_s, s)|^2\right) ds$$

$$+ p \int_{t_0}^{t} [1 + |x(s)|^2]^{\frac{p-2}{2}} x^T(s) g(x_s, s) dB(s)$$

$$\leq 2^{\frac{p-2}{2}} (1 + ||\xi||^p) + c_1 \int_{t_0}^{t} [1 + ||x_s||^2]^{\frac{p}{2}} ds$$

$$+ p \int_{t_0}^{t} [1 + |x(s)|^2]^{\frac{p-2}{2}} x^T(s) g(x_s, s) dB(s), \quad (4.4)$$

where $c_1 = p[\sqrt{K} + (p-1)K/2]$. Therefore

$$E\left(\sup_{t_0 \le s \le t} [1 + |x(s)|^2]^{\frac{p}{2}}\right)$$
$$\le 2^{\frac{p-2}{2}}(1 + E\|\xi\|^p) + c_1 E \int_{t_0}^t [1 + \|x_s\|^2]^{\frac{p}{2}} ds$$
$$+ pE\left(\sup_{t_0 \le s \le t} \int_{t_0}^s [1 + |x(r)|^2]^{\frac{p-2}{2}} x^T(r) g(x_r, r) dB(r)\right). \quad (4.5)$$

On the other hand, by the Burkholder–Davis–Gundy inequality (i.e. Theorem 1.7.3), we derive that

$$pE\left(\sup_{t_0 \le s \le t} \int_{t_0}^s [1 + |x(r)|^2]^{\frac{p-2}{2}} x^T(r) g(x_r, r) dB(r)\right)$$
$$\le 4\sqrt{2} pE \left(\int_{t_0}^t [1 + |x(s)|^2]^{p-2} |x^T(s) g(x_s, s)|^2 ds\right)^{\frac{1}{2}}$$
$$\le 4\sqrt{2} pE \left\{ \left(\sup_{t_0 \le s \le t} [1 + |x(s)|^2]^{\frac{p}{2}}\right) \int_{t_0}^t [1 + |x(s)|^2]^{\frac{p-4}{2}} |x(s)|^2 |g(x_s, s)|^2 ds \right\}^{\frac{1}{2}}$$
$$\le \frac{1}{2} E\left(\sup_{t_0 \le s \le t} [1 + |x(s)|^2]^{\frac{p}{2}}\right) + 16 p^2 K \, E \int_{t_0}^t [1 + \|x_s\|^2]^{\frac{p}{2}} ds. \quad (4.6)$$

Substituting this into (4.5) yields that

$$E\left(\sup_{t_0 \le s \le t} [1 + |x(s)|^2]^{\frac{p}{2}}\right)$$
$$\le 2^{\frac{p}{2}}(1 + E\|\xi\|^p) + CE \int_{t_0}^t [1 + \|x_s\|^2]^{\frac{p}{2}} ds, \quad (4.7)$$

where $C = 2c_1 + 32p^2 K = p[2\sqrt{K} + (33p-1)K]$ as defined before. Note that

$$E\left(\sup_{t_0-\tau \le s \le t} [1 + |x(s)|^2]^{\frac{p}{2}}\right)$$
$$\le E[1 + \|\xi\|^2]^{\frac{p}{2}} + E\left(\sup_{t_0 \le s \le t} [1 + |x(s)|^2]^{\frac{p}{2}}\right)$$
$$\le 2^{\frac{p-2}{2}}(1 + E\|\xi\|^p) + E\left(\sup_{t_0 \le s \le t} [1 + |x(s)|^2]^{\frac{p}{2}}\right).$$

It then follows from (4.7) that

$$E\left(\sup_{t_0-\tau \le s \le t} [1 + |x(s)|^2]^{\frac{p}{2}}\right)$$
$$\le \frac{3}{2} 2^{\frac{p}{2}}(1 + E\|\xi\|^p) + C \int_{t_0}^t E\left(\sup_{t_0-\tau \le r \le s} [1 + |x(r)|^2]^{\frac{p}{2}}\right) ds.$$

An application of the Gronwall inequality implies that

$$E\left(\sup_{t_0-\tau\leq s\leq t}[1+|x(s)|^2]^{\frac{p}{2}}\right) \leq \frac{3}{2}\, 2^{\frac{p}{2}}(1+E\|\xi\|^p)e^{C(t-t_0)}, \qquad (4.8)$$

and the desired assertion (4.3) follows. The proof is complete.
When $p = 2$, inequality (4.3) reduces to

$$E\left(\sup_{t_0-\tau\leq s\leq t}|x(s)|^2\right) \leq 3(1+E\|\xi\|^2)\exp\left[(4\sqrt{K}+130K)(t-t_0)\right].$$

On the other hand, Lemma 2.3 shows that

$$E\left(\sup_{t_0-\tau\leq s\leq t}|x(s)|^2\right) \leq (1+4E\|\xi\|^2)\exp\left[(3K(t-t_0+4)(t-t_0)\right].$$

Clearly, Theorem 4.1 gives a much better estimate for large t.

As an application of Theorem 4.1 we give an upper bound for the sample Lyapunov exponent.

Theorem 4.2 *Under the linear growth condition (4.2), we have*

$$\limsup_{t\to\infty}\frac{1}{t}\log|x(t)| \leq 2\sqrt{K}+65K \qquad \text{a.s.} \qquad (4.9)$$

In other words, the sample Lyapunov exponent of the solution should not be greater than $2\sqrt{K}+65K$.

Proof. For each $n = 1, 2, \cdots$, it follows from Theorem 4.1 (taking $p=2$) that

$$E\left(\sup_{t_0+n-1\leq t\leq t_0+n}|x(t)|^2\right) \leq \beta e^{\gamma n},$$

where $\beta = 3(1+E\|\xi\|^2)$ and $\gamma = 2[2\sqrt{K}+65K]$. Hence, for arbitrary $\varepsilon > 0$,

$$P\left\{\omega:\sup_{t_0+n-1\leq t\leq t_0+n}|x(t)|^2 > e^{(\gamma+\varepsilon)n}\right\} \leq \beta e^{-\varepsilon n}.$$

The Borel–Cantelli lemma now yields that for almost all $\omega \in \Omega$, there is a random integer $n_0 = n_0(\omega)$ such that

$$\sup_{t_0+n-1\leq t\leq t_0+n}|x(t)|^2 \leq e^{(\gamma+\varepsilon)n} \qquad \text{whenever } n \geq n_0.$$

Consequently, for almost all $\omega \in \Omega$, if $t_0+n-1 \leq t \leq t_0+n$ and $n \geq n_0$,

$$\frac{1}{t}\log|x(t)| \leq \frac{(\gamma+\varepsilon)n}{2(t_0+n-1)}.$$

Thus
$$\limsup_{t\to\infty} \frac{1}{t}\log|x(t)| \le \frac{\gamma+\varepsilon}{2} = 2\sqrt{K} + 65K + \frac{\varepsilon}{2} \quad a.s.$$

Since ε is arbitrary, the assertion (4.9) must hold.

As another application of Theorem 4.1 we now show the continuity of the pth moment of the solution.

Theorem 4.3 *Under the same conditions as Theorem 4.1, we have*
$$E|x(t) - x(s)|^p \le \beta(t)(t-s)^{\frac{p}{2}} \quad \text{for all } t_0 \le s < t < \infty, \qquad (4.10)$$

where
$$\beta(t) = \frac{3}{4}\, 2^p K^{\frac{p}{2}}(1 + E\|\xi\|^p)e^{C(t-t_0)}\left([2(t-t_0)]^{\frac{p}{2}} + [p(p-1)]^{\frac{p}{2}}\right)$$

and $C = p[2\sqrt{K} + (33p-1)K]$. In particular, the pth moment of the solution is continuous.

Proof. Note that
$$E|x(t) - x(s)|^p \le 2^{p-1} E\left|\int_s^t f(x_r, r)\,dr\right|^p + 2^{p-1} E\left|\int_s^t g(x_r, r)\,dB(r)\right|^p.$$

Using the Hölder inequality, Theorem 1.7.1 and the linear growth condition, we can then obtain that
$$E|x(t) - x(s)|^p \le [2(t-s)]^{p-1} E\int_s^t |f(x_r, r)|^p\, dr$$
$$+ \frac{1}{2}[2p(p-1)]^{\frac{p}{2}}(t-s)^{\frac{p-2}{2}} E\int_s^t |g(x_r, r)|^p\, dr$$
$$\le c_2(t-s)^{\frac{p-2}{2}}\int_s^t E(1 + \|x_r\|^2)^{\frac{p}{2}},$$

where $c_2 = 2^{\frac{p-2}{2}} K^{\frac{p}{2}}\left([2(t-t_0)]^{\frac{p}{2}} + [p(p-1)]^{\frac{p}{2}}\right)$. Applying (4.8) one sees that
$$E|x(t) - x(s)|^p \le c_2(t-s)^{\frac{p-2}{2}}\int_s^t \frac{3}{2}\, 2^{\frac{p}{2}}(1 + E\|\xi\|^p)e^{C(r-t_0)}\, dr$$
$$\le \frac{3}{2}\, c_2 2^{\frac{p}{2}}(1 + E\|\xi\|^p)e^{C(t-t_0)}(t-s)^{\frac{p}{2}},$$

which is the required inequality (4.10).

5.5 APPROXIMATE SOLUTIONS

In Chapter 2 we discussed Caratheodory's and Cauchy–Maruyama's approximate solutions to stochastic differential equations and we also pointed out the advantages of these approximation procedures in comparison with Picard's. In this section we shall establish Caratheodory's and Cauchy–Maruyama's approximate solutions to stochastic functional differential equations. To make the theory more understandable, we shall only discuss the case of stochastic differential delay equations but the reader will see that the theory can be extended to more general functional equations.

Let us first discuss the Caratheodory approximation procedure. Consider the stochastic differential delay equation

$$dx(t) = F(x(t), x(t - \delta(t)), t)dt + G(x(t), x(t - \delta(t)), t)dB(t) \quad (5.1)$$

on $t \in [t_0, T]$ with initial data (2.2), where $\delta : [t_0, T] \to [0, \tau]$, $F : R^d \times R^d \times [t_0, T] \to R^d$ and $G : R^d \times R^d \times [t_0, T] \to R^{d \times m}$ are all Borel measurable. We impose the uniform Lipschitz condition and the linear growth condition. That is, there exists a $\bar{K} > 0$ such that for all $t \in [t_0, T]$ and all $x, y, \bar{x}, \bar{y} \in R^d$

$$|F(x, y, t) - F(\bar{x}, \bar{y}, t)|^2 \bigvee |G(x, y, t) - G(\bar{x}, \bar{y}, t)|^2 \leq \bar{K}(|x - \bar{x}|^2 + |y - \bar{y}|^2); \quad (5.2)$$

and there is moreover a $K > 0$ such that for all $(x, y, t) \in R^d \times R^d \times [t_0, T]$,

$$|F(x, y, t)|^2 \bigvee |G(x, y, t)|^2 \leq K(1 + |x|^2 + |y|^2). \quad (5.3)$$

Recall that in Section 2.6 when we discussed the Caratheodory approximation for the stochastic differential equation

$$dx(t) = f(x(t), t)dt + g(x(t), t)dB(t),$$

the main idea was to replace the present state $x(t)$ with the past $x(t - 1/n)$ to obtain the delay equation

$$dx_n(t) = f(x_n(t - 1/n), t)dt + g(x_n(t - 1/n), t)dB(t)$$

and then showed that the solution $x_n(t)$ of this delay equation approximates the solution $x(t)$ of the original equation. When we try to carry over this procedure to the delay equation (5.1), we will naturally replace the present state $x(t)$ by its past $x(t - 1/n)$ as well but with what should we replace the past $x(t - \delta(t))$? In the first instance, one may be tempted to replace it with $x(t - \delta(t) - 1/n)$. However, on the second thought, one realizes that it is not necessary in the case of $\delta(t) \geq 1/n$. It is in this spirit we define the Caratheodory approximation as follows: For each integer $n \geq 2/\tau$, define $x_n(t)$ on $[t_0 - \tau, T]$ by

$$x_n(t_0 + \theta) = \xi(\theta) \quad \text{for } -\tau \leq \theta \leq 0$$

and

$$x_n(t) = \xi(0) + \int_{t_0}^t I_{D_n^c}(s)F(x_n(s-1/n), x_n(s-\delta(s)), s)ds$$

$$+ \int_{t_0}^t I_{D_n}(s)F(x_n(s-1/n), x_n(s-\delta(s))-1/n), s)ds$$

$$+ \int_{t_0}^t I_{D_n^c}(s)G(x_n(s-1/n), x_n(s-\delta(s)), s)dB(s)$$

$$+ \int_{t_0}^t I_{D_n}(s)G(x_n(s-1/n), x_n(s-\delta(s))-1/n), s)dB(s) \quad (5.4)$$

for $t_0 \leq t \leq T$, where

$$D_n = \{t \in [t_0, T] : \delta(t) < 1/n\} \quad \text{and} \quad D_n^c = [t_0, T] - D_n.$$

It is important to note that each $x_n(\cdot)$ can be determined explicitly by the stepwise iterated Itô integrals over the intervals $[t_0, t_0+1/n]$, $(t_0+1/n, t_0+2/n]$ etc. Let us now prepare a few lemmas in order to show the main result.

Lemma 5.1 *Let (5.3) hold. Then, for all $n \geq 2/\tau$,*

$$E\Big(\sup_{t_0-\tau \leq t \leq T} |x_n(s)|^2\Big) \leq \Big(\frac{1}{2} + 6E\|\xi\|^2\Big)e^{10K(T-t_0+4)(T-t_0)}. \quad (5.5)$$

Proof. By Hölder's inequality, Doob's martingale inequality and the linear growth condition (5.3), we can derive from (5.4) that for $t_0 \leq t \leq T$,

$$E\Big(\sup_{t_0 \leq s \leq t} |x_n(s)|^2\Big) \leq 5E|\xi(0)|^2$$

$$+ 5K(T-t_0+4)\int_{t_0}^t I_{D_n^c}(s)\big[1 + E|x_n(s-1/n)|^2 + E|x_n(s-\delta(s))|^2\big]ds$$

$$+ 5K(T-t_0+4)\int_{t_0}^t I_{D_n}(s)\big[1 + E|x_n(s-1/n)|^2 + E|x_n(s-\delta(s)-1/n)|^2\big]ds$$

$$\leq 5E\|\xi\|^2 + 5K(T-t_0+4)\int_{t_0}^t \Big[1 + 2E\Big(\sup_{t_0-\tau \leq r \leq s}|x_n(r)|^2\Big)\Big]ds.$$

Hence

$$\frac{1}{2} + E\Big(\sup_{t_0-\tau \leq s \leq t}|x_n(s)|^2\Big) \leq \frac{1}{2} + E\|\xi\|^2 + E\Big(\sup_{t_0 \leq s \leq t}|x_n(s)|^2\Big)$$

$$\leq \frac{1}{2} + 6E\|\xi\|^2 + 10K(T-t_0+4)\int_{t_0}^t \Big[\frac{1}{2} + E\Big(\sup_{t_0-\tau \leq r \leq s}|x_n(r)|^2\Big)\Big]ds.$$

The Gronwall inequality implies

$$\frac{1}{2} + E\left(\sup_{t_0-\tau\leq s\leq t} |x_n(s)|^2\right) \leq \left(\frac{1}{2} + 6E\|\xi\|^2\right)e^{10K(T-t_0+4)(t-t_0)}$$

and the required inequality (5.5) follows immediately.

Lemma 5.2 *Let (5.3) hold. Then the solution of equation (5.1) has the property*

$$E\left(\sup_{t_0-\tau\leq t\leq T} |x(t)|^2\right) \leq C_1 := \left(\frac{1}{2} + 4E\|\xi\|^2\right)e^{6K(T-t_0+4)(T-t_0)}. \qquad (5.6)$$

Moreover, for any $t_0 \leq s < t \leq T$ *with* $t - s < 1$,

$$E|x(t) - x(s)|^2 \leq C_2(t-s), \qquad (5.7)$$

where $C_2 = 4K(1 + 2C_1)$.

Proof. The proof of (5.6) is similar to that of Lemma 5.1 so we need only to show (5.7) but this is straightforward:

$$E|x(t) - x(s)|^2 \leq 2K(t - s + 1) \int_s^t \left[1 + E|x(r)|^2 + E|x(r - \delta(r))|^2\right] dr$$

$$\leq 4K(1 + 2C_1)(t - s)$$

as required.

We can now prove one of the main results in this section.

Theorem 5.3 *Let (5.2) and (5.3) hold. Then*

$$E\left(\sup_{t_0\leq t\leq T} |x(t) - x_n(t)|^2\right) \leq 4C_3 e^{5C_3(T-t_0)}$$

$$\times \left(\frac{6C_1 + TC_2}{n} + 2C_1\mu\{t \in [t_0, t_0 + \tau] : 0 < \delta(t) < 1/n\}\right), \qquad (5.8)$$

where C_1, C_2 *are defined in Lemma 5.2,* $C_3 = 4\bar{K}(T - t_0 + 4)$ *and* μ *stands for the Lebesgue measure on* $\mathbb{R}$.

Proof. By Hölder's inequality, Doob's martingale inequality and the Lipschitz condition (5.2), we can derive that, for $t_0 \leq t \leq T$,

$$E\left(\sup_{t_0\leq s\leq t} |x(s) - x_n(s)|^2\right)$$

$$\leq C_3 \int_{t_0}^t I_{D_n^c}(s)\left[E|x(s) - x_n(s - 1/n)|^2 + E|x(s - \delta(s)) - x_n(s - \delta(s))|^2\right] ds$$

$$+ C_3 \int_{t_0}^{t} I_{D_n}(s) \Big[E|x(s) - x_n(s - 1/n)|^2$$
$$+ E|x(s - \delta(s)) - x_n(s - \delta(s) - 1/n)|^2 \Big] ds$$
$$\leq 2C_3 \int_{t_0}^{t} \Big[E|x(s) - x(s - 1/n)|^2 + E|x(s - 1/n) - x_n(s - 1/n)|^2 \Big] ds$$
$$+ C_3 \int_{t_0}^{t} I_{D_n^c}(s) E|x(s - \delta(s)) - x_n(s - \delta(s))|^2 \Big] ds$$
$$+ 2C_3 \int_{t_0}^{t} I_{D_n}(s) \Big[E|x(s - \delta(s)) - x(s - \delta(s) - 1/n)|^2$$
$$+ E|x(s - \delta(s) - 1/n) - x_n(s - \delta(s) - 1/n)|^2 \Big] ds$$
$$\leq 5C_3 \int_{t_0}^{t} E\Big(\sup_{t_0 \leq r \leq s} |x(r) - x_n(r)|^2 \Big) ds + J_1 + J_2,$$

where
$$J_1 = 2C_3 \int_{t_0}^{T} E|x(s) - x(s - 1/n)|^2 ds$$

and
$$J_2 = 2C_3 \int_{t_0}^{T} I_{D_n}(s) E|x(s - \delta(s)) - x(s - \delta(s) - 1/n)|^2 ds.$$

Applying the Gronwall inequality we obtain that
$$E\Big(\sup_{t_0 \leq s \leq T} |x(s) - x_n(s)|^2 \Big) \leq (J_1 + J_2) e^{5C_3(T - t_0)}. \tag{5.9}$$

But, using Lemma 5.2, we can estimate
$$J_1 \leq 4C_3 \int_{t_0}^{t_0 + 1/n} (E|x(s)|^2 + E|x(s - 1/n)|^2) ds$$
$$+ 2C_3 \int_{t_0 + 1/n}^{T} E|x(s) - x(s - 1/n)|^2 ds$$
$$\leq 8C_1 C_3 / n + 2C_2 C_3 T / n = 2C_3(4C_1 + TC_2)/n. \tag{5.10}$$

Also, setting $D_0 = \{t \in [t_0, T] : \delta(t) = 0\}$,
$$J_2 = 2C_3 \int_{t_0}^{T} I_{D_0}(s) E|x(s) - x(s - 1/n)|^2 ds$$
$$+ 2C_3 \int_{t_0}^{T} I_{D_n - D_0}(s) E|x(s - \delta(s)) - x(s - \delta(s) - 1/n)|^2 ds$$
$$\leq 8C_1 C_3 \int_{t_0}^{t_0 + 1/n} I_{D_0}(s) ds + 2C_2 C_3 / n \int_{t_0 + 1/n}^{T} I_{D_0}(s) ds$$
$$+ 8C_1 C_3 \int_{t_0}^{t_0 + \tau + 1/n} I_{D_n - D_0}(s) ds + 2C_2 C_3 / n \int_{t_0 + \tau + 1/n}^{T} I_{D_n - D_0}(s) ds$$
$$\leq 2C_3(8C_1 + TC_2)/n + 8C_1 C_3 \mu([t_0, t_0 + \tau] \cap (D_n - D_0)). \tag{5.11}$$

Substituting (5.10) and (5.11) into (5.9) yields the required result (5.8). The proof is complete.

Let us now turn to the Cauchy–Maruyama approximation procedure. We first give the definition of the Cauchy–Maruyama approximation sequence. For each integer $n \geq 1$, define $x_n(t)$ on $[t_0 - \tau, T]$ as follows:

$$x_n(t_0 + \theta) = \xi(\theta) \quad \text{for } -\tau \leq \theta \leq 0$$

and

$$x_n(t) = x(t_0 + k/n)$$
$$+ \int_{t_0 + k/n}^{t} F(x_n(t_0 + k/n), x_n(t_0 + k/n - \delta(s)), s) ds$$
$$+ \int_{t_0 + k/n}^{t} G(x_n(t_0 + k/n), x_n(t_0 + k/n - \delta(s)), s) dB(s) \quad (5.12)$$

for $t_0 + k/n < t \leq [t_0 + (k+1)/n] \wedge T$, $k = 0, 1, 2, \cdots$. In the sequel of this section $x_n(t)$ always means the Cauchy–Maruyama approximation rather than the Caratheodory one. Clearly, $x_n(\cdot)$ can be determined explicitly by the stepwise iterated Itô integrals over the intervals $(t_0, t_0 + 1/n]$, $(t_0 + 1/n, t_0 + 2/n]$ etc. Moreover, if we define $\hat{x}_n(t_0) = x_n(t_0)$, $\tilde{x}_n(t_0) = x_n(t_0 - \delta(t_0))$,

$$\hat{x}_n(t) = x_n(t_0 + k/n) \quad \text{and} \quad \tilde{x}_n(t) = x_n(t_0 + k/n - \delta(t)) \quad (5.13)$$

for $t_0 + k/n < t \leq [t_0 + (k+1)/n] \wedge T$, $k = 0, 1, 2, \cdots$, it then follows from (5.12) that

$$x_n(t) = \xi(0) + \int_{t_0}^{t} F(\hat{x}_n(s), \tilde{x}_n(s), s) ds + \int_{t_0}^{t} G(\hat{x}_n(s), \tilde{x}_n(s), s) dB(s) \quad (5.14)$$

for $t_0 \leq t \leq T$. The following lemma shows that the Cauchy–Maruyama approximation sequence is bounded in L^2.

Lemma 5.4 *Let (5.3) hold. Then, for all $n \geq 1$,*

$$E\left(\sup_{t_0 - \tau \leq t \leq T} |x_n(t)|^2\right) \leq \left(\frac{1}{2} + 4E\|\xi\|^2\right) e^{6K(T - t_0 + 4)(T - t_0)}. \quad (5.15)$$

Proof. It is easy to show from (5.14) that, for $t_0 \leq t \leq T$,

$$E\left(\sup_{t_0 \leq s \leq t} |x_n(s)|^2\right) \leq 3E|\xi(0)|^2$$
$$+ 3K(T - t_0 + 4) \int_{t_0}^{t} \left[1 + E|\hat{x}_n(s)|^2 + E|\tilde{x}_n(s)|^2\right] ds.$$

Recalling the definition of $\hat{x}_n(t)$ and $\tilde{x}_n(t)$, we then see that

$$E\left(\sup_{t_0 \le s \le t} |x_n(s)|^2\right) \le 3E\|\xi\|^2$$
$$+3K(T-t_0+4)\int_{t_0}^{t}\left[1+2E\left(\sup_{t_0-\tau \le r \le s}|x_n(r)|^2\right)\right]ds.$$

Consequently

$$\frac{1}{2}+E\left(\sup_{t_0-\tau \le s \le t}|x_n(s)|^2\right) \le \frac{1}{2}+4E\|\xi\|^2$$
$$+6K(T-t_0+4)\int_{t_0}^{t}\left[\frac{1}{2}+E\left(\sup_{t_0-\tau \le r \le s}|x_n(r)|^2\right)\right]ds$$

and the required assertion (5.15) follows by applying the Gronwall inequality.

The following theorem shows that the Cauchy–Maruyama sequence converges to the unique solution of equation (5.1) and gives an estimate for the difference between the approximate solution $x_n(t)$ and the accurate solution $x(t)$.

Theorem 5.5 Let (5.2) and (5.3) hold. Assume that the initial data $\xi = \{\xi(\theta) : -\tau \le \theta \le 0\}$ is uniformly Lipschitz L^2-continuous, that is there is a positive constant β such that

$$E|\xi(\theta_1) - \xi(\theta_2)|^2 \le \beta(\theta_2 - \theta_1) \quad \text{if } -\tau \le \theta_1 < \theta_2 \le 0. \tag{5.16}$$

Then the difference between the Cauchy–Maruyama approximate solution $x_n(t)$ and the accurate solution $x(t)$ of equation (5.1) can be estimated as

$$E\left(\sup_{t_0 \le t \le T}|x(t) - x_n(t)|^2\right) \le \frac{4C_4}{n}[C_2(T-t_0) + \tau(\beta \vee C_2)]e^{4C_4(T-t_0)}, \tag{5.17}$$

where C_2 is defined in Lemma 5.2 and $C_4 = 2\bar{K}(T-t_0+4)$.

Proof. It is not difficult to show that, for $t_0 \le t \le T$,

$$E\left(\sup_{t_0 \le s \le t}|x(s) - x_n(s)|^2\right)$$
$$\le C_4 \int_{t_0}^{t}\left[E|x(s) - \hat{x}_n(s)|^2 + E|x(s-\delta(s)) - \tilde{x}_n(s)|^2\right]ds. \tag{5.18}$$

define $\hat{x}(t_0) = x(t_0)$, $\tilde{x}(t_0) = x(t_0 - \delta(t_0))$,

$$\hat{x}(t) = x(t_0 + k/n) \quad \text{and} \quad \tilde{x}(t) = x(t_0 + k/n - \delta(t)) \tag{5.19}$$

for $t_0 + k/n < t \leq [t_0 + (k+1)/n] \wedge T$, $k = 0, 1, 2, \cdots$. It then follows from (5.18) that

$$E\left(\sup_{t_0 \leq s \leq t} |x(s) - x_n(s)|^2\right)$$

$$\leq 2C_4 \int_{t_0}^{t} \left[E|\hat{x}(s) - \hat{x}_n(s)|^2 + E|\tilde{x}(s) - \tilde{x}_n(s)|^2\right] ds + J_3 + J_4$$

$$\leq 4C_4 \int_{t_0}^{t} E\left(\sup_{t_0 \leq r \leq s} |x(r) - x_n(r)|^2\right) ds + J_3 + J_4,$$

where

$$J_3 = 2C_4 \int_{t_0}^{T} E|x(s) - \hat{x}(s)|^2 ds$$

and

$$J_4 = 2C_4 \int_{t_0}^{T} E|x(s - \delta(s)) - \tilde{x}(s)|^2 ds.$$

By Gronwall's inequality, we obtain that

$$E\left(\sup_{t_0 \leq s \leq T} |x(s) - x_n(s)|^2\right) \leq (J_3 + J_4) e^{4C_4(T-t_0)}. \tag{5.20}$$

We now estimate J_3 and J_4. By Lemma 5.2,

$$J_3 = 2C_4 \sum_{k \geq 0} \int_{t_0+k/n}^{[t_0+(k+1)/n] \wedge T} E|x(s) - x(t_0 + k/n)|^2 ds$$

$$\leq \frac{2}{n} C_2 C_4 (T - t_0). \tag{5.21}$$

Also

$$J_4 = 2C_4 \sum_{k \geq 0} \int_{t_0+k/n}^{[t_0+(k+1)/n] \wedge T} E|x(s - \delta(s)) - x(t_0 + k/n - \delta(s))|^2 ds$$

$$\leq \frac{2}{n} C_2 C_4 (T - t_0) + 2C_4 \sum_{k} \int_{t_0+k/n}^{[t_0+(k+1)/n] \wedge \tau}$$

$$E|x(s - \delta(s)) - x(t_0 + k/n - \delta(s))|^2 ds. \tag{5.22}$$

It is easy to show, by condition (5.16) and Lemma 5.2, that

$$E|x(t) - x(s)|^2 \leq 2(\beta \vee C_2)(t - s) \quad \text{if } -\tau \leq s < t \leq \tau, \ t - s \leq 1.$$

We therefore see from (5.22) that

$$J_4 \leq \frac{2C_4}{n} [C_2(T - t_0) + 2\tau(\beta \vee C_2)]. \tag{5.23}$$

Substituting (5.21) and (5.23) into (5.20) yields the required assertion (5.17). The proof is complete.

In the case when the time delay function $\delta(\cdot)$ is Lipschitz continuous, the Cauchy–Maruyama approximate solutions can be defined by a simpler form, that is (5.12) can be replaced by

$$x_n(t) = x(t_0 + k/n)$$
$$+ \int_{t_0+k/n}^{t} F(x_n(t_0+k/n), x_n(t_0+k/n - \delta(t_0+k/n)), s)ds$$
$$+ \int_{t_0+k/n}^{t} G(x_n(t_0+k/n), x_n(t_0+k/n - \delta(t_0+k/n)), s)dB(s) \quad (5.24)$$

for $t_0 + k/n < t \leq [t_0 + (k+1)/n] \wedge T$, $k = 0, 1, 2, \cdots$, while $x_n(t_0 + \theta) = \xi(\theta)$ for $-\tau \leq \theta \leq 0$, the same as before. When both functions F and G are independent of t, this becomes even simpler, namely

$$x_n(t) = x(t_0 + k/n)$$
$$+ F(x_n(t_0+k/n), x_n(t_0+k/n - \delta(t_0+k/n)))(t - t_0 - k/n)ds$$
$$+ G(x_n(t_0+k/n), x_n(t_0+k/n - \delta(t_0+k/n)))[B(t) - B(t_0 + k/n)].$$

To be more precise, let us state this result to close this section.

Theorem 5.6 *In addition to the assumptions of Theorem 5.5, assume that $\delta(\cdot)$ is Lipschitz continuous, that is there is a positive constant α such that*

$$|\delta(t) - \delta(s)| \leq \alpha(t - s) \quad \text{if } t_0 \leq s < t \leq T. \quad (5.25)$$

Then, for every $n > 1 + \alpha$, the difference between the Cauchy–Maruyama approximate solution $x_n(t)$ defined by (5.24) and the accurate solution $x(t)$ of equation (5.1) can be estimated as

$$E\left(\sup_{t_0 \leq t \leq T} |x(t) - x_n(t)|^2 \right)$$
$$\leq \frac{4C_4}{n} \Big(C_2(1+\alpha)(T - t_0) + \tau[\beta \vee C_2(1+\alpha)] \Big) e^{4C_4(T-t_0)}. \quad (5.26)$$

where C_2 and C_4 are defined as before.

This theorem can be proved in the same way as in the proof of Theorem 5.5 with a little bit careful consideration on the estimation of J_4, but the details are left to the reader.

5.6 STABILITY THEORY—RAZUMIKHIN THEOREMS

Stochastic modelling has come to play an important role in many branches of science and industry. An area of particular interest has been the automatic control of stochastic systems, with consequent emphasis being placed on the analysis of stability in stochastic models. One of the most useful stochastic models which appear frequently in applications is the stochastic functional differential equation (2.10), namely

$$dx(t) = f(x_t, t)dt + g(x_t, t)dB(t) \quad \text{on } t \geq t_0. \tag{6.1}$$

In this section we shall discuss its stability property. Due to the page limit we shall only investigate the exponential stability—one of the most important stabilities. For this purpose we always assume that all of the assumptions of the existence-and-uniqueness Theorem 2.6 are fulfilled so equation (6.1) has a unique global solution for any given initial data $x_{t_0} = \xi$ satisfying (2.2), and we here denote the solution by $x(t; \xi)$. Assume furthermore that $f(0, t) \equiv 0$ and $g(0, t) \equiv 0$ so equation (6.1) has the solution $x(t) \equiv 0$ corresponding to the initial data $x_{t_0} = 0$. This solution is called the *trivial solution* or *equilibrium position*.

When we try to carry over the stability theory established in Chapter 4 to the stochastic functional differential equation, we will naturally employ the Lyapunov functionals rather than functions. For instance, it is not difficult to show the following result (cf. Kolmanovskii & Nosov (1986)):

Let $p \geq 2$ and c_1–c_3 be positive constants. Assume that there is a continuous functional $V : C([-\tau, 0]; R^d) \times [t_0, \infty) \to R$ such that

$$c_1 |\varphi(0)|^p \leq V(\varphi, t) \leq c_2 \|\varphi\|^p, \quad (\varphi, t) \in C([-\tau, 0]; R^d) \times [t_0, \infty) \tag{6.2}$$

and

$$EV(x_{t_2}, t_2) - EV(x_{t_1}, t_1) \leq -c_3 \int_{t_1}^{t_2} E|x(s)|^p ds, \quad t_0 \leq t_1 < t_2 < \infty. \tag{6.3}$$

Then the trivial solution of equation (6.1) is pth moment asymptotically stable.

This result is of course a natural generalization of the Lyapunov direct method but is somewhat inconvenient in applications. This is not only because condition (6.3) is not related to the coefficients f and g explicitly but also because it appears to be more difficult to construct the Lyapunov functionals than the Lyapunov functions. It is in this spirit that we would like to explore the possibility of using the rate of change of a function on R^d to determine sufficient conditions for stability.

To explain the idea, we need to introduce a few more new notations. Denote by $C^b_{\mathcal{F}_{t_0}}([-\tau, 0]; R^d)$ the family of all bounded, $\mathcal{F}_{t_0}$-measurable, $C([-\tau, 0]; R^d)$-valued random variables. For $p > 0$ and $t \geq 0$, denote by $L^p_{\mathcal{F}_t}([-\tau, 0]; R^d)$ the family of all $\mathcal{F}_t$-measurable $C([-\tau, 0]; R^d)$-valued random variables $\phi = \{\phi(\theta) : -\tau \leq \theta \leq 0\}$ such that $E\|\phi\|^p < \infty$. Moreover, for each function $V(x, t) \in$

$C^{2,1}(R^d \times [t_0 - \tau, \infty); R_+)$, define an operator $\mathcal{L}V$ from $C([-\tau, 0]; R^d) \times [t_0, \infty)$ to R by

$$\mathcal{L}V(\varphi, t) = V_t(\varphi(0), t) + V_x(\varphi(0), t)f(\varphi, t)$$
$$+ \frac{1}{2}\text{trace}[g^T(\varphi, t)V_{xx}(\varphi(0), t)g(\varphi, t)].$$

Then the expectation of the derivative of V along the solution $x(t; \xi) = x(t)$ of equation (6.1) is given by $E\mathcal{L}V(x_t, t)$. In order for $E\mathcal{L}V(x_t, t)$ to be negative for all initial data ξ and $t \geq t_0$, one would be forced to impose very severe restrictions on the functions $f(\varphi, t)$ and $g(\varphi, t)$ to the extent that the point $\varphi(0)$ plays a dominant role and, therefore, the results will apply only to equations that are very similar to stochastic differential equations. This seems to indicate that it is not good enough to use the Lyapunov functions. Fortunately, a few moments of reflection in the proper direction indicate that it is unnecessary to require $E\mathcal{L}V(x_t, t)$ be negative for all initial data and all $t \geq t_0$ in order to have asymptotic stability, and this is the basic idea exploited in this section. This idea originated with Razumikhin (1956, 1960) for the ordinary differential delay equation and was developed by several people to more general functional differential equations (cf. Hale & Lunel (1993) and the references therein) and by Mao (1996b) to stochastic functional differential equations. The results in this direction are generally referred to as theorems of Razumikhin type.

Let us now begin to establish the Razumikhin-type theorems on the exponential stability for the stochastic functional differential equation.

Theorem 6.1 *Let λ, p, c_1, c_2 all be positive constants and $q > 1$. Assume that there exists a function $V \in C^{2,1}(R^d \times [t_0 - \tau, \infty); R_+)$ such that*

$$c_1|x|^p \leq V(x, t) \leq c_2|x|^p \quad \text{for all } (x, t) \in R^d \times [t_0 - \tau, \infty) \quad (6.4)$$

and, moreover,

$$E\mathcal{L}V(\phi, t) \leq -\lambda EV(\phi(0), t) \quad (6.5)$$

for all $t \geq t_0$ and those $\phi \in L^p_{\mathcal{F}_t}([-\tau, 0]; R^d)$ satisfying

$$EV(\phi(\theta), t + \theta) < qEV(\phi(0), t) \quad \text{on } -\tau \leq \theta \leq 0.$$

Then for all $\xi \in C^b_{\mathcal{F}_{t_0}}([-\tau, 0]; R^d)$

$$E|x(t; \xi)|^p \leq \frac{c_2}{c_1} E\|\xi\|^p e^{-\gamma(t-t_0)} \quad \text{on } t \geq t_0, \quad (6.6)$$

where $\gamma = \min\{\lambda, \log(q)/\tau\}$.

Proof. Fix any initial data $\xi \in C^b_{\mathcal{F}_{t_0}}([-\tau, 0]; R^d)$ and write $x(t; \xi) = x(t)$ simply. Let $\varepsilon \in (0, \gamma)$ be arbitrary and set $\bar{\gamma} = \gamma - \varepsilon$. Define

$$U(t) = \max_{-\tau \leq \theta \leq 0}\left[e^{\bar{\gamma}(t+\theta)}EV(x(t+\theta), t+\theta)\right] \quad \text{for } t \geq t_0. \quad (6.7)$$

Noting that $E(\sup_{0\leq s\leq t}|x(s)|^r) < \infty$ for all $r > 0$ and, moreover, $x(t)$, $V(x,t)$ are continuous, we see that $EV(x(t),t)$ is continuous. Hence $U(t)$ is well define and is continuous. We claim that

$$D_+U(t) := \limsup_{h\to 0+} \frac{U(t+h) - U(t)}{t} \leq 0 \quad \text{for all } t \geq 0. \tag{6.8}$$

To show this, for each $t \geq t_0$ (fixed for the moment), define

$$\bar{\theta} = \max\{\theta \in [-\tau, 0] : e^{\bar{\gamma}(t+\theta)} EV(x(t+\theta), t+\theta) = U(t)\}.$$

Obviously, $\bar{\theta}$ is well defined, $\bar{\theta} \in [-\tau, 0]$ and

$$U(t) = e^{\bar{\gamma}(t+\bar{\theta})} EV(x(t+\bar{\theta}), t+\bar{\theta}).$$

If $\bar{\theta} < 0$, then

$$e^{\bar{\gamma}(t+\theta)} EV(x(t+\theta), t+\theta) < e^{\bar{\gamma}(t+\bar{\theta})} EV(x(t+\bar{\theta}), t+\bar{\theta}) \quad \text{for all } \bar{\theta} < \theta \leq 0.$$

It is therefore easy to observe that for all $h > 0$ sufficiently small

$$e^{\bar{\gamma}(t+h)} EV(x(t+h), t+h) \leq e^{\bar{\gamma}(t+\bar{\theta})} EV(x(t+\bar{\theta}), t+\bar{\theta}),$$

hence
$$U(t+h) \leq U(t) \quad \text{and} \quad D_+U(t) \leq 0.$$

If $\bar{\theta} = 0$, then

$$e^{\bar{\gamma}(t+\theta)} EV(x(t+\theta), t+\theta) \leq e^{\bar{\gamma}t} EV(x(t), t) \quad \text{for all } -\tau \leq \theta \leq 0.$$

So

$$EV(x(t+\theta), t+\theta) \leq e^{-\bar{\gamma}\theta} EV(x(t), t)$$
$$\leq e^{\bar{\gamma}\tau} EV(x(t), t) \quad \text{for all } -\tau \leq \theta \leq 0. \tag{6.9}$$

Note that either $EV(x(t), t) = 0$ or $EV(x(t), t) > 0$. If $EV(x(t), t) = 0$, then (6.9) and (6.4) yield that $x(t+\theta) = 0$ a.s. for all $-\tau \leq \theta \leq 0$. Recalling the fact that $f(0,t) \equiv 0$ and $g(0,t) \equiv 0$, one sees that $x(t+h) = 0$ a.s. for all $h > 0$ hence $U(t+h) = 0$ and $D_+U(t) = 0$. On the other hand, in the case when $EV(x(t), t) > 0$, (6.9) implies

$$EV(x(t+\theta), t+\theta) < qEV(x(t), t) \quad \text{for all } -\tau \leq \theta \leq 0$$

since $e^{\bar{\gamma}\tau} < q$. Thus, by condition (6.5),

$$E\mathcal{L}V(x_t, t) \leq -\lambda EV(x(t), t).$$

However, by Itô's formula, one can derive that for all $h > 0$

$$e^{\bar{\gamma}(t+h)} EV(x(t+h), t+h) - e^{\bar{\gamma}t} EV(x(t), t)$$
$$= \int_t^{t+h} e^{\bar{\gamma}s} [\bar{\gamma} EV(x(s), s) + E\mathcal{L}V(x_s, s)] ds.$$

Note that

$$\bar{\gamma} EV(x(t), t) + E\mathcal{L}V(x_t, t) \leq -(\lambda - \bar{\gamma}) EV(x(t), t) < 0.$$

One sees from the continuity of V etc. that for all $h > 0$ sufficiently small

$$\bar{\gamma} EV(x(s), s) + E\mathcal{L}V(x_s, s) \leq 0 \quad \text{if } t \leq s \leq t + h,$$

and, consequently,

$$e^{\bar{\gamma}(t+h)} EV(x(t+h), t+h) \leq e^{\bar{\gamma}t} EV(x(t), t).$$

Therefore, $U(t+h) = U(t)$ for all $h > 0$ sufficiently small and then $D_+U(t) = 0$. Inequality (6.8) has been proved. It now follows from (6.8) immediately that

$$U(t) \leq U(0) \quad \text{for all } t \geq t_0.$$

By the definition of $U(t)$ and condition (6.4) one sees

$$E|x(t)|^p \leq \frac{c_2}{c_1} E\|\xi\|^p e^{-\bar{\gamma}(t-t_0)} = \frac{c_2}{c_1} E\|\xi\|^p e^{-(\gamma-\varepsilon)(t-t_0)}.$$

Since ε is arbitrary, the required assertion (6.6) must hold. The proof is complete.

As pointed out in Chapter 4, the pth moment exponential stability and almost sure exponential stability do not imply each other in general. However, we are going to show that under an irrestrictive condition the pth moment exponential stability implies almost sure exponential stability

Theorem 6.2 *Let $p \geq 1$. Assume that there is a constant $K > 0$ such that for every solution $x(t)$ of equation (6.1)*

$$E(|f(x_t, t)|^p + |g(x_t, t)|^p) \leq K \sup_{-\tau \leq \theta \leq 0} E|x(t+\theta)|^p \quad \text{on } t \geq 0. \tag{6.10}$$

Then (6.6) implies

$$\limsup_{t \to \infty} \frac{1}{t} \log |x(t; \xi)| \leq -\frac{\gamma}{p} \quad a.s. \tag{6.11}$$

In particular, if, in addition to the above conditions, all of the assumptions of Theorem 6.1 are satisfied, then the trivial solution of equation (6.1) is almost surely exponentially stable.

Proof. Fix any $\xi \in C^b_{\mathcal{F}_{t_0}}([-\tau, 0]; R^d)$ and write $x(t; \xi) = x(t)$ again. For each integer $k \geq 2$,

$$E\|x_{t_0+k\tau}\|^p = E\Big(\sup_{0\leq h \leq \tau} |x(t_0 + (k-1)\tau + h)|^p\Big)$$

$$\leq 3^{p-1}\bigg[E|x(t_0 + (k-1)\tau)|^p + E\Big(\int_{t_0+(k-1)\tau}^{t_0+k\tau} |f(x_t,t)|dt\Big)^p$$

$$+ E\Big(\sup_{0\leq h \leq \tau} \Big|\int_{t_0+(k-1)\tau}^{t_0+(k-1)\tau+h} g(x_t,t)dB(t)\Big|^p\Big)\bigg]. \tag{6.12}$$

But, by Hölder's inequality, condition (6.10) and Theorem 6.1, one can derive that

$$E\Big(\int_{t_0+(k-1)\tau}^{t_0+k\tau} |f(x_t,t)|dt\Big)^p \leq \tau^{p-1}\int_{t_0+(k-1)\tau}^{t_0+k\tau} E|f(x_t,t)|^p dt$$

$$\leq K\tau^{p-1}\int_{t_0+(k-1)\tau}^{t_0+k\tau} \Big(\sup_{-\tau \leq \theta \leq 0} E|x(t+\theta)|^p\Big) dt$$

$$\leq \frac{Kc_2\tau^{p-1}}{c_1} E\|\xi\|^p \int_{t_0+(k-1)\tau}^{t_0+k\tau} e^{-\gamma(t-\tau-t_0)} dt$$

$$\leq \frac{Kc_2\tau^p}{c_1} E\|\xi\|^p e^{-(k-2)\tau\gamma}. \tag{6.13}$$

On the other hand, by the Burkholder–Davis–Gundy inequality,

$$J := E\Big(\sup_{0\leq h \leq \tau} \Big|\int_{t_0+(k-1)\tau}^{t_0+(k-1)\tau+h} g(x_t,t)dB(t)\Big|^p\Big)$$

$$\leq C_p E\Big(\int_{t_0+(k-1)\tau}^{t_0+k\tau} |g(x_t,t)|^2 dt\Big)^{\frac{p}{2}}, \tag{6.14}$$

where C_p is a positive constant dependent of p only. Note from condition (6.10) that

$$|g(\varphi, t)|^p \leq K\|\varphi\|^p \quad \text{for } (\varphi, t) \in C([-\tau, 0]; R^d) \times [t_0, \infty).$$

Let $\sigma \in (0, 1/3^{p-1}K)$ be sufficiently small for

$$\frac{3^{p-1}K\sigma}{1 - 3^{p-1}K\sigma} < e^{-\gamma\tau}. \tag{6.15}$$

One can then derives from (6.14) that

$$J \leq C_p E\bigg[\Big(\sup_{t_0+(k-1)\tau \leq t \leq t_0+k\tau} |g(x_t,t)|\Big) \int_{t_0+(k-1)\tau}^{t_0+k\tau} |g(x_t,t)|dt\bigg]^{\frac{p}{2}}$$

$$\leq \sigma E\Big(\sup_{t_0+(k-1)\tau \leq t \leq t_0+k\tau} |g(x_t,t)|^p\Big) + \frac{C_p^2}{4\sigma} E\bigg[\int_{t_0+(k-1)\tau}^{t_0+k\tau} |g(x_t,t)|dt\bigg]^p$$

$$\le K\sigma E\left(\sup_{t_0+(k-1)\tau \le t \le t_0+k\tau} \|x_t\|^p\right) + \frac{C_p^2 \tau^{p-1}}{4\sigma} \int_{t_0+(k-1)\tau}^{t_0+k\tau} E|g(x_t,t)|^p dt$$

$$\le K\sigma\left(E\|x_{t_0+k\tau}\|^p + E\|x_{t_0+(k-1)\tau}\|^p\right)$$

$$+ \frac{Kc_2 C_p^2 \tau^p}{4\sigma c_1} E\|\xi\|^p e^{-(k-2)\tau\gamma}. \tag{6.16}$$

Substituting (6.6), (6.13) and (6.16) into (6.12) and then making use of (6.15) we obtain that

$$E\|x_{t_0+k\tau}\|^p \le e^{-\tau\gamma} E\|x_{t_0+(k-1)\tau}\|^p + Ce^{-(k-2)\tau\gamma}, \tag{6.17}$$

where C is a constant independent of k. By induction, one can easily show from (6.17) that

$$E\|x_{t_0+k\tau}\|^p \le e^{-k\tau\gamma} E\|x_{t_0}\|^p + kCe^{-(k-2)\tau\gamma}$$
$$\le (Ce^{2\tau\gamma} + E\|\xi\|^p)(k+1)e^{-k\tau\gamma}. \tag{6.18}$$

We shall now show that (6.18) implies the required assertion (6.11). Let $\varepsilon \in (0,\gamma)$ be arbitrary. By (6.18),

$$P\left\{\omega : \|x_{t_0+k\tau}\| > e^{-(\gamma-\varepsilon)k\tau/p}\right\}$$
$$\le e^{(\gamma-\varepsilon)k\tau} E\|x_{t_0+k\tau}\|^p \le (Ce^{2\tau\gamma} + E\|\xi\|^p)(k+1)e^{-\varepsilon k\tau}.$$

In view of the well-known Borel–Cantelli lemma, one sees that for almost all $\omega \in \Omega$

$$\|x_{t_0+k\tau}\| \le e^{-(\gamma-\varepsilon)k\tau/p} \tag{6.19}$$

holds for all but finitely many k. Hence there exists a $k_0(\omega)$, for all $\omega \in \Omega$ excluding a P-null set, for which (6.19) holds whenever $k \ge k_0$. Consequently, for almost all $\omega \in \Omega$,

$$\frac{1}{t}\log|x(t)| \le -\frac{k\tau(\gamma-\varepsilon)}{p[t_0+(k-1)\tau]}.$$

if $t_0+(k-1)\tau \le t \le t_0+k\tau$, $k \ge k_0$. Therefore

$$\limsup_{t\to\infty} \frac{1}{t}\log|x(t)| \le -\frac{\gamma-\varepsilon}{p} \quad a.s.$$

and the required (6.11) follows by letting $\varepsilon \to 0$. The proof is complete.

In the case of $p \in (0,1)$, we need a slightly stronger condition than (6.10) in order to imply the almost sure exponential stability from the pth moment exponential stability.

Theorem 6.3 Let $p \in (0,1)$. Assume that there is a constant $K > 0$ such that for every solution $x(t)$ of equation (6.1)

$$E\Big(\sup_{-\tau \leq \theta \leq 0}\big[|f(x_{t+\theta}, t+\theta)|^p + |g(x_{t+\theta}, t+\theta)|^p\big]\Big)$$
$$\leq K \sup_{-2\tau \leq r \leq 0} E|x(t+r)|^p \quad \text{on } t \geq \tau. \qquad (6.20)$$

Then (6.6) implies

$$\limsup_{t \to \infty} \frac{1}{t} \log|x(t;\xi)| \leq -\frac{\gamma}{p} \quad \text{a.s.} \qquad (6.21)$$

Proof. Fix any $\xi \in C_{\mathcal{F}_{t_0}}^b([-\tau, 0]; R^d)$ and write $x(t;\xi) = x(t)$. Noting that, for any $a, b, c \geq 0$,

$$(a+b+c)^p \leq [3(a \vee b \vee c)]^p \leq 3^p(a^p \vee b^p \vee c^p) \leq 3^p(a^p + b^p + c^p),$$

we have, for each integer $k \geq 2$,

$$E\|x_{t_0+k\tau}\|^p = E\Big(\sup_{0 \leq h \leq \tau} |x(t_0 + (k-1)\tau + h)|^p\Big)$$
$$\leq 3^p \Big[E|x(t_0 + (k-1)\tau)|^p + E\Big(\int_{t_0+(k-1)\tau}^{t_0+k\tau} |f(x_t, t)| dt\Big)^p$$
$$+ E\Big(\sup_{0 \leq h \leq \tau} \Big|\int_{t_0+(k-1)\tau}^{t_0+(k-1)\tau+h} g(x_t, t) dB(t)\Big|^p\Big)\Big]. \qquad (6.22)$$

By conditions (6.20) and (6.6),

$$E\Big(\int_{t_0+(k-1)\tau}^{t_0+k\tau} |f(x_t, t)| dt\Big)^p \leq \tau^p E\Big(\sup_{t_0+(k-1)\tau \leq t \leq t_0+k\tau} |f(x_t, t)|^p\Big)$$
$$\leq K\tau^p \Big(\sup_{t_0+(k-2)\tau \leq t \leq t_0+k\tau} E|x(t)|^p\Big) \leq \frac{Kc_2\tau^p}{c_1} E\|\xi\|^p e^{-(k-2)\tau\gamma}. \qquad (6.23)$$

Also, by the Burkholder–Davis–Gundy inequality etc.,

$$E\Big(\sup_{0 \leq h \leq \tau} \Big|\int_{t_0+(k-1)\tau}^{t_0+(k-1)\tau+h} g(x_t, t) dB(t)\Big|^p\Big)$$
$$\leq C_p E\Big(\int_{t_0+(k-1)\tau}^{t_0+k\tau} |g(x_t, t)|^2 dt\Big)^{\frac{p}{2}}$$
$$\leq C_p \tau^{\frac{p}{2}} E\Big[\sup_{t_0+(k-1)\tau \leq t \leq t_0+k\tau} |g(x_t, t)|^p\Big]$$
$$\leq K C_p \tau^{\frac{p}{2}} \Big[\sup_{t_0+(k-2)\tau \leq t \leq t_0+k\tau} E|x(t)|^p\Big]$$
$$\leq \frac{Kc_2 C_p \tau^{\frac{p}{2}}}{c_1} E\|\xi\|^p e^{-(k-2)\tau\gamma}, \qquad (6.24)$$

where C_p is a positive constant dependent of p only. Substituting (6.6), (6.23) and (6.24) into (6.22) yields

$$E\|x_{t_0+k\tau}\|^p \leq Ce^{-(k-2)\tau\gamma}, \qquad (6.25)$$

where C is a constant independent of k. It is now the same as the proof of Theorem 6.2 to derive the required assertion (6.21) from (6.25). The proof is complete.

Let us now apply the above Razumikhin-type theorems to some special cases of equation (6.1).

(i) *Stochastic Differential Delay Equations*

First of all, consider a stochastic differential delay equation

$$\begin{aligned} dx(t) = {} & F(x(t), x(t-\delta_1(t)), \cdots, x(t-\delta_k(t)), t)dt \\ & + G(x(t), x(t-\delta_1(t)), \cdots, x(t-\delta_k(t)), t)dB(t) \end{aligned} \qquad (6.26)$$

on $t \geq t_0$ with initial data $x_0 = \xi$ satisfying (2.2), where $\delta_i : [t_0, \infty) \to [0, \tau]$, $1 \leq i \leq k$, are all continuous, and

$$F: R^d \times R^{n \times k} \times [t_0, \infty) \to R^d, \qquad G: R^d \times R^{n \times k} \times [t_0, \infty) \to R^{n \times m}.$$

We assume that both F and G satisfy the local Lipschitz condition and the linear growth condition (cf. Theorem 3.2). We also assume that $F(0, \cdots, 0, t) \equiv 0$ and $G(0, \cdots, 0, t) \equiv 0$.

Theorem 6.4 *Let $\lambda, \lambda_1, \cdots, \lambda_k, p, c_1, c_2$ all be positive numbers. Assume that there exists a function $V(x,t) \in C^{2,1}(R^d \times [t_0 - \tau, \infty); R_+)$ such that*

$$c_1|x|^p \leq V(x,t) \leq c_2|x|^p \quad \text{for all } (x,t) \in R^d \times [t_0 - \tau, \infty), \qquad (6.27)$$

and

$$\begin{aligned} & V_t(x,t) + V_x(x,t)F(x, y_1, \cdots, y_k, t) \\ & + \frac{1}{2}\text{trace}\left[G^T(x, y_1, \cdots, y_k, t)V_{xx}(x,t)G(x, y_1, \cdots, y_k, t)\right] \\ & \leq -\lambda V(x,t) + \sum_{i=1}^k \lambda_i V(y_i, t - \delta_i(t)) \end{aligned} \qquad (6.28)$$

for all $(x, y_1, \cdots, y_k, t) \in R^d \times R^{n \times k} \times [t_0, \infty)$. If $\lambda > \sum_{i=1}^k \lambda_i$, then the trivial solution of equation (6.26) is pth moment exponentially stable and its pth moment Lyapunov exponent should not be greater than $-(\lambda - q\sum_{i=1}^k \lambda_i)$, where $q \in (1, \lambda/\sum_{i=1}^k \lambda_i)$ is the unique root of $\lambda - q\sum_{i=1}^k \lambda_i = \log(q)/\tau$. If, in addition, $p \geq 1$ and there is a $K > 0$ such that

$$|F(x, y_1, \cdots, y_k, t)| \vee |G(x, y_1, \cdots, y_k, t)| \leq K\left(|x| + \sum_{i=1}^k |y_i|\right) \qquad (6.29)$$

for all $(x, y_1, \cdots, y_k, t) \in R^d \times R^{n \times k} \times [t_0, \infty)$, then the trivial solution of equation (6.26) is also almost surely exponentially stable and its sample Lyapunov exponent should not be greater than $-(\lambda - q\sum_{i=1}^{k} \lambda_i)/p$.

Proof. Define, for $(\varphi, t) \in C([-\tau, 0]; R^d) \times [t_0, \infty)$,

$$f(\varphi, t) = F(\varphi(0), \varphi(-\delta_1(t)), \cdots, \varphi(-\delta_k(t)), t)$$

and

$$g(\varphi, t) = G(\varphi(0), \varphi(-\delta_1(t)), \cdots, \varphi(-\delta_k(t)), t).$$

Then equation (6.26) becomes equation (6.1). Moreover, the operator $\mathcal{L}V$ becomes

$$\mathcal{L}V(\varphi, t) = V_t(\varphi(0), t) + V_x(\varphi(0), t) F(\varphi(0), \varphi(-\delta_1(t)), \cdots, \varphi(-\delta_k(t)), t)$$
$$+ \frac{1}{2} \text{trace}\Big[G^T(\varphi(0), \varphi(-\delta_1(t)), \cdots, \varphi(-\delta_k(t)), t)$$
$$\times V_{xx}(\varphi(0), t) G(t, \varphi(0), \varphi(-\delta_1(t)), \cdots, \varphi(-\delta_k(t)), t)\Big].$$

If $t \geq t_0$ and $\phi \in L^p_{\mathcal{F}_t}([-\tau, 0]; R^d)$ satisfying

$$EV(\phi(\theta), t + \theta) < qEV(\phi(0), t) \quad \text{for all } -\tau \leq \theta \leq 0,$$

then by condition (6.28)

$$E\mathcal{L}V(\phi, t) \leq -\lambda EV(\phi(0), t) + \sum_{i=1}^{k} \lambda_i EV(\phi(-\delta_i(t)), t - \delta_i(t))$$
$$\leq -\left(\lambda - q\sum_{i=1}^{k} \lambda_i\right) EV(\phi(0), t).$$

So, by Theorem 6.1, the trivial solution of equation (6.26) is pth moment exponentially stable and, moreover, its pth moment Lyapunov exponent should not be greater than $-(\lambda - q\sum_{i=1}^{k} \lambda_i)$. If furthermore $p \geq 1$ and (6.29) holds, then for all $t \geq t_0$ and $\phi \in L^p_{\mathcal{F}_t}([-\tau, 0]; R^d)$,

$$E\Big(|f(\phi, t)|^p + |g(\phi, t)|^p\Big)$$
$$\leq 2E\bigg(K\Big[|\phi(0)| + \sum_{i=1}^{k} |\phi(-\delta_i(t))|\Big]\bigg)^p$$
$$\leq 2K^p(1+k)^{p-1} E\Big[|\phi(0)|^p + \sum_{i=1}^{k} |\phi(-\delta_i(t))|^p\Big]$$
$$\leq 2K^p(1+k)^p \sup_{-\tau \leq \theta \leq 0} E|\phi(\theta)|^p.$$

Therefore, by Theorem 6.2, the trivial solution of equation (6.26) is almost surely exponentially stable and its sample Lyapunov exponent should not be greater than $-(\lambda - q\sum_{i=1}^{k}\lambda_i)/p$. The proof is therefore complete.

Corollary 6.5 *Assume that there is a $\lambda > 0$ such that*

$$x^T F(x, 0, \cdots, 0, t) \leq -\lambda|x|^2 \quad \text{for all } (x, t) \in R^d \times [t_0, \infty). \tag{6.30}$$

Assume also that there are nonnegative numbers α_i, β_i, $0 \leq i \leq k$ such that

$$|F(x, 0, \cdots, 0, t) - F(\bar{x}, y_1, \cdots, y_k, t)| \leq \alpha_0|x - \bar{x}| + \sum_{i=1}^{k} \alpha_i|y_i| \tag{6.31}$$

and

$$|G(x, y_1, \cdots, y_k, t)|^2 \leq \beta_0|x|^2 + \sum_{i=1}^{k} \beta_i|y_i|^2 \tag{6.32}$$

for all $t \geq t_0$ and $x, \bar{x}, y_1, \cdots, y_k \in R^d$. If $p \geq 2$ and

$$\lambda > \sum_{i=1}^{k} \alpha_i + \frac{p-1}{2}\sum_{i=0}^{k} \beta_i, \tag{6.33}$$

then the trivial solution of equation (6.26) is pth moment exponentially stable and is also almost surely exponentially stable.

Proof. Note first that (6.29) follows from (6.31), (6.32) and $F(0, \cdots, 0, t) \equiv 0$. Let $V(x, t) = |x|^p$ and verify (6.28) as follows: For all $(x, y_1, \cdots, y_k, t) \in R^d \times R^{n \times k} \times [t_0, \infty)$,

$$V_t(x, t) + V_x(x, t)F(x, y_1, \cdots, y_k, t)$$
$$+ \frac{1}{2}\text{trace}\left[G^T(x, y_1, \cdots, y_k, t)V_{xx}(x, t)G(x, y_1, \cdots, y_k, t)\right]$$
$$= p|x|^{p-2}x^T F(x, 0, \cdots, 0, t)$$
$$+ p|x|^{p-2}x^T\left[F(x, y_1, \cdots, y_k, t) - F(x, 0, \cdots, 0, t)\right]$$
$$+ \frac{p}{2}|x|^{p-2}|G(x, y_1, \cdots, y_k, t)|^2$$
$$+ \frac{p(p-2)}{2}|x|^{p-4}|x^T G(x, y_1, \cdots, y_k, t)|^2$$
$$\leq -\left(p\lambda - \frac{p(p-1)\beta_0}{2}\right)|x|^p + p\sum_{i=1}^{k}\alpha_i|x|^{p-1}|y_i|$$
$$+ \frac{p(p-1)}{2}\sum_{i=1}^{k}\beta_i|x|^{p-2}|y_i|^2. \tag{6.34}$$

Using the elementary inequality

$$u^\alpha v^{1-\alpha} \leq \alpha u + (1-\alpha)v \quad \text{for } u, v \geq 0, \ 0 \leq \alpha < 1, \tag{6.35}$$

we have
$$|x|^{p-1}|y_i| = (|x|^p)^{\frac{p-1}{p}} (|y_i|^p)^{\frac{1}{p}} \le \frac{p-1}{p}|x|^p + \frac{1}{p}|y_i|^p$$

and
$$|x|^{p-2}|y_i|^2 \le \frac{p-2}{p}|x|^p + \frac{2}{p}|y_i|^p.$$

Substituting these into (6.34) yields that

the left-hand side of (6.34)
$$\le -\left(p\lambda - \frac{p(p-1)}{2}\beta_0 - (p-1)\sum_{i=1}^{k}\alpha_i - \frac{(p-1)(p-2)}{2}\sum_{i=1}^{k}\beta_i\right)|x|^p$$
$$+ \sum_{i=1}^{k}(\alpha_i + (p-1)\beta_i)|y_i|^p.$$

Now the conclusions follow from Theorem 6.4 immediately and the proof is complete.

The conditions of Corollary 6.5 are delay-independent and so the conclusions. However, (6.30) may not hold sometimes and, instead, one may have $x^T F(x,x,\cdots,x,t) \le -\lambda|x|^2$. For example, $F(x,y_1,\cdots,y_k,t) = ax - \sum_{i=1}^{k} b_i y_i$ with $0 \le a < \sum_{i=1}^{k} b_i$. In this case, the delay effect plays the main role in stabilizing the system. The following Corollary deals with this case.

Corollary 6.6 *Assume that there is a $\lambda > 0$ such that*

$$x^T F(x,x,\cdots,x,t) \le -\lambda|x|^2 \quad \text{for all } (x,t) \in R^d \times [t_0,\infty). \tag{6.36}$$

Let $p \ge 2$ and assume furthermore that there are nonnegative numbers α_i, β_i, $0 \le i \le k$ such that

$$|F(x,x,\cdots,x,t) - F(\bar{x},y_1,\cdots,y_k,t)|^p$$
$$\le \alpha_0|x-\bar{x}|^p + \sum_{i=1}^{k}\alpha_i|x-y_i|^p \tag{6.37}$$

and

$$|G(x,y_1,\cdots,y_k,t)|^p \le \beta_0|x|^p + \sum_{i=1}^{k}\beta_i|y_i|^p \tag{6.38}$$

for all $t \ge t_0$ and $x, \bar{x}, y_1, \cdots, y_k \in R^d$. If

$$\lambda > (K\hat{\alpha})^{\frac{1}{p}} + \frac{1}{2}(p-1)\hat{\beta}^{\frac{2}{p}}, \tag{6.39}$$

where
$$K = 2^{p-1}\left[\tau^p(\alpha_0 + \hat{\alpha}) + \bar{C}_p \tau^{\frac{p}{2}}\hat{\beta}\right], \qquad \bar{C}_p = \left[\frac{p(p-1)}{2}\right]^{\frac{p}{2}},$$

$$\hat{\alpha} = \sum_{i=1}^{k}\alpha_i, \qquad \hat{\beta} = \sum_{i=0}^{k}\beta_i,$$

then the trivial solution of equation (6.26) is pth moment exponentially stable and is also almost surely exponentially stable.

Proof. Regard equation (6.26) as a delay equation on $t \geq t_0 + \tau$ with initial data on $[t_0 - \tau, t_0 + \tau]$, that is, consider the delay interval of length 2τ instead of τ. It is easy to show that for $t \geq \tau$,

$$E\mathcal{L}|x_t|^p \leq -p\lambda E|x(t)|^p$$
$$+ pE\Big[|x(t)|^{p-1}|F(x(t),\cdots,x(t),t)$$
$$\quad - F(x(t), x(t-\delta_1(t)), \cdots, x(t-\delta_k(t)), t)|\Big]$$
$$+ \frac{p(p-1)}{2}E\Big[|x(t)|^{p-2}|G(x(t), x(t-\delta_1(t)), \cdots, x(t-\delta_k(t)), t)|^2\Big],$$

where condition (6.36) has been used. Note from inequality (6.35) that for $u, v \geq 0$ and $\varepsilon_1 > 0$,

$$u^{p-1}v \leq (\varepsilon_1 u^p)^{\frac{p-1}{p}}\left(\frac{v^p}{\varepsilon_1^{p-1}}\right)^{\frac{1}{p}} \leq \frac{\varepsilon_1(p-1)}{p}u^p + \frac{1}{p\varepsilon_1^{p-1}}v^p.$$

Applying this and condition (6.37) one obtains that
$$pE\Big[|x(t)|^{p-1}|F(x(t),\cdots,x(t),t) - F(x(t), x(t-\delta_1(t)), \cdots, x(t-\delta_k(t)), t)|\Big]$$
$$\leq \varepsilon_1(p-1)E|x(t)|^p + \frac{1}{\varepsilon_1^{p-1}}\sum_{i=1}^{k}\alpha_i E|x(t) - x(t-\delta_i(t))|^p.$$

Similarly one can show that
$$\frac{p(p-1)}{2}E\Big[|x(t)|^{p-2}|G(x(t), x(t-\delta_1(t)), \cdots, x(t-\delta_k(t)), t)|^2\Big]$$
$$\leq \frac{1}{2}\varepsilon_2(p-1)(p-2)E|x(t)|^p$$
$$\quad + \frac{(p-1)}{\varepsilon_2^{(p-2)/2}}\left(\beta_0 E|x(t)|^p + \sum_{i=1}^{k}\beta_i E|x(t-\delta_i(t))|^p\right)$$
$$\leq \frac{1}{2}\varepsilon_2(p-1)(p-2)E|x(t)|^p + \frac{(p-1)\hat{\beta}}{\varepsilon_2^{(p-2)/2}}\sup_{-\tau \leq \theta \leq 0} E|x(t+\theta)|^p,$$

where $\varepsilon_2 > 0$, like ε_1, is to be determined. Summarising the aboves one obtains that
$$E\mathcal{L}|x_t|^p \leq -p\lambda E|x(t)|^p + \varepsilon_1(p-1)E|x(t)|^p$$
$$+ \frac{1}{\varepsilon_1^{p-1}} \sum_{i=1}^{k} \alpha_i E|x(t) - x(t - \delta_i(t))|^p$$
$$+ \frac{1}{2}\varepsilon_2(p-1)(p-2)E|x(t)|^p$$
$$+ \frac{(p-1)\hat{\beta}}{\varepsilon_2^{(p-2)/2}} \sup_{-\tau \leq \theta \leq 0} E|x(t+\theta)|^p. \qquad (6.40)$$

On the other hand, by Hölder's inequality, Theorem 1.7.1 and the assumptions one can derive that
$$E|x(t) - x(t - \delta_i(t))|^p$$
$$\leq 2^{p-1} E\left|\int_{t-\delta_i(t)}^{t} F(x(s), x(s-\delta_1(s)), \cdots, x(s-\delta_k(s)), s)ds\right|^p$$
$$+ 2^{p-1} E\left|\int_{t-\delta_i(t)}^{t} G(x(s), x(s-\delta_1(s)), \cdots, x(s-\delta_k(s)), s)dB(s)\right|^p$$
$$\leq (2\tau)^{p-1} \int_{t-\delta_i(t)}^{t} E|F(x(s), x(s-\delta_1(s)), \cdots, x(s-\delta_k(s)), s)|^p ds$$
$$+ 2^{p-1}\bar{C}_p \tau^{(p-2)/2} \int_{t-\delta_i(t)}^{t} E|G(x(s), x(s-\delta_1(s)), \cdots, x(s-\delta_k(s)), s)|^p ds$$
$$\leq (2\tau)^{p-1} \int_{t-\tau}^{t} \left(\alpha_0 E|x(s)|^p + \sum_{i=1}^{k} \alpha_i E|x(s-\delta_i(s))|^p\right) ds$$
$$+ 2^{p-1}\bar{C}_p \tau^{(p-2)/2} \int_{t-\tau}^{t} \left(\beta_0 E|x(s)|^p + \sum_{i=1}^{k} \beta_i E|x(s-\delta_i(s))|^p\right) ds$$
$$\leq 2^{p-1}\left[\tau^p(\alpha_0 + \hat{\alpha}) + \bar{C}_p \tau^{p/2} \hat{\beta}\right] \sup_{-2\tau \leq \theta \leq 0} E|x(t+\theta)|^p$$
$$= K \sup_{-2\tau \leq \theta \leq 0} E|x(t+\theta)|^p \qquad (6.41)$$

for $t \geq \tau$, $1 \leq i \leq k$, where $K, \bar{C}_p$ etc. have been defined above. Substituting (6.41) into (6.40) and choosing $\varepsilon_1 = (K\hat{\alpha})^{1/p}$, $\varepsilon_2 = \hat{\beta}^{2/p}$ one then obtains
$$E\mathcal{L}|x_t|^p \leq -p\lambda E|x(t)|^p$$
$$+ \left(p(K\hat{\alpha})^{\frac{1}{p}} + \frac{1}{2}p(p-1)\hat{\beta}^{\frac{2}{p}}\right) \sup_{-2\tau \leq \theta \leq 0} E|x(t+\theta)|^p. \qquad (6.42)$$

By (6.39), one can choose $q > 1$ such that
$$\lambda > q\left((K\hat{\alpha})^{\frac{1}{p}} + \frac{1}{2}(p-1)\hat{\beta}^{\frac{2}{p}}\right).$$

Therefore, if $E|x(t+\theta)|^p < qE|x(t)|^p$ for $-2\tau \leq \theta \leq 0$, (6.42) implies

$$E\mathcal{L}|x_t|^p \leq -p\left(\lambda - q(K\hat{a})^{\frac{1}{p}} - \frac{1}{2}q(p-1)\hat{\beta}^{\frac{2}{p}}\right)E|x(t)|^p.$$

Finally the conclusions follow from Theorems 6.1 and 6.2. The proof is complete.

(ii) *Stochastically Perturbed Equations*

Let us now turn to consider a stochastic equation of the form

$$dx(t) = [\psi(x(t),t) + F(x_t,t)]dt + g(x_t,t)dB(t) \quad \text{on } t \geq t_0 \quad (6.43)$$

with initial data $x_{t_0} = \xi \in C^b_{\mathcal{F}_{t_0}}([-\tau,0];R^d)$. Here g is as defined in Section 5.2, $\psi : R^d \times [t_0,\infty) \to R^d$ and $F : C([-\tau,0];R^d) \times [t_0,\infty) \to R^d$. As before, assume that ψ, F, g satisfy the local Lipschitz condition and the linear growth condition (cf. Theorem 2.6), moreover, $\psi(0,t) = F(0,t) \equiv 0$ and $g(0,t) \equiv 0$. Under these conditions, equation (6.43) has a unique global solution. Equation (6.43) can be regarded as the stochastically perturbed equation of the ordinary differential equation

$$\dot{x}(t) = \psi(x(t),t). \quad (6.44)$$

To a certain degree it has been known that if equation (6.44) is exponentially stable and the stochastic perturbation is sufficiently small, then the perturbed equation (6.43) will remain exponentially stable (cf. Mao, 1994a). The critical research in this direction is to give better bound for the stochastic perturbation. We now apply the Razumikhin-type theorems to establish a number of new results.

Theorem 6.7 *Let $\lambda, c_1, c_2, \beta_1, \cdots, \beta_4$ all be positive numbers and $p \geq 2$, $q > 1$. Assume that there exists a function $V(x,t) \in C^{2,1}(R^d \times [-\tau,\infty); R_+)$ such that*

$$c_1|x|^p \leq V(x,t) \leq c_2|x|^p \quad \text{for all } (x,t) \in R^d \times [t_0 - \tau, \infty),$$

and

$$V_t(x,t) + V_x(x,t)\psi(x,t) \leq -\lambda V(x,t),$$

$$|V_x(x,t)| \leq \beta_1[V(x,t)]^{\frac{p-1}{p}}, \quad \|V_{xx}(x,t)\| \leq \beta_2[V(x,t)]^{\frac{p-2}{p}}$$

for all $(x,t) \in R^d \times [t_0,\infty)$. Assume also that

$$E|F(\phi,t)|^p \leq \beta_3 EV(\phi(0),t) \quad \text{and} \quad E|g(\phi,t)|^p \leq \beta_4 EV(\phi(0),t)$$

for all $t \geq t_0$ and those $\phi \in L^p_{\mathcal{F}_t}([-\tau,0];R^d)$ satisfying

$$EV(\phi(\theta), t+\theta) < qEV(\phi(0),t) \quad \text{for all } -\tau \leq \theta \leq 0. \quad (6.45)$$

If
$$\lambda > \beta_1 \beta_3^{\frac{1}{p}} + \frac{1}{2}\beta_2 \beta_4^{\frac{2}{p}}, \qquad (6.46)$$

then the trivial solution of equation (6.43) is pth moment exponentially stable. In addition, if there is a constant $K > 0$ such that, for all $t \geq t_0$ and $\phi \in L^p_{\mathcal{F}_t}([-\tau, 0]; R^d)$,

$$E|\psi(\phi(0),t)|^p + E|F(\phi,t)|^p + E|g(\phi,t)|^p \leq K \sup_{-\tau \leq \theta \leq 0} E|\phi(\theta)|^p,$$

then the trivial solution of equation (6.43) is also almost surely exponentially stable.

Proof. Define $f(\varphi,t) = \psi(\varphi(0),t) + F(\varphi,t)$ so that equation (6.43) becomes equation (6.1). Moreover

$$\mathcal{L}V(\varphi,t) = V_t(\varphi(0),t) + V_x(\varphi(0),t)[\psi(\varphi(0),t) + F(\varphi,t)]$$
$$+ \frac{1}{2}\text{trace}\big[g^T(\varphi,t)V_{xx}(\varphi(0),t)g(\varphi,t)\big].$$

Hence for $t \geq t_0$ and those $\phi \in L^p_{\mathcal{F}_t}([-\tau,0]; R^d)$ satisfying (6.45) one can derive from the assumptions that

$$E\mathcal{L}V(\phi,t) \leq -\lambda EV(\phi(0),t) + \beta_1 E\Big([V(\phi(0),t)]^{\frac{p-1}{p}}|F(\phi,t)|\Big)$$
$$+ \frac{\beta_2}{2} E\Big([V(\phi(0),t)]^{\frac{p-2}{p}}|g(\phi,t)|^2\Big). \qquad (6.47)$$

But for any $\varepsilon > 0$

$$E\Big([V(\phi(0),t)]^{\frac{p-1}{p}}|F(\phi,t)|\Big) = E\left[\big(\varepsilon V(\phi(0),t)\big)^{\frac{p-1}{p}}\left(\frac{|F(\phi,t)|^p}{\varepsilon^{p-1}}\right)^{\frac{1}{p}}\right]$$
$$\leq \frac{\varepsilon(p-1)}{p} EV(\phi(0),t) + \frac{1}{p\varepsilon^{p-1}} E|F(\phi,t)|^p$$
$$\leq \Big(\frac{\varepsilon(p-1)}{p} + \frac{\beta_3}{p\varepsilon^{p-1}}\Big) EV(\phi(0),t),$$

where the elementary inequality (6.35) has been used once again. In particular, if choose $\varepsilon = \beta_3^{1/p}$, then

$$E\Big([V(\phi(0),t)]^{\frac{p-1}{p}}|F(\phi,t)|\Big) \leq \beta_3^{\frac{1}{p}} EV(\phi(0),t).$$

Similarly, one can show

$$E\Big([V(\phi(0),t)]^{\frac{p-2}{p}}|g(\phi,t)|^2\Big) \leq \beta_4^{\frac{2}{p}} EV(\phi(0),t).$$

Substituting these into (6.47) yields

$$E\mathcal{L}V(\phi,t) \leq -\left(\lambda - \beta_1\beta_3^{\frac{1}{p}} - \frac{1}{2}\beta_2\beta_4^{\frac{2}{p}}\right)EV(\phi(0),t).$$

Now the conclusions follow from Theorems 6.1 and 6.2 immediately. The proof is complete.

Corollary 6.8 *Assume that there is a $\lambda > 0$ such that*

$$x^T\psi(x,t) \leq -\lambda|x|^2 \quad \text{for all } (x,t) \in R^d \times [t_0,\infty).$$

Assume also that there are two functions $\alpha_1(\cdot), \alpha_2(\cdot) \in C([-\tau,0]; R_+)$ such that

$$|F(\varphi,t)| \leq \int_{-\tau}^{0} \alpha_1(\theta)|\varphi(\theta)|d\theta \quad \text{and} \quad |g(\varphi,t)|^2 \leq \int_{-\tau}^{0} \alpha_2(\theta)|\varphi(\theta)|^2 d\theta$$

for all $t \geq t_0$ and $\varphi \in C([-\tau,0]; R^d)$. If $p \geq 2$ and

$$\lambda > (\tau\bar{\alpha}_1)^{\frac{1}{p}} + \frac{p-1}{2}(\tau\bar{\alpha}_2)^{\frac{2}{p}}, \tag{6.48}$$

where

$$\bar{\alpha}_1 = \left(\int_{-\tau}^{0} |\alpha_1(\theta)|^{\frac{p}{p-1}} d\theta\right)^{p-1},$$

$$\bar{\alpha}_2 = \begin{cases} \max_{-\tau \leq \theta \leq 0} \alpha_2(\theta), & \text{if } p = 2, \\ \left(\int_{-\tau}^{0} |\alpha_2(\theta)|^{\frac{p}{p-2}} d\theta\right)^{\frac{p-2}{2}}, & \text{if } p > 2, \end{cases}$$

then the trivial solution of equation (6.43) is pth moment exponentially stable. In addition, if there is a $K > 0$ such that $|\psi(x,t)| \leq K|x|$ for all $(x,t) \in R^d \times [t_0,\infty)$, then the trivial solution of equation (6.43) is also almost surely exponentially stable.

Proof. Let $V(x,t) = |x|^p$. Then

$$V_t(x,t) + V_x(x,t)\psi(x,t) \leq -p\lambda|x|^p,$$
$$|V_x(x,t)| \leq p|x|^{p-1}, \quad \|V_{xx}(x,t)\| \leq p(p-1)|x|^{p-2}$$

for all $(x,t) \in R^d \times [t_0,\infty)$. By (6.48) one can choose $q > 1$ such such that

$$\lambda > (q\tau\bar{\alpha}_1)^{\frac{1}{p}} + \frac{p-1}{2}(q\tau\bar{\alpha}_2)^{\frac{2}{p}}. \tag{6.49}$$

Now for $t \geq t_0$ and $\phi \in L^p_{\mathcal{F}_t}([-\tau,0]; R^d)$ satisfying

$$E|\phi(\theta)|^p < qE|\phi(0)|^p \quad \text{for all } -\tau \leq \theta \leq 0,$$

one can easily show that

$$E|F(\phi,t)|^p \leq q\tau\bar{a}_1 E|\phi(0)|^p$$

and

$$E|g(\phi,t)|^p \leq q\tau\bar{a}_2 E|\phi(0)|^p.$$

So the conclusions follow from Theorem 6.7 and the proof is complete.

(iii) *Examples*

In the following two examples we shall omit mentioning the initial data which are always assumed to be in $C^b_{\mathcal{F}_{t_0}}([-\tau,0]; R^d)$ anyway.

Example 6.9 Consider a linear stochastic differential delay equation

$$dx(t) = -[Ax(t) + Bx(t-\delta(t))]dt + Cx(t-\delta(t))dB(t) \quad \text{on } t \geq t_0, \quad (6.50)$$

where A, B, C all are $d \times d$ constant matrices, $B(t)$ is a one-dimensional Brownian motion and $\delta : [t_0, \infty) \to [-\tau, 0]$ is continuous.

Case (i). Assume that $A + A^T$ is positive definite and its smallest eigenvalue is denoted by $\lambda_{\min}(A + A^T)$. In this case, one can easily conclude by Corollary 6.5 that if $p \geq 2$ and

$$\frac{1}{2}\lambda_{\min}(A + A^T) > \|B\| + \frac{p-1}{2}\|C\|^2, \quad (6.51)$$

then the trivial solution of equation (6.50) is both pth moment and almost surely exponentially stable.

Case (ii). Assume that $A + A^T + B + B^T$ is positive definite. To apply Corollary 6.5, write equation (6.50) as

$$dx(t) = -[(A+B)x(t) + Bx(t-\delta(t)) - Bx(t-\delta_2(t))]dt + Cx(t-\delta(t))dB(t) \quad (6.52)$$

with $\delta_2(t) \equiv 0$. By Corollary 6.5, one then sees that if $p \geq 2$ and

$$\frac{1}{2}\lambda_{\min}(A + A^T + B + B^T) > 2\|B\| + \frac{p-1}{2}\|C\|^2, \quad (6.53)$$

then the trivial solution of equation (6.52), i.e. (6.50) is pth moment as well as almost surely exponentially stable. Of course, in this case one may also apply Corollary 6.6 to obtain a delay-dependent result. For simplicity, choose $p = 2$. Note that for any $\rho > 0$

$$|Ax + By - A\bar{x} - B\bar{y}|^2 \leq (1+\rho^{-1})\|A\|^2|x-\bar{x}|^2 + (1+\rho)\|B\|^2|y-\bar{y}|^2.$$

One can then apply Corollary 6.6 (with $p = 2$) to conclude that if

$$\frac{1}{2}\lambda_{\min}(A + A^T + B + B^T)$$
$$> \frac{1}{2}\|C\|^2 + \inf_{\rho>0}\left\{\|B\|\left[2(1+\rho)\right.\right.$$
$$\left.\left.\times \left(\tau^2[(1+\rho^{-1})\|A\|^2 + (1+\rho)\|B\|^2] + \tau\|C\|^2\right)\right]^{\frac{1}{2}}\right\}, \quad (6.54)$$

then the trivial solution of equation (5.1) is 2nd moment as well as almost surely exponentially stable.

As a special case, let us look at a one-dimensional linear delay equation

$$dx(t) = -bx(t - \delta(t))dt + cx(t - \delta(t))dB(t) \quad (6.55)$$

with $b > c^2/2$. In this case, criteria (6.51) and (6.53) do not work but (6.54) reduces to

$$b > \frac{c^2}{2} + b\sqrt{2(\tau^2 b^2 + \tau c^2)}.$$

Hence, if

$$\tau < \frac{1}{2b^2}\left(\sqrt{c^4 + \frac{1}{2}(2b - c^2)^2} - c^2\right)$$

then the trivial solution of equation (6.55) is both 2nd moment and almost surely exponentially stable.

Example 6.10 Consider a stochastic oscillator described by a semi-linear stochastic functional differential equation

$$\ddot{z}(t) + 3\dot{z}(t) + 2z(t) = \sigma_1(z_t, \dot{z}_t) + \sigma_2(z_t, \dot{z}_t)\dot{B}(t) \quad (6.56)$$

on $t \geq t_0$, where $\dot{B}(t)$ is a one-dimensional white noise, i.e. $B(t)$ a Brownian motion, both $\sigma_1, \sigma_2 : C([-\tau, 0]; R^2) \to R$ are locally Lipschitz continuous and, moreover,

$$|\sigma_1(\varphi)| \vee |\sigma_2(\varphi)| \leq \int_{-\tau}^0 |\varphi(\theta)|d\theta, \qquad \varphi \in C([-\tau, 0]; R^2).$$

We claim that if

$$\tau < \frac{\sqrt{42} - \sqrt{14}}{14} \quad (6.57)$$

then the trivial solution of equation (6.56) is 2nd moment and almost exponentially stable. To show this, introduce a new variable $x = (z, \dot{z})^T$ and write equation (6.56) as a two-dimensional stochastic functional differential equation

$$dx(t) = [Ax(t) + F(x_t)]dt + G(x_t)dB(t), \quad (6.58)$$

where

$$A = \begin{pmatrix} 0 & 1 \\ -2 & -3 \end{pmatrix}, \quad F(\varphi) = \begin{pmatrix} 0 \\ \sigma_1(\varphi) \end{pmatrix}, \quad G(\varphi) = \begin{pmatrix} 0 \\ \sigma_2(\varphi) \end{pmatrix}.$$

It is easy to find

$$H = \begin{pmatrix} 1 & 1 \\ -1 & -2 \end{pmatrix}, \quad \text{and hence} \quad H^{-1} = \begin{pmatrix} 2 & 1 \\ -1 & -1 \end{pmatrix},$$

such that

$$H^{-1}AH = \begin{pmatrix} -1 & 0 \\ 0 & -2 \end{pmatrix}.$$

Set

$$Q = (H^{-1})^T H^{-1} = \begin{pmatrix} 5 & 3 \\ 3 & 2 \end{pmatrix}$$

and define $V(x,t) = x^T Q x$ for $x \in R^2$. It is easy to verify

$$\frac{1}{7}|x|^2 \le V(x) \le 7|x|^2.$$

We further compute

$$\mathcal{L}V(\varphi,t) = 2\varphi^T(0)Q[A\varphi(0) + F(\varphi)] + G^T(\varphi)QG(\varphi)$$
$$\le -2V(\varphi(0)) + 2|\varphi^T(0)(H^{-1})^T| \, |H^{-1}F(\varphi)| + 2|\sigma_2(\varphi)|^2$$
$$\le -2V(\varphi(0)) + \sqrt{14\tau}V(\varphi(0)) + \frac{2}{\sqrt{14\tau}}|\sigma_1(\varphi)|^2 + 2|\sigma_2(\varphi)|^2$$
$$\le -(2 - \sqrt{14\tau})V(\varphi(0)) + (\sqrt{14} + 14\tau)\int_{-\tau}^{0} V(\varphi(\theta))d\theta. \tag{6.59}$$

By condition (6.57) one can find $q > 1$ such that

$$2 - \sqrt{14}(1+q)\tau - 14q\tau^2 > 0.$$

Therefore, for any $\phi \in L^2_{\mathcal{F}_t}([-\tau,0]; R^d)$ satisfying $EV(\phi(\theta)) < qEV(\phi(0))$ on $-\tau \le \theta \le 0$, (6.59) yields

$$E\mathcal{L}V(\phi,t) \le -(2 - \sqrt{14}(1+q)\tau - 14q\tau^2)EV(\phi(0)).$$

Thus the conclusions follow from Theorems 6.1 and 6.2.

5.7 STOCHASTIC SELF-STABILIZATION

We consider the following problem of stochastic self-stabilization in this section. Suppose we are given a nonlinear Itô equation in R^d, namely

$$dx(t) = f(x(t),t)dt + ug(x(t),t)dB(t) \tag{7.1}$$

on $t \geq t_0 = 0$ with initial value $x(0) = x_0 \in R^d$ (it is just for convenience to set $t_0 = 0$ and the theory clearly works for general $t_0 \geq 0$). Here $u > 0$ is the noise intensity parameter, $B(t)$ is an m-dimensional Brownian motion, both $f : R^d \times R_+ \to R^d$ and $g : R^d \times R_+ \to R^{d \times m}$ are locally Lipschitz continuous. We impose the standing hypothesis:

(H7.1) There exists a symmetric positive-definite $d \times d$-matrix Q, and three positive constants K, α, β with $2\beta > \alpha$, such that

$$|x^T Q f(x,t)| \leq K|x|^2,$$
$$\text{trace}(g^T(x,t)Qg(x,t)) \leq \alpha x^T Q x,$$
$$|x^T Q g(x,t)|^2 \geq \beta |x^T Q x|^2$$

for all $t \geq 0$ and $x \in R^d$.

By Theorem 4.3.3, we know that this hypothesis guarantees that for all sufficiently large u the trivial solution of equation (7.1) is almost surely exponentially stable. Equation (7.1) is regarded as a stochastically stabilized system of an ordinary differential equation $\dot{x}(t) = f(x(t),t)$, which is unstable in general. In other words, equation (7.1) is stabilized by white noise provided that the noise intensity is large enough. We ask whether it is true that equation (7.1) stabilizes itself if the intensity parameter u is replaced by e.g. $\int_0^t |x(s)| ds$. That is, is the trivial solution of the stochastic integrodifferential equation

$$dx(t) = f(x(t),t)dt + \left(\int_0^t |x(s)| ds\right) g(x(t),t) dB(t) \tag{7.3}$$

almost surely $L^1(R_+; R^d)$-stable (i.e. $\int_0^\infty |x(t)| dt < \infty$ a.s.)? The main aim of this section is to give a positive answer. A rough argument goes as follows. If equation (7.3) is not almost surely $L^1(R_+; R^d)$-stable, then for some $\omega \in \Omega$ (with positive probability) $\int_0^\infty |x(t,\omega)| dt = \infty$. Consequently for all large t, $\int_0^t |x(s,\omega)| ds$ will be sufficiently large. Therefore, by the property of equation (7.1), one should have $\int_0^\infty |x(t,\omega)| dt < \infty$, and this yields a contradiction.

This argument is, of course, not a mathematical proof, but indicates that it is possible to replace the noise intensity parameter u in various ways in order to stabilize the system more precisely. One may replace u by $\int_0^t |r(s)x(s)|^p ds$, where $p > 0$ and $r(\cdot)$ is a continuous $R^{n \times d}$-valued function defined on R_+ satisfying $||r(t)|| \leq M e^{\gamma t}$ for all $t \geq 0$, which shall be called a *convergence rate function*. In this section we shall show that the stochastic integrodifferential equation

$$dx(t) = f(x(t),t)dt + \left(\int_0^t |r(s)x(s)|^p ds\right) g(x(t),t) dB(t) \tag{7.4}$$

has the property

$$\int_0^\infty |r(t)x(t)|^p dt < \infty \quad \text{a.s.} \tag{7.5}$$

Before proving this result, let us point out that by choosing various convergence rate functions, one can stabilize the system in different ways. For example, in order to stabilize the i-th component of the solution in the sense $\int_0^\infty |x_i(t)|^p dt < \infty$ a.s., one can choose the convergence rate function

$$r(t) = (0, \underbrace{\cdots, 0, 1}_{i \text{ times}}, 0, \cdots, 0)_{1 \times d}.$$

To stabilize the difference between the i-th and j-th component of the solution, in the sense $\int_0^\infty |x_i(t) - x_j(t)|^p dt < \infty$ a.s., one can choose

$$r(t) = (0, \underbrace{\cdots, 0, 1}_{i \text{ times}}, 0, \underbrace{\cdots, 0, -1}_{j \text{ times}}, 0, \cdots, 0)_{1 \times d}.$$

Moreover, to stabilize the system in the sense $\int_0^\infty e^{\gamma t} |x(t)|^p dt < \infty$ a.s., one can choose $r(t) = e^{\gamma t/p} I_{d \times d}$ where $I_{d \times d}$ is the $d \times d$ identity matrix.

Let us now turn to prove property (7.5). We first point out that the local Lipschitz continuity of the coefficients f and g as well as the standing hypothesis (H7.1) guarantee the existence and uniqueness of the global solution, denoted by $x(t; x_0)$, of equation (7.4) (cf. Theorem 2.7 and the detailed proof can be found in Mao, 1996c). It is also clear that equation (7.4) admits a trivial solution $x(t; 0) = 0$ for hypothesis (H7.1) implies that $f(0, t) \equiv 0$ and $g(0, t) \equiv 0$. To prove the main result, we need prepare a lemma which shows that under hypothesis (H7.1) the solution will never reach zero if it starts from a non-zero point.

Lemma 7.1 *Let hypothesis (H7.1) hold. Then the solution of equation (7.4) has the property that*

$$P\{x(t; x_0) \neq 0 \text{ for all } t \geq 0\} = 1$$

provided $x_0 \neq 0$.

Proof. Suppose the assertion is false. Then there exists some $x_0 \neq 0$ such that $P(\tau < \infty) > 0$, where τ is the time of first reaching state zero, i.e.

$$\tau = \inf\{t \geq 0 : x(t) = 0\}.$$

Here we write $x(t; x_0) = x(t)$. Hence one can find $\bar{t} > 0$ and $\theta > 0$ large enough to ensure that $P(B) > 0$, where

$$B = \{\omega : \tau \leq \bar{t} \text{ and } |x(t)| \leq \theta - 1 \text{ for all } 0 \leq t \leq \tau\}.$$

For each $0 < \varepsilon < |x_0|$, define

$$\tau_\varepsilon = \inf\{t \geq 0 : |x(t)| \leq \varepsilon \text{ or } |x(t)| \geq \theta\}.$$

Then by Itô's formula, for $0 \leq t \leq \bar{t}$,

$$E[|x^T(t \wedge \tau_\varepsilon)Qx(t \wedge \tau_\varepsilon)|^{-1}]$$

$$\leq |x_0^T Q x_0|^{-1} + 2E \int_0^{t \wedge \tau_\varepsilon} |x^T(s)Qx(s)|^{-2}|x^T(s)Qf(x(s),s)|ds$$

$$+ 4E \int_0^{t \wedge \tau_\varepsilon} |x^T(s)Qx(s)|^{-3}|x^T(s)Qg(x(s),s)|^2 \left(\int_0^s |r(u)x(u)|^p du\right)^2 ds.$$

By (H7.1), one can derive that

$$E[|x^T(t \wedge \tau_\varepsilon)Qx(t \wedge \tau_\varepsilon)|^{-1}]$$

$$\leq |x_0^T Q x_0|^{-1} + \mu E \int_0^{t \wedge \tau_\varepsilon} |x^T(s)Qx(s)|^{-1} ds$$

$$\leq |x_0^T Q x_0|^{-1} + \mu \int_0^t E[|x^T(s \wedge \tau_\varepsilon)Qx(s \wedge \tau_\varepsilon)|^{-1}] ds$$

where μ is a constant dependent of $K, \alpha, \beta, \bar{t}, \theta, Q$ but independent of ε. An application of the Gronwall inequality yields

$$E[|x^T(\bar{t} \wedge \tau_\varepsilon)Qx(\bar{t} \wedge \tau_\varepsilon)|^{-1}] \leq |x_0^T Q x_0|^{-1} e^{\mu \bar{t}}.$$

Note that if $\omega \in B$, then $\tau_\varepsilon \leq \bar{t}$ and $|x(\tau_\varepsilon)| = \varepsilon$. It therefore follows from the above inequality that

$$(\varepsilon^2 ||Q||)^{-1} P(B) \leq |x_0^T Q x_0|^{-1} e^{\mu \bar{t}}.$$

Letting $\varepsilon \to 0$ one obtains $P(B) = 0$, but this contradicts the definition of B. The proof is therefore complete.

We can now establish the main results of this section. To make the statement more clear, we state the condition on the convergence rate function $r(t)$ as another hypothesis:

(H7.2) There exists a pair of constants $M > 0$ and $\gamma \geq 0$ such that

$$||r(t)|| \leq M e^{\gamma t} \quad \text{for all } t \geq 0.$$

Theorem 7.2 Let (H7.1) and (H7.2) hold. Then the solution of equation (7.4) has the property that

$$\int_0^\infty |r(t)x(t;x_0)|^p dt < \infty \quad a.s. \tag{7.5}$$

for all $x_0 \in R^d$.

Proof. Since hypothesis (H7.1) guarantees $x(t;0) \equiv 0$, one only needs to show that (7.5) holds for $x_0 \neq 0$. For any $x_0 \neq 0$, by Lemma 7.1, the solution

$x(t; x_0) \neq 0$ for all $t \geq 0$ almost surely. Suppose (7.5) is false, then there exists some $x_0 \neq 0$ for which $P(\Omega^*) > 0$, where

$$\Omega^* = \{\omega \in \Omega : \int_0^\infty |r(t)x(t;x_0)|^p dt = \infty\}.$$

For convenience, again write $x(t;x_0) = x(t)$. By Itô's formula and hypothesis (H7.1), one can show that for any $t \geq 0$

$$\log(x^T(t)Qx(t)) \leq \log(x_0^T Q x_0) + \frac{2Kt}{\lambda_{\min}(Q)} + \alpha \int_0^t \left(\int_0^s |r(v)x(v)|^p dv\right)^2 ds$$
$$- 2\int_0^t \left(\int_0^s |r(v)x(v)|^p dv\right)^2 \frac{|x^T(s)Qg(x(s),s)|^2}{(x^T(s)Qx(s))^2} ds + M(t), \quad (7.6)$$

where

$$M(t) = 2\int_0^t \left(\int_0^s |r(v)x(v)|^p dv\right) \frac{x^T(s)Qg(x(s),s)}{x^T(s)Qx(s)} dB(s)$$

is a continuous martingale vanishing at $t = 0$. Let $k = 1, 2, \cdots$. Then by the exponential martingale inequality (i.e. Theorem 1.7.4)

$$P\left(\omega : \sup_{0 \leq t \leq k}\left[M(t) - \frac{2\beta - \alpha}{8\beta}\langle M(t), M(t)\rangle\right] > \frac{8\beta \log k}{2\beta - \alpha}\right) \leq \frac{1}{k^2},$$

where

$$\langle M(t), M(t)\rangle = 4\int_0^t \left(\int_0^s |r(v)x(v)|^p dv\right)^2 \frac{|x^T(s)Qg(x(s),s)|^2}{(x^T(s)Qx(s))^2} ds.$$

Hence the well-known Borel–Cantelli lemma yields that for almost all $\omega \in \Omega$ there exists a random integer $k_1(\omega)$ such that for all $k \geq k_1$

$$\sup_{0 \leq t \leq k}\left[M(t) - \frac{2\beta - \alpha}{8\beta}\langle M(t), M(t)\rangle\right] \leq \frac{8\beta \log k}{2\beta - \alpha},$$

that is, for $0 \leq t \leq k$,

$$M(t) \leq \frac{8\beta \log k}{2\beta - \alpha} + \frac{2\beta - \alpha}{8\beta}\langle M(t), M(t)\rangle$$
$$\leq \frac{8\beta \log k}{2\beta - \alpha} + \frac{2\beta - \alpha}{2\beta}\int_0^t \left(\int_0^s |r(v)x(v)|^p dv\right)^2 \frac{|x^T(s)Qg(x(s),s)|^2}{(x^T(s)Qx(s))^2} ds. \quad (7.7)$$

Substituting (7.7) into (7.6) and then applying (H7.1) one obtains that

$$\log(x^T(t)Qx(t)) \leq \log(x_0^T Q x_0) + \frac{2Kt}{\lambda_{\min}(Q)} + \frac{8\beta \log k}{2\beta - \alpha}$$
$$- \frac{2\beta - \alpha}{2}\int_0^t \left(\int_0^s |r(v)x(v)|^p dv\right)^2 ds \quad (7.8)$$

for all $0 \leq t \leq k$, $k \geq k_1$ almost surely. Recalling the definition of Ω^*, one sees that for every $\omega \in \Omega^*$, there exists a random integer $k_2(\omega)$ such that

$$\int_0^t |r(s)x(s)|^p ds \geq \sqrt{\frac{4K/\lambda_{\min}(Q) + 4\gamma + 8}{2\beta - \alpha}} \qquad \text{for all } t \geq k_2. \tag{7.9}$$

It then follows from (7.8) and (7.9) that for almost all $\omega \in \Omega^*$, if $k - 1 \leq t \leq k$, $k \geq k_1 \vee (k_2 + 1)$,

$$\log(x^T(t)Qx(t))$$
$$\leq \log(x_0^T Q x_0) + \frac{2Kk}{\lambda_{\min}(Q)} + \frac{8\beta \log k}{2\beta - \alpha} - \frac{2\beta - \alpha}{2} \int_{k_2}^t \left(\int_0^s |r(v)x(v)|^p dv \right)^2 ds$$
$$\leq \log(x_0^T Q x_0) + \frac{2Kk}{\lambda_{\min}(Q)} + \frac{8\beta \log k}{2\beta - \alpha} - (2K/\lambda_{\min}(Q) + 2\gamma + 4)(k - 1 - k_2)$$
$$= \log(x_0^T Q x_0) + \frac{2K(k_2 + 1)}{\lambda_{\min}(Q)} + \frac{8\beta \log k}{2\beta - \alpha} - 2(\gamma + 2)(k - 1 - k_2).$$

Consequently,

$$\frac{1}{t} \log(x^T(t)Qx(t))$$
$$\leq \frac{1}{k-1} \left(\log(x_0^T Q x_0) + \frac{2K(k_2 + 1)}{\lambda_{\min}(Q)} + \frac{8\beta \log k}{2\beta - \alpha} - 2(\gamma + 2)(k - 1 - k_2) \right).$$

It then follows that

$$\limsup_{t \to \infty} \frac{1}{t} \log(x^T(t)Qx(t)) \leq -2(\gamma + 2) \qquad \text{for almost all } \omega \in \Omega^*. \tag{7.10}$$

Thus, for almost all $\omega \in \Omega^*$, there exists a random number $k_3(\omega)$ such that

$$\frac{1}{t} \log(x^T(t)Qx(t)) \leq -2(\gamma + 2) \qquad \text{for all } t \geq k_3,$$

and hence

$$|x(t)| \leq \frac{e^{-(\gamma+2)t}}{\sqrt{\lambda_{\min}(Q)}} \qquad \text{for all } t \geq k_3.$$

Therefore, by hypothesis (H7.2), for almost all $\omega \in \Omega^*$

$$\int_0^\infty |r(t)x(t)|^p dt \leq \int_0^{k_3} M^p e^{p\gamma t} |x(t)|^p dt + \int_{k_3}^\infty \frac{M^p e^{-pt}}{[\lambda_{\min}(Q)]^{p/2}} dt < \infty.$$

However, this contradicts the definition of Ω^*. So (7.5) must hold, and the proof is complete.

The following Theorem gives more precise estimates for the solution.

Theorem 7.3 *Let (H7.1) and (H7.2) hold. Then for every $x_0 \in R^d$, either*

$$\int_0^\infty |r(t)x(t;x_0)|^p dt \leq \sqrt{\frac{2K}{(2\beta-\alpha)\lambda_{\min}(Q)}} \qquad (7.11)$$

or

$$\limsup_{t\to\infty} \frac{1}{t} \log(|x(t;x_0)|) < 0 \qquad (7.12)$$

holds for almost all $\omega \in \Omega$.

Proof. Again one only needs to show the conclusions for all $x_0 \neq 0$. Fix $x_0 \neq 0$ arbitrarily, and write $x(t;x_0) = x(t)$. Define

$$\bar{\Omega} = \left\{ \omega \in \Omega : \int_0^\infty |r(t)x(t)|^p dt > \sqrt{\frac{2K}{(2\beta-\alpha)\lambda_{\min}(Q)}} \right\}.$$

Clearly one only needs to show that (7.12) holds for almost all $\omega \in \bar{\Omega}$. For each $i = 1, 2, \cdots$, define

$$\bar{\Omega}_i = \left\{ \omega \in \bar{\Omega} : \int_0^\infty |r(t)x(t)|^p dt > (1+i^{-1})\sqrt{\frac{2K}{(2\beta-\alpha)\lambda_{\min}(Q)}} \right\}.$$

Now, $\bar{\Omega} = \bigcup_{i=1}^\infty \bar{\Omega}_i$, and hence one only needs to show that for each $i \geq 1$, (7.12) holds for almost all $\omega \in \bar{\Omega}_i$. Fix any $i \geq 1$. Then in the same way as (7.8) one can derive that for each $\omega \in \Omega - \hat{\Omega}$, with $\hat{\Omega}$ a P-null set, there exists a random integer $k_4(\omega)$ such that

$$\log(x^T(t)Qx(t)) \leq \log(x_0^T Qx_0) + \frac{2Kt}{\lambda_{\min}(Q)} + \frac{4\beta(1+i^{-1})\log k}{2\beta-\alpha}$$
$$- \frac{2\beta-\alpha}{1+i^{-1}} \int_0^t \left(\int_0^s |r(v)x(v)|^p dv \right)^2 ds \qquad (7.13)$$

for all $0 \leq t \leq k$, $k \geq k_4$. On the other hand, for every $\omega \in \bar{\Omega}_i$ there exists a random number $k_5(\omega)$ such that

$$\int_0^t |r(s)x(s)|^p ds \geq (1+i^{-1})\sqrt{\frac{2K}{(2\beta-\alpha)\lambda_{\min}(Q)}} \quad \text{for all } t \geq k_5. \qquad (7.14)$$

It then follows from (7.13) and (7.14) that for all $\omega \in \bar{\Omega}_i - \hat{\Omega}$, if $k-1 \leq t \leq k$, $k \geq k_4 \vee (k_5+1)$,

$$\log(x^T(t)Qx(t)) \leq \log(x_0^T Qx_0) + \frac{2K(k_5+1)}{\lambda_{\min}(Q)}$$
$$+ \frac{4\beta(1+i^{-1})\log k}{2\beta-\alpha} - \frac{2K}{i\lambda_{\min}(Q)}(k-1-k_5).$$

This implies that

$$\limsup_{t\to\infty} \frac{1}{t} \log(x^T(t)Qx(t)) \leq -\frac{2K}{i\lambda_{\min}(Q)} \quad \text{for all } \omega \in \bar{\Omega}_i - \hat{\Omega}.$$

Consequently,

$$\limsup_{t\to\infty} \frac{1}{t} \log(|x(t)|) \leq -\frac{K}{i\lambda_{\min}(Q)} < 0 \quad \text{for all } \omega \in \bar{\Omega}_i - \hat{\Omega}.$$

The proof is now complete.

Theorem 7.3 shows that if

$$\int_0^\infty |r(t)x(t;x_0)|^p dt > \sqrt{\frac{2K}{(2\beta-\alpha)\lambda_{\min}(Q)}}$$

then the solution $x(t;x_0)$ tends to zero exponentially. This is quite natural since one can show by Theorem 4.3.3 that under hypothesis (H7.1) equation (7.1) is almost surely exponentially stable provided the noise intensity parameter

$$u > \sqrt{\frac{2K}{(2\beta-\alpha)\lambda_{\min}(Q)}}.$$

On the other hand, if

$$\int_0^\infty |r(t)x(t;x_0)|^p dt \leq \sqrt{\frac{2K}{(2\beta-\alpha)\lambda_{\min}(Q)}}$$

then the noise intensity may not be large enough to stabilize the system exponentially. In other words, if the noise intensity parameter u is replaced by $\int_0^t |r(s)x(s)|^p ds$, then equation (7.1) may not always stabilize itself exponentially. But we now start to discuss how to stabilize equation (7.1) in the sense of almost sure asymptotic stability or exponential stability. By replacing the noise intensity parameter u with $\sup_{0\leq s\leq t} |r(s)x(s)|$, equation (7.1) becomes a stochastic functional differential equation

$$dx(t) = f(x(t),t)dt + \left(\sup_{0\leq s\leq t} |r(s)x(s)|\right) g(x(t),t)dB(t) \tag{7.15}$$

on $t \geq 0$ with initial value $x(0) = x_0 \in R^d$. It can be shown in the same way as Lemma 7.1 that under hypothesis (H7.1) the unique global solution of equation (7.15), again denoted by $x(t;x_0)$, will never reach zero if it starts from a non-zero point. Moreover, we have the following result.

Theorem 7.4 *Let (H7.1) and (H7.2) hold. Then for all $x_0 \in R^d$, the solution of equation (7.15) has the property*

$$\sup_{0\leq t<\infty} |r(t)x(t;x_0)| < \infty \quad a.s. \tag{7.16}$$

Furthermore,

(i) *if* $\lambda_{\min}(r^T(t)r(t)) \to \infty$ *as* $t \to \infty$, *then*

$$\lim_{t \to \infty} |x(t; x_0)| = 0 \quad a.s. \tag{7.17}$$

(ii) *if* $\liminf_{t \to \infty} \log[\lambda_{\min}(r^T(t)r(t))]/t \geq \lambda > 0$, *then*

$$\limsup_{t \to \infty} \frac{1}{t} \log(|x(t; x_0)|) \leq -\frac{\lambda}{2} \quad a.s. \tag{7.18}$$

Proof. Suppose (7.16) is false, then there exists some $x_0 \neq 0$ for which $P(\Omega^*) > 0$, where

$$\Omega^* = \{\omega \in \Omega : \sup_{0 \leq t < \infty} |r(t)x(t; x_0)| = \infty\}.$$

Again write $x(t; x_0) = x(t)$. In the same way as the proof of (7.8), one can show that there exists a finite random integer $k_6(\omega)$ such that

$$\log(x^T(t)Qx(t)) \leq \log(x_0^T Q x_0) + \frac{2Kt}{\lambda_{\min}(Q)} + \frac{8\beta \log k}{2\beta - \alpha}$$
$$- \frac{2\beta - \alpha}{2} \int_0^t \left(\sup_{0 \leq v \leq s} |r(v)x(v)| \right)^2 ds$$

for all $0 \leq t \leq k$, $k \geq k_6$ almost surely. By the definition of Ω^*, for every $\omega \in \Omega$ there exists a random integer $k_7(\omega)$ such that

$$\sup_{0 \leq s \leq t} |r(s)x(s)| \geq \sqrt{\frac{2(2K/\lambda_{\min}(Q) + 2(\gamma + 2))}{2\beta - \alpha}} \quad \text{for all } t \geq k_7.$$

One can then derive from these two inequalities that for almost all $\omega \in \Omega^*$, there exists a random number $k_8(\omega)$ such that

$$|x(t)| \leq \frac{e^{-(\gamma+1)t}}{\sqrt{\lambda_{\min}(Q)}} \quad \text{for all } t \geq k_8.$$

Therefore, by hypothesis (H7.2), for almost all $\omega \in \Omega^*$

$$\sup_{0 \leq t < \infty} |r(t)x(t)| \leq \sup_{0 \leq t \leq k_8} Me^{\gamma t}|x(t)| + \sup_{k_8 \leq t < \infty} \frac{Me^{-t}}{\sqrt{\lambda_{\min}(Q)}} < \infty.$$

However this contradicts the definition of Ω^*. So (7.16) must hold, whence both (7.17) and (7.18) follow from (7.16) immediately. The proof is complete.

Before discussing specific examples, let us point out that it is possible to extend Theorem 7.3 to the above case to obtain even more precise estimates for the solutions, but the details are left to the reader.

Let us now illustrate our theory by discussing several examples.

Example 7.5 First, consider the one-dimensional equation

$$dx(t) = f(x(t),t)dt + \left(\int_0^t e^s |x(s)|^p ds\right) \sigma x(t) dB(t) \qquad (7.19)$$

on $t \geq 0$ with $x(0) = x_0 \in R$. Here $B(t)$ is a one-dimensional Brownian motion, $\sigma \neq 0$, $p > 0$ and $f : R \times R_+ \to R$ satisfies

$$|f(x,t)| \leq K|x| \qquad \text{for all } x \in R,\ t \geq 0 \qquad (7.20)$$

and some $K > 0$. It is easy to see that hypothesis (H7.1) holds with $Q = 1$ and $\alpha = \beta = \sigma^2$. Hence, by Theorem 7.2, the solution of equation (7.19) satisfies

$$\int_0^\infty e^t |x(t;x_0)|^p dt < \infty \qquad \text{a.s.}$$

Example 7.6 Consider the d-dimensional semi-linear equation

$$dx(t) = f(x(t),t)dt + \left(\int_0^t |x(s)|ds\right) \sum_{i=1}^m G_i x(s) dB_i(s) \qquad (7.21)$$

on $t \geq 0$ with $x(0) = x_0 \in R^d$. Here $f(x,t)$ is as defined before and G_i ($1 \leq i \leq m$) are all symmetric positive definite $d \times d$-matrices. Note that for all $x \in R^d$

$$\sum_{i=1}^m |G_i x|^2 \leq \sum_{i=1}^m \|G_i\|^2 |x|^2$$

and

$$\sum_{i=1}^m |x^T G_i x|^2 \geq \sum_{i=1}^m \lambda_{\min}^2(G_i) |x|^4.$$

Assume that

$$2 \sum_{i=1}^m \lambda_{\min}^2(G_i) > \sum_{i=1}^m \|G_i\|^2, \qquad (7.22)$$

and also that there exists a positive constant K such that

$$|x^T f(x,t)| \leq K|x|^2 \qquad \text{for all } x \in R^d,\ t \geq 0. \qquad (7.23)$$

Then hypothesis (H7.1) is satisfied with Q the identity matrix and

$$\alpha = \sum_{i=1}^m \|G_i\|^2, \qquad \beta = \sum_{i=1}^m \lambda_{\min}^2(G_i).$$

Therefore, by Theorem 7.2, equation (7.21) is almost surely $L(R_+; R^d)$-stable, that is, $\int_0^\infty |x(t;x_0)|dt < \infty$ a.s. Moreover, applying Theorem 7.3 one can obtain even more precise estimates for the solution: for every $x_0 \in R^d$, either

$$\int_0^\infty |x(t;x_0)|dt \leq \sqrt{\frac{2K}{\sum_{i=1}^m (2\lambda_{\min}^2(G_i) - ||G_i||^2)}}$$

or

$$\limsup_{t\to\infty} \frac{1}{t} \log(|x(t;x_0)|) < 0$$

holds for almost all $\omega \in \Omega$.

Example 7.7 Finally, let us consider the two-dimensional nonlinear equation

$$dx(t) = f(x(t))dt + \left(\sup_{0\leq s\leq t} e^s |x_1(s) - x_2(s)|\right) g(x(t))dB(t) \quad (7.24)$$

on $t \geq 0$ with $x(0) = x_0 \in R^2$. Here $B(t)$ is a one-dimensional Brownian motion and

$$f(x) = \begin{pmatrix} x_1 \sin x_2 - x_2^2 \\ x_2 \cos x_1 + x_1 x_2 \end{pmatrix}, \quad g(x) = \begin{pmatrix} 8x_1 + \cos x_2 \\ 9x_2 + \sin x_1 \end{pmatrix}$$

for $x = (x_1, x_2)^T \in R^2$. It is easy to verify that

$$|x^T f(x)| \leq |x|^2, \quad |g(x)|^2 \leq 91.6|x|^2, \quad |x^T g(x)|^2 \geq 54.4|x|^4.$$

So hypothesis (H7.1) is satisfied. Also hypothesis (H7.2) holds since $r(t) = e^t(1,-1)$. By Theorem 7.4, one therefore sees that the solution of equation (7.24) satisfies

$$\sup_{0\leq t<\infty} e^t |x_1(t;x_0) - x_2(t;x_0)| < \infty \quad a.s.,$$

which implies

$$\limsup_{t\to\infty} \frac{1}{t} \log|x_1(t;x_0) - x_2(t;x_0)| \leq -1 \quad a.s.$$

That is, the first and second component of the solution will tend to each other almost surely exponentially fast.

6

Stochastic Equations of Neutral Type

6.1 INTRODUCTION

In this chapter we introduce another class of stochastic equations depending on past and present values but that involve derivatives with delays as well as the function itself. Such equations historically have been referred to as *neutral stochastic functional differential equations*, or *neutral stochastic differential delay equations*. Such equations are more difficult to motivate but often arise in the study of two or more simple oscillatory systems with some interconnections between them. For example, Brayton (1976) considered the problem of lossless transmission. This problem may be described by the following system of partial differential equations

$$L\frac{\partial i}{\partial t} = -\frac{\partial v}{\partial x}, \quad C\frac{\partial v}{\partial t} = -\frac{\partial i}{\partial x}, \quad 0 < x < 1, \ t > 0,$$

with the boundary conditions

$$E - v(0,t) - Ri(0,t) = 0, \quad C_1 \frac{dv(1,t)}{dt} = i(1,t) - g(v(1,t)).$$

We now indicate how one can transform this problem into a neutral differential delay equation. If $s = (LC)^{-1/2}$ and $z = (L/C)^{1/2}$, then the general solution of the partial differential equation is given by

$$v(x,t) = \phi(x - st) + \psi(x + st), \quad i(x,t) = \frac{1}{z}[\phi(x - st) - \psi(x + st)]$$

or
$$2\phi(x-st) = v(x,t) + zi(x,t), \quad 2\psi(x+st) = v(x,t) - zi(x,t).$$

This implies
$$2\phi(-st) = v\left(1, t+\frac{1}{z}\right) + zi\left(1, t+\frac{1}{s}\right),$$
$$2\psi(st) = v\left(1, t-\frac{1}{z}\right) - zi\left(1, t-\frac{1}{s}\right).$$

Using these expressions in the general solution and using the first boundary condition at $t - 1/s$, one obtains that

$$i(1,t) - Ki\left(1, t-\frac{2}{s}\right) = \alpha - \frac{1}{z}v(1,t) - \frac{K}{z}v\left(1, t-\frac{2}{s}\right),$$

where $K = (z-R)/(z+R)$, $\alpha = 2E/(z+R)$. Inserting the second boundary condition and letting $u(t) = v(1+t)$, we obtain the equation

$$\frac{d}{dt}\left[u(t) - Ku\left(t-\frac{2}{s}\right)\right] = f\left(u(t), u\left(t-\frac{2}{s}\right)\right), \tag{1.1}$$

where
$$C_1 f(u, \bar{u}) = \alpha - \frac{1}{z}u - \frac{K}{z}\bar{u} - g(u) + Kg(\bar{u}).$$

Another similar equation encountered by Rubanik (1969) in his study of vibrating masses attached to an elastic bar is

$$\begin{cases} \ddot{x}(t) + \omega_1^2 x(t) = \varepsilon f_1(x(t), \dot{x}(t), y(t), \dot{y}(t)) + \gamma_1 \ddot{y}(t-\tau), \\ \ddot{y}(t) + \omega_2^2 x(t) = \varepsilon f_2(x(t), \dot{x}(t), y(t), \dot{y}(t)) + \gamma_2 \ddot{x}(t-\tau). \end{cases} \tag{1.2}$$

In studying the collision problem in electrodynamics, Driver (1963) considered the system of neutral type

$$\dot{x}(t) = f_1(x(t), x(\delta(t))) + f_2(x(t), x(\delta(t)))\dot{x}(\delta(t)), \tag{1.3}$$

where $\delta(t) \leq t$. Generally, a neutral functional differential equation has the form

$$\frac{d}{dt}[x(t) - D(x_t)] = f(x_t, t). \tag{1.4}$$

Taking into account stochastic perturbations, we are led to a neutral stochastic functional differential equation

$$d[x(t) - D(x_t)] = f(x_t, t)dt + g(x_t, t)dB(t). \tag{1.5}$$

In this chapter we shall discuss various properties of this stochastic equation of neutral type. However, the presentation will not be as detailed as the one for the stochastic functional differential equations of the previous chapter. We concentrate only on those proofs that are significantly different from the ones for functional equations.

6.2 NEUTRAL STOCHASTIC FUNCTIONAL DIFFERENTIAL EQUATIONS

As usual, we are still working on the given complete probability space $(\Omega, \mathcal{F}, P)$ with the filtration $\{\mathcal{F}_t\}_{t\geq 0}$ satisfying the usual conditions, and $B(t)$ is the given m-dimensional Brownian motion defined on the space. Let $\tau > 0$ and $0 \leq t_0 < T < \infty$. Let

$$D : C([-\tau, 0]; R^d) \to R^d,$$
$$f : C([-\tau, 0]; R^d) \times [t_0, T] \to R^d,$$
$$g : C([-\tau, 0]; R^d) \times [t_0, T] \to R^{d \times m}$$

all be Borel measurable. Consider the d-dimensional neutral stochastic functional differential equation

$$d[x(t) - D(x_t)] = f(x_t, t)dt + g(x_t, t)dB(t) \quad \text{on } t_0 \leq t \leq T. \quad (2.1)$$

By the definition of Itô's stochastic differential, equation (2.1) means that for every $t_0 \leq t \leq T$,

$$x(t) - D(x_t) = x(t_0) - D(x_{t_0}) + \int_{t_0}^t f(x_s, s)ds + \int_{t_0}^t g(x_s, s)dB(s). \quad (2.2)$$

For the initial-value problem of this equation, we must specify the initial data on the entire interval $[t_0 - \tau, t_0]$, and hence we impose the initial condition:

$$x_{t_0} = \xi = \{\xi(\theta) : -\tau \leq \theta \leq 0\} \in L^2_{\mathcal{F}_{t_0}}([-\tau, 0]; R^d), \quad (2.3)$$

that is, ξ is an $\mathcal{F}_{t_0}$-measurable $C([-\tau, 0]; R^d)$-valued random variable such that $E\|\xi\|^2 < \infty$. The initial-value problem for equation (2.1) is to find the solution of equation (2.1) satisfying the initial data (2.3). To be more precise, we give the definition of the solution.

Definition 2.1 An R^d-valued stochastic process $x(t)$ on $t_0 - \tau \leq t \leq T$ is called a solution to equation (2.1) with initial data (2.3) if it has the following properties:

(i) it is continuous and $\{x_t\}_{t_0 \leq t \leq T}$ is $\mathcal{F}_t$-adapted;
(ii) $\{f(x_t, t)\} \in \mathcal{L}^1([t_0, T]; R^d)$ and $\{g(x_t, t)\} \in \mathcal{L}^2([t_0, T]; R^{d \times m})$;
(iii) $x_{t_0} = \xi$ and (2.2) holds for every $t_0 \leq t \leq T$.

A solution $x(t)$ is said to be unique if any other solution $\bar{x}(t)$ is indistinguishable from it, that is

$$P\{x(t) = \bar{x}(t) \text{ for all } t_0 - \tau \leq t \leq T\} = 1.$$

Let us now begin to establish the theory of the existence and uniqueness of the solution. Obviously, the Lipschitz condition as well as the linear growth condition on the functionals f and g are required, for equation (2.1) reduces to the stochastic functional differential equation discussed in the previous chapter if $D(\cdot) \equiv 0$. The question is: What condition should be imposed on the functional D? It turns out that D should be uniformly Lipschitz continuous with the Lipschitz coefficient less than 1.

Theorem 2.2 *Assume that there exist two positive constants $\bar{K}$ and K such that for all $\varphi, \phi \in C([-\tau, 0]; R^d)$ and $t \in [t_0, T]$,*

$$|f(\varphi, t) - f(\phi, t)|^2 \bigvee |g(\varphi, t) - g(\phi, t)|^2 \leq \bar{K}\|\varphi - \phi\|^2; \quad (2.4)$$

and for all $(\varphi, t) \in C([-\tau, 0]; R^d) \times [t_0, T]$,

$$|f(\varphi, t)|^2 \bigvee |g(\varphi, t)|^2 \leq K(1 + \|\varphi\|^2). \quad (2.5)$$

Assume also that there is a $\kappa \in (0, 1)$ such that for all $\varphi, \phi \in C([-\tau, 0]; R^d)$,

$$|D(\varphi) - D(\phi)| \leq \kappa\|\varphi - \phi\|. \quad (2.6)$$

Then there exists a unique solution $x(t)$ to equation (2.1) with initial data (2.3). Moreover, the solution belongs to $\mathcal{M}^2([t_0 - \tau, T]; R^d)$.

In order to prove this theorem, let us present two useful lemmas.

Lemma 2.3 *For any $a, b \geq 0$ and $0 < \alpha < 1$ we have*

$$(a+b)^2 \leq \frac{a^2}{\alpha} + \frac{b^2}{1-\alpha}.$$

Proof. Note that for any $\varepsilon > 0$

$$(a+b)^2 = a^2 + 2ab + b^2 \leq (1+\varepsilon)a^2 + (1+\varepsilon^{-1})b^2.$$

Letting $\varepsilon = (1-\alpha)/\alpha$ we obtain the required inequality.

Lemma 2.4 *Let (2.5) and (2.6) hold. Let $x(t)$ be a solution to equation (2.1) with initial data (2.3). Then*

$$E\left(\sup_{t_0-\tau \leq t \leq T} |x(t)|^2\right) \leq \left(1 + \frac{4 + \kappa\sqrt{\kappa}}{(1-\kappa)(1-\sqrt{\kappa})} E\|\xi\|^2\right)$$

$$\times \exp\left[\frac{3K(T-t_0)(T-t_0+4)}{(1-\kappa)(1-\sqrt{\kappa})}\right]. \quad (2.7)$$

Sec.6.2] Neutral Stochastic Functional Differential Equations

In particular, $x(t)$ belongs to $\mathcal{M}^2([t_0 - \tau, T]; R^d)$.

Proof. For every integer $n \geq 1$, define the stopping time

$$\tau_n = T \wedge \inf\{t \in [t_0, T] : ||x_t|| \geq n\}.$$

Clearly, $\tau_n \uparrow T$ a.s. Set $x^n(t) = x(t \wedge \tau_n)$ for $t \in [t_0 - \tau, T]$. Then, for $t_0 \leq t \leq T$,

$$x^n(t) = D(x_t^n) - D(\xi) + J^n(t),$$

where

$$J^n(t) = \xi(0) + \int_{t_0}^t f(x_s^n, s) I_{[[t_0, \tau_n]]}(s) ds + \int_{t_0}^t g(x_s^n, s) I_{[[t_0, \tau_n]]}(s) dB(s).$$

Applying Lemma 2.3 twice one derives that

$$|x^n(t)|^2 \leq \frac{1}{\kappa}|D(x_t^n) - D(\xi)|^2 + \frac{1}{1-\kappa}|J^n(t)|^2$$

$$\leq \kappa ||x_t^n - \xi||^2 + \frac{1}{1-\kappa}|J^n(t)|^2$$

$$\leq \sqrt{\kappa}||x_t^n||^2 + \frac{\kappa}{1-\sqrt{\kappa}}||\xi||^2 + \frac{1}{1-\kappa}|J^n(t)|^2,$$

where condition (2.6) has also been used. Hence

$$E\left(\sup_{t_0 \leq s \leq t} |x^n(s)|^2\right) \leq \sqrt{\kappa} E\left(\sup_{t_0 - \tau \leq s \leq t} |x^n(s)|^2\right)$$

$$+ \frac{\kappa}{1-\sqrt{\kappa}} E||\xi||^2 + \frac{1}{1-\kappa} E\left(\sup_{t_0 \leq s \leq t} |J^n(s)|^2\right).$$

Noting that $\sup_{t_0 - \tau \leq s \leq t} |x^n(s)|^2 \leq ||\xi||^2 + \sup_{t_0 \leq s \leq t} |x^n(s)|^2$, one sees that

$$E\left(\sup_{t_0 - \tau \leq s \leq t} |x^n(s)|^2\right) \leq \sqrt{\kappa} E\left(\sup_{t_0 - \tau \leq s \leq t} |x^n(s)|^2\right)$$

$$+ \frac{1 + \kappa - \sqrt{\kappa}}{1 - \sqrt{\kappa}} E||\xi||^2 + \frac{1}{1-\kappa} E\left(\sup_{t_0 \leq s \leq t} |J^n(s)|^2\right).$$

Consequently

$$E\left(\sup_{t_0 - \tau \leq s \leq t} |x^n(s)|^2\right) \leq \frac{1 + \kappa - \sqrt{\kappa}}{(1 - \sqrt{\kappa})^2} E||\xi||^2$$

$$+ \frac{1}{(1-\kappa)(1-\sqrt{\kappa})} E\left(\sup_{t_0 \leq s \leq t} |J^n(s)|^2\right). \tag{2.8}$$

On the other hand, by Hölder's inequality, Doob's martingale inequality and the linear growth condition (2.5), one can show that

$$E\left(\sup_{t_0 \leq s \leq t} |J^n(s)|^2\right) \leq 3E\|\xi\|^2 + 3K(T - t_0 + 4)\int_{t_0}^t (1 + E\|x_s^n\|^2)ds.$$

Substituting this into (2.8) yields that

$$E\left(\sup_{t_0-\tau \leq s \leq t} |x^n(s)|^2\right) \leq \frac{4 + \kappa\sqrt{\kappa}}{(1-\kappa)(1-\sqrt{\kappa})} E\|\xi\|^2$$
$$+ \frac{3K(T - t_0 + 4)}{(1-\kappa)(1-\sqrt{\kappa})} \int_{t_0}^t (1 + E\|x_s^n\|^2)ds.$$

Therefore

$$1 + E\left(\sup_{t_0-\tau \leq s \leq t} |x^n(s)|^2\right) \leq 1 + \frac{4 + \kappa\sqrt{\kappa}}{(1-\kappa)(1-\sqrt{\kappa})} E\|\xi\|^2$$
$$+ \frac{3K(T - t_0 + 4)}{(1-\kappa)(1-\sqrt{\kappa})} \int_{t_0}^t \left[1 + E\left(\sup_{t_0-\tau \leq r \leq s} |x^n(r)|^2\right)\right]ds.$$

Now the Gronwall inequality yields that

$$1 + E\left(\sup_{t_0-\tau \leq t \leq T} |x^n(t)|^2\right)$$
$$\leq \left(1 + \frac{4 + \kappa\sqrt{\kappa}}{(1-\kappa)(1-\sqrt{\kappa})} E\|\xi\|^2\right) \exp\left[\frac{3K(T - t_0)(T - t_0 + 4)}{(1-\kappa)(1-\sqrt{\kappa})}\right].$$

Consequently

$$E\left(\sup_{t_0-\tau \leq t \leq T_n} |x(t)|^2\right)$$
$$\leq \left(1 + \frac{4 + \kappa\sqrt{\kappa}}{(1-\kappa)(1-\sqrt{\kappa})} E\|\xi\|^2\right) \exp\left[\frac{3K(T - t_0)(T - t_0 + 4)}{(1-\kappa)(1-\sqrt{\kappa})}\right].$$

Finally the required inequality (2.7) follows by letting $n \to \infty$. The proof is complete.

Proof of Theorem 2.2 <u>Uniqueness</u>. Let $x(t)$ and $\bar{x}(t)$ be the two solutions. By Lemma 2.4, both of them belong to $\mathcal{M}^2([t_0 - \tau, T]; R^d)$. Note that

$$x(t) - \bar{x}(t) = D(x_t) - D(\bar{x}_t) + J(t),$$

where

$$J(t) = \int_{t_0}^t [f(x_s, s) - f(\bar{x}_s, s)]ds + \int_{t_0}^t [g(x_s, s) - g(\bar{x}_s, s)]dB(s).$$

Neutral Stochastic Functional Differential Equations

By Lemma 2.3 and condition (2.6), one sees easily that

$$|x(t) - \bar{x}(t)|^2 \leq \kappa ||x_t - \bar{x}_t||^2 + \frac{1}{1-\kappa}|J(t)|^2.$$

Therefore

$$E\left(\sup_{t_0 \leq s \leq t} |x(s) - \bar{x}(s)|^2\right)$$

$$\leq \kappa E\left(\sup_{t_0 \leq s \leq t} |x(s) - \bar{x}(s)|^2\right) + \frac{1}{1-\kappa} E\left(\sup_{t_0 \leq s \leq t} |J(s)|^2\right),$$

which implies

$$E\left(\sup_{t_0 \leq s \leq t} |x(s) - \bar{x}(s)|^2\right) \leq \frac{1}{(1-\kappa)^2} E\left(\sup_{t_0 \leq s \leq t} |J(s)|^2\right).$$

On the other hand, one can easily show that

$$E\left(\sup_{t_0 \leq s \leq t} |J(s)|^2\right) \leq 2\bar{K}(T - t_0 + 4) \int_{t_0}^{t} E||x_s - \bar{x}_s||^2 ds$$

$$\leq 2\bar{K}(T - t_0 + 4) \int_{t_0}^{t} E\left(\sup_{t_0 \leq r \leq s} |x(r) - \bar{x}(r)|^2\right) ds.$$

Therefore

$$E\left(\sup_{t_0 \leq s \leq t} |x(s) - \bar{x}(s)|^2\right) \leq \frac{2\bar{K}(T - t_0 + 4)}{(1-\kappa)^2} \int_{t_0}^{t} E\left(\sup_{t_0 \leq r \leq s} |x(r) - \bar{x}(r)|^2\right) ds.$$

The Gronwall inequality then yields that

$$E\left(\sup_{t_0 \leq t \leq T} |x(t) - \bar{x}(t)|^2\right) = 0.$$

This implies that $x(t) = \bar{x}(t)$ for $t_0 \leq t \leq T$, hence for all $t_0 - \tau \leq t \leq T$, almost surely. The uniqueness has been proved.

Existence. We divide the whole proof of the existence into two steps:

Step 1. We impose an additional condition: $T - t_0$ is sufficiently small so that

$$\delta := \kappa + \frac{2\bar{K}(T - t_0 + 4)(T - t_0)}{1 - \kappa} < 1. \tag{2.9}$$

Define $x_{t_0}^0 = \xi$ and $x^0(t) = \xi(0)$ for $t_0 \leq t \leq T$. For each $n = 1, 2, \cdots$, set $x_{t_0}^n = \xi$ and define, by the Picard iterations,

$$x^n(t) - D(x_t^{n-1}) = \xi(0) - D(\xi)$$

$$+ \int_{t_0}^{t} f(x_s^{n-1}, s) ds + \int_{t_0}^{t} g(x_s^{n-1}, s) dB(s) \tag{2.10}$$

for $t \in [t_0, T]$. It is not difficult to show that $x^n(\cdot) \in \mathcal{M}^2([t_0 - \tau, T]; R^d)$ (the details are left to the reader). Note that for $t_0 \leq t \leq T$,

$$x^1(t) - x^0(t) = x^1(t) - \xi(0) = D(x_t^0) - D(\xi)$$
$$+ \int_{t_0}^t f(x_s^0, s)ds + \int_{t_0}^t g(x_s^0, s)dB(s).$$

In the similar way as in the proof of the uniqueness one derives that

$$E\left(\sup_{t_0 \leq t \leq T} |x^1(t) - x^0(t)|^2\right)$$
$$\leq \kappa E\left(\sup_{t_0 \leq t \leq T} ||x_t^0 - \xi||^2\right) + \frac{2K(T - t_0 + 4)}{1 - \kappa} E \int_{t_0}^T (1 + ||x_t^0||^2)dt$$
$$\leq 2\kappa E||\xi||^2 + \frac{2K(T - t_0 + 4)}{1 - \kappa}(1 + E||\xi||^2)(T - t_0) := C. \qquad (2.11)$$

Note also that for $n \geq 1$ and $t_0 \leq t \leq T$,

$$x^{n+1}(t) - x^n(t) = D(x_t^n) - D(x_t^{n-1})$$
$$+ \int_{t_0}^t [f(x_s^n, s) - f(x_s^{n-1}, s)]ds + \int_{t_0}^t [g(x_s^n, s) - g(x_s^{n-1}, s)]dB(s).$$

In the same way as in the proof of the uniqueness one derives that

$$E\left(\sup_{t_0 \leq t \leq T} |x^{n+1}(t) - x^n(t)|^2\right) \leq \kappa E\left(\sup_{t_0 \leq t \leq T} |x^n(t) - x^{n-1}(t)|^2\right)$$
$$+ \frac{2\bar{K}(T - t_0 + 4)}{1 - \kappa} \int_{t_0}^T E\left(\sup_{t_0 \leq s \leq t} |x^n(s) - x^{n-1}(s)|^2\right)dt$$
$$\leq \delta E\left(\sup_{t_0 \leq t \leq T} |x^n(t) - x^{n-1}(t)|^2\right)$$
$$\leq \delta^n E\left(\sup_{t_0 \leq t \leq T} |x^1(t) - x^0(t)|^2\right)$$
$$\leq C\delta^n, \qquad (2.12)$$

where (2.11) has been used. Using the additional condition (2.9), one can show from (2.12) that there is a solution to equation (2.1) in the same way as in the proof of Theorem 2.3.1.

Step 2. We need to remove the additional condition (2.9). Let $\sigma > 0$ be sufficiently small for

$$\kappa + \frac{2\bar{K}\sigma(\sigma + 4)}{1 - \kappa} < 1.$$

By step 1, there is a solution to equation (2.1) on $[t_0 - \tau, t_0 + \sigma]$. Now consider equation (2.1) on $[t_0 + \sigma, t_0 + 2\sigma]$ with initial data $x_{t_0+\sigma}$. By step 1 again, there is a solution to equation (2.1) on $[t_0 + \sigma, t_0 + 2\sigma]$. Repeating this procedure we see that there is a solution to equation (2.1) on the entire interval $[t_0 - \tau, T]$. The proof is complete.

As in the theory of stochastic functional differential equations, we can replace the uniform Lipschitz condition (2.4) with the local Lipschitz condition.

Theorem 2.5 *Let (2.5) and (2.6) hold, but replace condition (2.4) with the following local Lipschitz condition: For every integer $n \geq 1$, there exists a positive constant K_n such that, for all $t \in [t_0, T]$ and those $\varphi, \phi \in C([-\tau, 0]; R^d)$ with $||\varphi|| \vee ||\phi|| \leq n$,*

$$|f(\varphi, t) - f(\phi, t)|^2 \bigvee |g(\varphi, t) - g(\phi, t)|^2 \leq K_n ||\varphi - \phi||^2. \qquad (2.13)$$

Then there exists a unique solution $x(t)$ to the initial-value problem (2.1) and (2.3), and the solution belongs to $M^2([t_0 - \tau, T]; R^d)$.

This theorem can be proved by a truncation procedure but the details are left to the reader.

In what follows we often discuss the neutral stochastic functional differential equation on $[t_0, \infty)$, namely

$$d[x(t) - D(x_t)] = f(x_t, t)dt + g(x_t, t)dB(t) \qquad \text{on } t \in [t_0, \infty) \qquad (2.14)$$

with initial data (2.3), where f and g are of course now the mappings from $C([-\tau, 0]; R^d) \times [t_0, \infty)$ to R^d and $R^{d \times m}$, respectively. If the assumptions of the existence-and-uniqueness theorem hold on every finite subinterval $[t_0, T]$ of $[t_0, \infty)$, then equation (2.14) has a unique solution $x(t)$ on the entire interval $[t_0 - \tau, \infty)$. Such a solution is called a *global* solution.

6.3 NEUTRAL STOCHASTIC DIFFERENTIAL DELAY EQUATIONS

An important class of neutral stochastic functional differential equations is the neutral stochastic differential delay equations. Let us begin with the discussion of the following neutral delay equation

$$d[x(t) - \tilde{D}(x(t - \tau))] = F(x(t), x(t - \tau), t)dt + G(x(t), x(t - \tau), t)dB(t) \quad (3.1)$$

on $t \in [t_0, T]$ with initial data (2.3), where $F : R^d \times R^d \times [t_0, T] \to R^d$, $G : R^d \times R^d \times [t_0, T] \to R^{d \times m}$ and $\tilde{D} : R^d \to R^d$. If we define

$$f(\varphi, t) = F(\varphi(0), \varphi(-\tau), t), \qquad g(\varphi, t) = G(\varphi(0), \varphi(-\tau), t)$$
$$\text{and} \quad D(\varphi) = \tilde{D}(\varphi(-\tau))$$

for $\varphi \in C([-\tau, 0]; R^d)$ and $t \in [t_0, T]$, then equation (3.1) can be written as equation (2.1). Therefore, we can apply the existence-and-uniqueness theorems established in the previous section to the delay equation (3.1). For example, let F and G satisfy the local Lipschitz condition (5.3.2) and the linear growth condition (5.3.3); moreover, let $\tilde{D}$ be Lipschitz continuous with the Lipschitz coefficient less than 1, that is there is a $\kappa \in (0,1)$ such that

$$|\tilde{D}(x) - \tilde{D}(y)| \leq \kappa |x-y| \quad \text{for all } x, y \in R^d, \tag{3.2}$$

then there is a unique solution to the neutral delay equation (3.1). However, in the same spirit as explained in Section 5.3, we can do considerably better.

Theorem 3.1 *Assume that there is a $K > 0$ such that for all $x, y \in R^d \times R^d$ and $t \in [t_0, T]$,*

$$|F(x,y,t)|^2 \bigvee |G(x,y,t)|^2 \bigvee |\tilde{D}(x)|^2 \leq K(1 + |x|^2 + |y|^2). \tag{3.3}$$

Assume also that both $F(x,y,t)$ and $G(x,y,t)$ are locally Lipschitz continuous in x only, that is, for every integer $n \geq 1$, there exists a positive constant K_n such that for all $t \in [t_0, T]$, $y \in R^d$ and $x, \bar{x} \in R^d$ with $|x| \vee |\bar{x}| \leq n$,

$$|F(x,y,t) - F(\bar{x},y,t)|^2 \bigvee |G(x,y,t) - G(\bar{x},y,t)|^2 \leq K_n |x - \bar{x}|^2. \tag{3.4}$$

Then there exists a unique solution to the delay equation (3.1).

Proof. On $[t_0, t_0 + \tau]$, equation (3.1) becomes

$$x(t) = \xi(0) + D(\xi(t - t_0 - \tau)) - D(\xi(-\tau))$$
$$+ \int_{t_0}^{t} F(x(s), \xi(s - t_0 - \tau), s) ds + \int_{t_0}^{t} G(x(s), \xi(s - t_0 - \tau), s) dB(s).$$

But this is a stochastic integral equation (not neutral and without delay), and conditions (3.3)-(3.4) guarantee the existence and uniqueness of the solution on $[t_0, t_0 + \tau]$ (the reader can verify this in the same way as in the proof of Theorem 2.3.1). Proceeding this argument on $[t_0 + \tau, t_0 + 2\tau], [t_0 + 2\tau, t_0 + 3\tau]$ etc., we obtain the unique solution on the entire interval $[t_0 - \tau, T]$.

Let us proceed to discuss the equations in which the delay is time dependent. Let $\delta : [t_0, T] \to [0, \tau]$ be a Borel measurable function. Consider the neutral stochastic differential delay equation

$$d[x(t) - \tilde{D}(x(t - \delta(t)))]$$
$$= F(x(t), x(t - \delta(t)), t) dt + G(x(t), x(t - \delta(t)), t) dB(t) \tag{3.5}$$

on $t \in [t_0, T]$ with initial data (2.3). This is again a special case of equation (2.1) if define

$$f(\varphi, t) = F(\varphi(0), \varphi(-\delta(t)), t), \quad g(\varphi, t) = G(\varphi(0), \varphi(-\delta(t)), t)$$
$$\text{and} \quad D(\varphi) = \tilde{D}(\varphi(-\delta(t)))$$

for $\varphi \in C([-\tau, 0]; R^d)$ and $t \in [t_0, T]$. Hence, conditions (5.3.2), (5.3.3) and (3.2) will guarantee the existence and uniqueness of the solution to this delay equation. On the other hand, if the delay is really "true" in the sense that $\sup_{t_0 \leq t \leq T} \delta(t) > 0$, then in the same way as in the proof of theorem 3.1 we can show that conditions (3.3) and (3.4) will guarantee the existence and uniqueness of the solution to equation (3.5). We summarize these results as a theorem.

Theorem 3.2 *If conditions (5.3.2), (5.3.3) and (3.2) are fulfilled, then there is a unique solution to equation (3.5). On the other hand, if the time lag $\delta(t)$ is positive everywhere, that is $\sup_{t_0 \leq t \leq T} \delta(t) > 0$, then conditions (3.3) and (3.4) are sufficient to guarantee the existence and uniqueness of the solution to equation (3.5).*

There is no difficulty to extend this result to a more general neutral stochastic differential equation with several time-varying delays but the details are left to the reader.

6.4 MOMENT AND PATHWISE ESTIMATES

In this section we shall establish the exponential estimates for the solution of equation (2.14), namely

$$d[x(t) - D(x_t)] = f(x_t, t)dt + g(x_t, t)dB(t) \quad \text{on } t \in [t_0, \infty) \quad (4.1)$$

with initial data (2.3). Let $x(t)$ be the unique global solution of the equation. To give the exponential estimates, we need to impose the linear growth condition: There is a constant $K > 1$ such that for all $(\varphi, t) \in C([-\tau, 0]; R^d) \times [t_0, T]$,

$$|f(\varphi, t)|^2 \bigvee |g(\varphi, t)|^2 \leq K(1 + ||\varphi||^2). \quad (4.2)$$

In addition, we assume that there is a constant $\kappa \in (0, 1)$ such that

$$|D(\varphi)| \leq \kappa ||\varphi|| \quad (4.3)$$

for all $\varphi \in C([-\tau, 0]; R^d)$. Note that (4.3) follows from (2.6) if in addition $D(0) = 0$. It is much more technical to establish L^p-estimates for the solution of the neutral stochastic functional differential equation than for the solution of a stochastic functional differential equation. We need to prepare several lemmas.

Lemma 4.1 *Let $p > 1$, $\varepsilon > 0$ and $a, b \in R$. Then*

$$|a + b|^p \leq \left[1 + \varepsilon^{\frac{1}{p-1}}\right]^{p-1} \left(|a|^p + \frac{|b|^p}{\varepsilon}\right).$$

Proof. By the Hölder's inequality, we have that

$$|a+b|^p = \left|a + \varepsilon^{\frac{1}{p}}\frac{b}{\varepsilon^{\frac{1}{p}}}\right|^p \le \left[1+\varepsilon^{\frac{1}{p-1}}\right]^{p-1}\left(|a|^p + \frac{|b|^p}{\varepsilon}\right)$$

as required.

Lemma 4.2 *Let $p \ge 2$ and $\varepsilon, a, b > 0$. Then*

$$a^{p-1}b \le \frac{(p-1)\varepsilon a^p}{p} + \frac{b^p}{p\varepsilon^{p-1}}$$

and

$$a^{p-2}b^2 \le \frac{(p-2)\varepsilon a^p}{p} + \frac{2b^p}{p\varepsilon^{(p-2)/2}}.$$

Proof. Using the elementary inequality $a^r b^{1-r} \le ra + (1-r)b$ for any $r \in [0,1]$, we derive that

$$a^{p-1}b = (\varepsilon a^p)^{\frac{p-1}{p}}\left(\frac{b^p}{\varepsilon^{p-1}}\right)^{\frac{1}{p}} \le \frac{(p-1)\varepsilon a^p}{p} + \frac{b^p}{p\varepsilon^{p-1}},$$

which is the first inequality required. Using this inquality, we derive that

$$a^{p-2}b^2 = (a^2)^{\frac{p}{2}-1}b^2 \le \frac{(p-2)\varepsilon a^p}{p} + \frac{2b^p}{p\varepsilon^{(p-2)/2}},$$

which is the second inequality required.

Lemma 4.3 *Let $p \ge 1$ and let (4.3) hold. Then*

$$|\varphi(0) - D(\varphi)|^p \le (1+\kappa)^p\|\varphi\|^p$$

for all $\varphi \in C([-\tau,0];R^n)$.

Proof. The required inequality follows from (4.3) directly when $p = 1$ so we only need to show the lemma for $p > 1$. Let $\varepsilon > 0$ be arbitrary. By Lemma 4.1 and condition (4.3), one derives that

$$|\varphi(0) - D(\varphi)|^p \le \left[1+\varepsilon^{\frac{1}{p-1}}\right]^{p-1}\left(|\varphi(0)|^p + \frac{|D(\varphi)|^p}{\varepsilon}\right)$$

$$\le \left[1+\varepsilon^{\frac{1}{p-1}}\right]^{p-1}\left(1 + \frac{\kappa^p}{\varepsilon}\right)\|\varphi\|^p.$$

The required inequality now follows by letting $\varepsilon = \kappa^{p-1}$.

Lemma 4.4 *Let $p > 1$ and (4.3) hold. Then*

$$\sup_{t_0 \le s \le t}|x(s)|^p \le \frac{\kappa}{1-\kappa}\|\xi\|^p + \frac{1}{(1-\kappa)^p}\sup_{t_0 \le s \le t}|x(s) - D(x_s)|^p.$$

Proof. For any $\varepsilon > 0$, by Lemma 4.1, we have that

$$|x(s)|^p = |D(x_s) + x(s) - D(x_s)|^p$$
$$\le \left[1 + \varepsilon^{\frac{1}{p-1}}\right]^{p-1} \left(\frac{|D(x_s)|^p}{\varepsilon} + |x(s) - D(x_s)|^p\right)$$
$$\le \left[1 + \varepsilon^{\frac{1}{p-1}}\right]^{p-1} \left(\frac{\kappa^p ||x_s||^p}{\varepsilon} + |x(s) - D(x_s)|^p\right).$$

Letting $\varepsilon = [\frac{\kappa}{1-\kappa}]^{p-1}$ yields that

$$|x(s)|^p \le \kappa ||x_s||^p + \frac{1}{(1-\kappa)^{p-1}} |x(s) - D(x_s)|^p.$$

Therefore

$$\sup_{t_0 \le s \le t} |x(s)|^p \le \kappa \sup_{t_0 \le s \le t} ||x_s||^p + \frac{1}{(1-\kappa)^{p-1}} \sup_{t_0 \le s \le t} |x(s) - D(x_s)|^p$$
$$\le \kappa ||\xi||^p + \kappa \sup_{t_0 \le s \le t} ||x(s)||^p + \frac{1}{(1-\kappa)^{p-1}} \sup_{t_0 \le s \le t} |x(s) - D(x_s)|^p,$$

and the required assertion follows immediately.

We can now begin to establish the main results in this section.

Theorem 4.5 *Let $p \ge 2$ and $E||\xi||^p < \infty$. Let (4.2) and (4.3) hold. Then*

$$E\left(\sup_{t_0 - \tau \le s \le t} |x(s)|^p\right) \le (1 + \bar{C} E||\xi||^p) e^{C(t-t_0)}, \tag{4.4}$$

where

$$C = \frac{2p(1+\kappa)^{p-2}}{(1-\kappa)^p} \left[\sqrt{2K}(1+\kappa) + K(33p-1)\right]$$

and

$$\bar{C} = \frac{1}{1-\kappa} + \frac{2(1+\kappa)^p}{(1-\kappa)^p}.$$

Proof. By Itô's formula, we can show that

$$|x(t) - D(x_t)|^p \le |\xi(0) - D(\xi)|^p$$
$$+ \int_{t_0}^t \left[p|x(s) - D(x_s)|^{p-1}|f(x_s, s)|\right.$$
$$+ \frac{p(p-1)}{2} |x(s) - D(x_s)|^{p-2} |g(x_s, s)|^2 \bigg] ds$$
$$+ p \int_{t_0}^t |x(s) - D(x_s)|^{p-2} (x(s) - D(x_s))^T g(x_s, s) dB(s). \tag{4.5}$$

Applying Lemmas 4.2 and 4.3 along with condition (4.2), we easily see that for any $\varepsilon > 0$,

$$|x(s) - D(x_s)|^{p-1}|f(x_s,s)| \leq \frac{(p-1)\varepsilon(1+\kappa)^p}{p}||x_s||^p + \frac{K^{\frac{p}{2}}}{p\varepsilon^{p-1}}(1+||x_s||^2)^{\frac{p}{2}}$$

$$\leq \left[\frac{(p-1)\varepsilon(1+\kappa)^p}{p} + \frac{(2K)^{\frac{p}{2}}}{p\varepsilon^{p-1}}\right](1+||x_s||^p).$$

Letting $\varepsilon = \sqrt{2K}/(1+\kappa)$ yields that

$$|x(s) - D(x_s)|^{p-1}|f(x_s,s)| \leq \sqrt{2K}(1+\kappa)^{p-1}(1+||x_s||^p).$$

Similarly, we can show that

$$|x(s) - D(x_s)|^{p-2}|g(x_s,s)|^2 \leq 2K(1+\kappa)^{p-2}(1+||x_s||^p).$$

Also, by Lemma 4.3,

$$|\xi(0) - D(\xi)|^p \leq (1+\kappa)^p ||\xi||^p.$$

We therefore obtain from (4.5) that

$$E\left(\sup_{t_0 \leq s \leq t} |x(s) - D(x_s)|^p\right)$$

$$\leq (1+\kappa)^p E||\xi||^p + C_1 \int_{t_0}^t (1+E||x_s||^p)ds$$

$$+ pE\left(\sup_{t_0 \leq s \leq t}\left|\int_{t_0}^s |x(r) - D(x_r)|^{p-2}(x(r) - D(x_r))^T g(x_r,r)dB(r)\right|\right), \quad (4.6)$$

where $C_1 = p\sqrt{2K}(1+\kappa)^{p-1} + p(p-1)K(1+\kappa)^{p-2}$. On the other hand, by the Burkholder–Davis–Gundy inequality (i.e. Theorem 1.7.3) and the assumptions, we derive that

$$pE\left(\sup_{t_0 \leq s \leq t}\left|\int_{t_0}^s |x(r) - D(x_r)|^{p-2}(x(r) - D(x_r))^T g(x_r,r)dB(r)\right|\right)$$

$$\leq 4p\sqrt{2}E\left(\int_{t_0}^t |x(s) - D(x_s)|^{2p-2}|g(x_s,s)|^2 ds\right)^{\frac{1}{2}}$$

$$\leq 4p\sqrt{2}E\left\{\left(\sup_{t_0 \leq s \leq t}|x(s) - D(x_s)|^p\right)\int_{t_0}^t |x(s) - D(x_s)|^{p-2}|g(x_s,s)|^2 ds\right\}^{\frac{1}{2}}$$

$$\leq \frac{1}{2}E\left(\sup_{t_0 \leq s \leq t}|x(s) - D(x_s)|^p\right) + 16p^2 E\int_{t_0}^t |x(s) - D(x_s)|^{p-2}|g(x_s,s)|^2 ds$$

$$\leq \frac{1}{2}E\left(\sup_{t_0 \leq s \leq t}|x(s) - D(x_s)|^p\right) + 32Kp^2(1+\kappa)^{p-2}\int_{t_0}^t (1+E||x_s||^p)ds. \quad (4.7)$$

Substituting this into (4.6) yields that

$$E\left(\sup_{t_0 \leq s \leq t} |x(s) - D(x_s)|^p\right) \leq 2(1+\kappa)^p E\|\xi\|^p + C_2 \int_{t_0}^t (1 + E\|x_s\|^p) ds, \quad (4.8)$$

where $C_2 = 2C_1 + 64Kp^2(1+\kappa)^{p-2}$. Applying Lemma 4.4 we then see that

$$E\left(\sup_{t_0 \leq s \leq t} |x(s)|^p\right) \leq C_3 E\|\xi\|^p + \frac{C_2}{(1-\kappa)^p} \int_{t_0}^t (1 + E\|x_s\|^p) ds,$$

where $C_3 = \kappa/(1-\kappa) + 2(1+\kappa)^p/(1-\kappa)^p$. Consequently

$$1 + E\left(\sup_{t_0-\tau \leq s \leq t} |x(s)|^p\right) \leq 1 + E\|\xi\|^p + E\left(\sup_{t_0 \leq s \leq t} |x(s)|^p\right)$$

$$\leq 1 + (1+C_3)E\|\xi\|^p + \frac{C_2}{(1-\kappa)^p} \int_{t_0}^t \left[1 + E\left(\sup_{t_0-\tau \leq r \leq s} |x(r)|^p\right)\right] ds.$$

Finally, by the Gronwall inequality, we obtain that

$$1 + E\left(\sup_{t_0-\tau \leq s \leq t} |x(s)|^p\right) \leq [1 + (1+C_3)E\|\xi\|^p] \exp\left[\frac{C_2(t-t_0)}{(1-\kappa)^p}\right]$$

and the required assertion (4.4) follows since $C = C_2/(1-\kappa)^p$ and $\bar{C} = 1 + C_3$. The proof is complete.

In the above proof, we have made several enlargements when use the linear growth condition (4.2) to estimate some terms. To obtain a more precise result, we can pool these terms together and estimate them as a whole.

Theorem 4.6 *Let $p \geq 2$ and $E\|\xi\|^p < \infty$. Let (4.3) hold. Assume that there is a constant $\lambda > 0$ such that*

$$2p|\varphi(0) - D(\varphi)|^{p-1}|f(\varphi,t)| + p(33p-1)|\varphi(0) - D(\varphi)|^{p-2}|g(\varphi,t)|^2$$
$$\leq \lambda(1 + \|\varphi\|^p) \quad (4.9)$$

for all $(\varphi, t) \in C([-\tau, 0]; R^d) \times [t_0, \infty)$. Then

$$E\left(\sup_{t_0-\tau \leq s \leq t} |x(s)|^p\right) \leq (1 + \bar{C}E\|\xi\|^p) \exp\left[\frac{\lambda(t-t_0)}{(1-\kappa)^p}\right], \quad (4.10)$$

where $\bar{C}$ is the same as defined in Theorem 4.5.

Proof. It is not difficult to derive from (4.5), (4.7) and condition (4.9) that

$$E\left(\sup_{t_0 \leq s \leq t} |x(s) - D(x_s)|^p\right) \leq 2(1+\kappa)^p E\|\xi\|^p$$
$$+ E\int_{t_0}^t \Big[2p|x(s) - D(x_s)|^{p-1}|f(x_s,s)|$$
$$+ p(33p-1)|x(s) - D(x_s)|^{p-2}|g(x_s,s)|^2\Big] ds$$
$$\leq 2(1+\kappa)^p E\|\xi\|^p + \lambda \int_{t_0}^t (1 + E\|x_s\|^p) ds, \quad (4.11)$$

which is similar to (4.8). From here, it is the same as in the proof of Theorem 4.5 to show that

$$1 + E\left(\sup_{t_0-\tau \leq s \leq t} |x(s)|^p\right) \leq \left[1 + \bar{C}E\|\xi\|^p\right]\exp\left[\frac{\lambda(t-t_0)}{(1-\kappa)^p}\right],$$

and the desired assertion (4.10) follows.

From the conclusion of Theorem 4.6 we can obtain the pathwise asymptotic estimate for the solution.

Theorem 4.7 *Let (4.3) hold. Assume that there is a constant $\lambda > 0$ such that*

$$4|\varphi(0) - D(\varphi)\|f(\varphi,t)| + 130|g(\varphi,t)|^2 \leq \lambda(1 + \|\varphi\|^2) \tag{4.12}$$

for all $(\varphi, t) \in C([-\tau, 0]; R^d) \times [t_0, \infty)$. Then

$$\limsup_{t \to \infty} \frac{1}{t} \log |x(s)| \leq \frac{\lambda}{2(1-\kappa)^2} \qquad a.s.$$

The proof is the same as that of Theorem 5.4.2 by using the result of Theorem 4.6 with $p = 2$.

We now apply this theorem to obtain the pathwise asymptotic estimate for the solution under conditions (4.2) and (4.3).

Corollary 4.8 *Let (4.2) and (4.3) hold. Then*

$$\limsup_{t \to \infty} \frac{1}{t} \log |x(s)| \leq \frac{1}{(1-\kappa)^2}\left[2\sqrt{K}(1+\kappa) + 65K\right] \qquad a.s. \tag{4.13}$$

Proof. By conditions (4.2) and (4.3) we estimate that

$$4|\varphi(0) - D(\varphi)\|f(\varphi,t)| + 130|g(\varphi,t)|^2$$
$$\leq 4(1+\kappa)\|\varphi\|\sqrt{K(1+\|\varphi\|^2)} + 130K(1+\|\varphi\|^2)$$
$$\leq \left[4\sqrt{K}(1+\kappa) + 130K\right](1+\|\varphi\|^2).$$

Hence the conclusion follows from Theorem 4.7.

To close this section, let us point out that if we apply Theorem 4.5 we can only obtain that

$$\limsup_{t \to \infty} \frac{1}{t} \log |x(s)| \leq \frac{1}{(1-\kappa)^2}\left[2\sqrt{2K}(1+\kappa) + 130K\right] \qquad a.s.$$

which is worse than (4.13), and this shows the advantage of Theorem 4.6.

6.5 L^p-CONTINUITY

Let us proceed to discuss the L^p-continuity of the solution $x(t)$ of equation (4.1). Bearing in mind that almost all sample paths of the solution are continuous, we can easily conclude from Theorem 4.5 that the solution is continuous in L^p by applying the dominated convergence theorem (i.e. Theorem 1.2.2). On the other hand, with a little more effort, we can more precisely estimate the L^p-difference between $x(t+\delta)$ and $x(t)$. Let $E||\xi||^p < \infty$. Since all the sample paths of $\xi(\cdot)$ are continuous on $[-\tau, 0]$, the dominated convergence theorem implies that $\xi(\cdot)$ is L^p-continuous, hence uniformly L^p-continuous on $[-\tau, 0]$. Therefore, for any $0 < \delta < \tau$, there is a $\beta_\delta > 0$ such that

$$E|\xi(\theta_1) - x(\theta_2)|^p \leq \beta_\delta \quad \text{if } \theta_1, \theta_2 \in [-\tau, 0] \text{ and } |\theta_1 - \theta_2| \leq \delta. \tag{5.1}$$

Moreover, we introduce a new notation $L^p_\mathcal{F}([-\tau, 0]; R^d)$ which denotes the family of all $C([-\tau, 0]; R^d)$-valued $\mathcal{F}$-measurable random variables ϕ such that $E||\phi||^p < \infty$.

Theorem 5.1 *Let $p \geq 2$ and $E||\xi||^p < \infty$. Let (4.2) hold. Assume that $D(0) = 0$ and, moreover, there is a constant $\kappa \in (0, 1)$ such that*

$$E|D(\phi) - D(\psi)|^p \leq \kappa^p \sup_{-\tau \leq \theta \leq 0} E|\phi(\theta) - \psi(\theta)|^p \tag{5.2}$$

for all $\phi, \psi \in L^p_\mathcal{F}([-\tau, 0]; R^d)$. Then for any $T > t_0$ and $0 < \delta < \tau$,

$$E|x(t+\delta) - x(t)|^p \leq \frac{\kappa}{1-\kappa}(1 + 2^{p-1})\beta_\delta + \frac{H_2 \delta^{\frac{p}{2}}}{(1-\kappa)^p}$$
$$+ \frac{\kappa 2^{p-1}}{(1-\kappa)(1-\sqrt{\kappa})}\left(\kappa H_4 \beta_\delta + \frac{H_3 \delta^{\frac{p}{2}}}{(1-\kappa)^{p-1}}\right), \tag{5.3}$$

whenever $t_0 \leq t \leq T$, where β_δ has been defined above, H_2–H_4 are constants dependent of $K, \kappa, p, \tau, T, \xi$ only and will be defined in the proof below.

Proof. First, let us show that condition (4.3) is fulfilled. Since $C([-\tau, 0]; R^d) \in L^p_\mathcal{F}([-\tau, 0]; R^d)$, we see from the fact $D(0) = 0$ and condition (5.2) that for any $\varphi \in C([-\tau, 0]; R^d)$,

$$|D(\varphi)|^p = E|D(\varphi)|^p = E|D(\varphi) - D(0)|^p \leq k^p E||\varphi||^p = k^p ||\varphi||^p,$$

and (4.3) follows. Therefore, by Theorem 4.5, we have that

$$E\left(\sup_{t_0-\tau \leq s \leq T+\tau} |x(s)|^p\right) \leq H_1 := (1 + \bar{C}E||\xi||^p)e^{C(T+\tau-t_0)}, \tag{5.4}$$

where C and $\bar{C}$ are defined in Theorem 4.5. In the same way as in the proof of Theorem 5.4.3, we derive that for $t_0 \le t \le T$,

$$E|x(t+\delta) - D(x_{t+\delta}) - x(t) + D(x_t)|^p$$
$$\le (2\delta)^{p-1} E \int_t^{t+\delta} |f(x_s,s)|^p ds + \frac{1}{2}[2p(p-1)]^{\frac{p}{2}} \delta^{\frac{p-2}{2}} E \int_t^{t+\delta} |g(x_s,s)|^p ds$$
$$\le \left[(2\delta)^{p-1} + \frac{1}{2}[2p(p-1)]^{\frac{p}{2}} \delta^{\frac{p-2}{2}}\right] 2^{\frac{p-2}{2}} K^{\frac{p}{2}}(1+H_1)\delta \le H_2 \delta^{\frac{p}{2}}, \quad (5.5)$$

where

$$H_2 = 2^{\frac{p-2}{2}} K^{\frac{p}{2}}(1+H_1)\left[2^{p-1}\tau^{\frac{p}{2}} + \frac{1}{2}[2p(p-1)]^{\frac{p}{2}}\right].$$

On the other hand, by Lemma 4.1 and condition (5.2), we can derive that for any $\varepsilon > 0$,

$$E|x(t+\delta) - x(t)|^p$$
$$= E|D(x_{t+\delta}) - D(x_t) + x(t+\delta) - D(x_{t+\delta}) - x(t) + D(x_t)|^p$$
$$\le \left[1 + \varepsilon^{\frac{1}{p-1}}\right]^{p-1} \bigg(\frac{1}{\varepsilon} E|D(x_{t+\delta}) - D(x_t)|^p$$
$$\quad + E|x(t+\delta) - D(x_{t+\delta}) - x(t) + D(x_t)|^p\bigg)$$
$$\le \left[1 + \varepsilon^{\frac{1}{p-1}}\right]^{p-1} \bigg(\frac{\kappa^p}{\varepsilon} \sup_{-\tau \le \theta \le 0} E|x(t+\delta+\theta) - x(t+\theta)|^p$$
$$\quad + E|x(t+\delta) - D(x_{t+\delta}) - x(t) + D(x_t)|^p\bigg).$$

Letting $\varepsilon = \left[\frac{\kappa}{1-\kappa}\right]^{p-1}$ and using (5.5) we see that

$$E|x(t+\delta) - x(t)|^p \le \kappa \sup_{-\tau \le \theta \le 0} E|x(t+\delta+\theta) - x(t+\theta)|^p + \frac{H_2 \delta^{\frac{p}{2}}}{(1-\kappa)^{p-1}}$$

holds for all $t_0 \le t \le T$. Consequently,

$$\sup_{t_0 \le t \le T} E|x(t+\delta) - x(t)|^p$$
$$\le \kappa \sup_{t_0-\tau \le t \le T} E|x(t+\delta) - x(t)|^p + \frac{H_2 \delta^{\frac{p}{2}}}{(1-\kappa)^{p-1}}$$
$$\le \kappa \sup_{t_0 \le t \le T} E|x(t+\delta) - x(t)|^p$$
$$\quad + \kappa \sup_{t_0-\tau \le t \le t_0} E|x(t+\delta) - x(t)|^p + \frac{H_2 \delta^{\frac{p}{2}}}{(1-\kappa)^{p-1}}.$$

This implies that

$$\sup_{t_0 \le t \le T} E|x(t+\delta) - x(t)|^p$$
$$\le \frac{\kappa}{1-\kappa} \sup_{t_0-\tau \le t \le t_0} E|x(t+\delta) - x(t)|^p + \frac{H_2 \delta^{\frac{p}{2}}}{(1-\kappa)^p}. \quad (5.6)$$

But, by (5.1), we derive that

$$\sup_{t_0-\tau\leq t\leq t_0} E|x(t+\delta) - x(t)|^p$$
$$\leq \beta_\delta + \sup_{t_0-\delta\leq t\leq t_0} E|x(t+\delta) - x(t)|^p$$
$$\leq \beta_\delta + 2^{p-1} \sup_{t_0-\delta\leq t\leq t_0} \left[E|x(t_0) - x(t)|^p + E|x(t+\delta) - x(t_0)|^p \right]$$
$$\leq (1+2^{p-1})\beta_\delta + 2^{p-1} \sup_{t_0\leq t\leq t_0+\delta} E|x(t) - \xi(0)|^p.$$

Substituting this into (5.6) yields that

$$\sup_{t_0\leq t\leq T} E|x(t+\delta) - x(t)|^p \leq \frac{\kappa}{1-\kappa}(1+2^{p-1})\beta_\delta + \frac{H_2 \delta^{\frac{p}{2}}}{(1-\kappa)^p}$$
$$+ \frac{\kappa 2^{p-1}}{1-\kappa} \sup_{t_0\leq t\leq t_0+\delta} E|x(t) - \xi(0)|^p. \qquad (5.7)$$

The assertion of the theorem follows now from the following lemma.

Lemma 5.2 *Under the same assumptions of Theorem 5.1,*

$$\sup_{t_0\leq t\leq t_0+\delta} E|x(t) - \xi(0)|^p \leq \frac{1}{1-\sqrt{\kappa}}\left(\kappa H_4 \beta_\delta + \frac{H_3 \delta^{\frac{p}{2}}}{(1-\kappa)^{p-1}}\right), \qquad (5.8)$$

where H_3 and H_4 are constants dependent of K, κ, p, τ, ξ only and will be defined in the proof below.

Proof. In the same way as in the above proof, we can show that for $t_0 \leq t \leq t_0 + \delta$,

$$E|x(t) - D(x_t) - \xi(0) + D(\xi)|^p \leq H_3 \delta^{\frac{p}{2}}$$

and

$$E|x(t) - \xi(0)|^p \leq \kappa \sup_{-\tau\leq\theta\leq 0} E|x(t+\theta) - x(\theta)|^p$$
$$+ \frac{1}{(1-\kappa)^{p-1}} E|x(t) - D(x_t) - \xi(0) + D(\xi)|^p,$$

where

$$H_3 = 2^{\frac{p-2}{2}} K^{\frac{p}{2}} \left[1 + (1+\bar{C}E||\xi||^p)e^{C\tau} \right] \left[2^{p-1}\tau^{\frac{p}{2}} + \frac{1}{2}[2p(p-1)]^{\frac{p}{2}} \right],$$

C and $\bar{C}$ are defined in Theorem 4.5. Thus

$$E|x(t) - \xi(0)|^p \leq \kappa \sup_{-\tau\leq\theta\leq 0} E|x(t+\theta) - \xi(\theta)|^p + \frac{H_3 \delta^{\frac{p}{2}}}{(1-\kappa)^{p-1}}. \qquad (5.9)$$

On the other hand, by (5.1), we can derive that

$$\sup_{-\tau \leq \theta \leq 0} E|x(t+\theta) - \xi(\theta)|^p$$
$$\leq \sup_{-\tau \leq \theta \leq -(t-t_0)} E|\xi(t-t_0+\theta) - \xi(\theta)|^p + \sup_{-(t-t_0) \leq \theta \leq 0} E|x(t+\theta) - \xi(\theta)|^p$$
$$\leq \beta_\delta + \sup_{-(t-t_0) \leq \theta \leq 0} E|x(t+\theta) - \xi(0) + \xi(0) - \xi(\theta)|^p. \tag{5.10}$$

But, using Lemma 4.1, we can show that

$$\sup_{-(t-t_0) \leq \theta \leq 0} E|x(t+\theta) - \xi(0) + \xi(0) - \xi(\theta)|^p$$
$$\leq \sup_{-(t-t_0) \leq \theta \leq 0} \left(\frac{1}{\sqrt{\kappa}} E|x(t+\theta) - \xi(0)|^p + \frac{1}{[1-\kappa^{1/2(p-1)}]^{p-1}} E|\xi(0) - \xi(\theta)|^p \right)$$
$$\leq \frac{1}{\sqrt{\kappa}} \sup_{t_0 \leq s \leq t_0+\delta} E|x(s) - \xi(0)|^p + \frac{\beta_\delta}{[1-\kappa^{1/2(p-1)}]^{p-1}}.$$

Substituting this into (5.10) gives

$$\sup_{-\tau \leq \theta \leq 0} E|x(t+\theta) - \xi(\theta)|^p$$
$$\leq H_4 \beta_\delta + \frac{1}{\sqrt{\kappa}} \sup_{t_0 \leq s \leq t_0+\delta} E|x(s) - \xi(0)|^p,$$

where $H_4 = 1 + [1-\kappa^{1/2(p-1)}]^{-(p-1)}$. Putting this into (5.9) yields

$$E|x(t) - \xi(0)|^p \leq \sqrt{\kappa} \sup_{t_0 \leq s \leq t_0+\delta} E|x(s) - \xi(0)|^p$$
$$+ \kappa H_4 \beta_\delta + \frac{H_3 \delta^{\frac{p}{2}}}{(1-\kappa)^{p-1}},$$

Since this holds for all $t_0 \leq t \leq t_0 + \delta$, we must have

$$\sup_{t_0 \leq t \leq t_0+\delta} E|x(t) - \xi(0)|^p \leq \sqrt{\kappa} \sup_{t_0 \leq s \leq t_0+\delta} E|x(s) - \xi(0)|^p$$
$$+ \kappa H_4 \beta_\delta + \frac{H_3 \delta^{\frac{p}{2}}}{(1-\kappa)^{p-1}},$$

and the required assertion (5.3) follows. The proof is now complete.

To close this section let us point out that although condition (5.2) is stronger than the Lipschitz condition (2.6), it is satisfied in many important cases. For example, if $D(\varphi) = \tilde{D}(\varphi(-\tau))$ for $\varphi \in C([-\tau, 0]; R^d)$ as in Section 3 and condition (3.2) is satisfied, then

$$E|D(\phi) - D(\psi)|^p \leq \kappa^p E|\phi(-\tau) - \psi(-\tau)|^p \leq \kappa^p \sup_{-\tau \leq \theta \leq 0} E|\phi(\theta) - \psi(\theta)|^p$$

for $\phi, \psi \in L^p_{\mathcal{F}}([-\tau,0]; R^d)$. Also, if D is defined by

$$D(\varphi) = \frac{1}{\tau}\int_{-\tau}^0 \Psi(\varphi(\theta))d\theta,$$

where $\Psi: R^d \to R^d$ satisfying $|\Psi(x) - \Psi(y)| \leq \kappa |x-y|$ with $\kappa \in (0,1)$. Then

$$E|D(\phi) - D(\psi)|^p \leq \frac{1}{\tau^p} E\left|\int_{-\tau}^0 [\Psi(\phi(\theta)) - \Psi(\psi(\theta))]d\theta\right|^p$$

$$\leq \frac{1}{\tau} E\int_{-\tau}^0 |\Psi(\phi(\theta)) - \Psi(\psi(\theta))|^p d\theta \leq \frac{\kappa^p}{\tau}\int_{-\tau}^0 E|\phi(\theta) - \psi(\theta)|^p d\theta$$

$$\leq \kappa^p \sup_{-\tau \leq \theta \leq 0} E|\phi(\theta) - \psi(\theta)|^p.$$

6.6 EXPONENTIAL STABILITY

In this section we shall study the stability problem for the neutral stochastic functional differential equation

$$d[x(t) - D(x_t)] = f(x_t, t)dt + g(x_t, t)dB(t) \quad \text{on } t \geq t_0. \tag{6.1}$$

For this purpose, we assume that f, g and D are smooth enough (e.g. continuous) so that the equation has a unique global solution for any given initial data $x_{t_0} = \xi \in L^2_{\mathcal{F}_{t_0}}([-\tau, 0]; R^d)$, and the solution is denoted by $x(t;\xi)$. We have already shown that almost all the sample paths of the solution are continuous and, moreover, the 2nd moment of the solution is continuous. We furthermore assume that $f(0,t) \equiv 0$, $g(0,t) \equiv 0$ and $D(0) = 0$. Therefore, the equation admits a trivial solution $x(t;0) \equiv 0$ corresponding to the initial data $x_{t_0} = 0$. Due to the page limit we shall only discuss the mean square and almost sure exponential stability of the trivial solution. The main technique used in this section is the Razumikhin argument (cf. Section 5.6). Let us first establish a result on the exponential stability in mean square.

Theorem 6.1 *Assume that there is a constant $\kappa \in (0,1)$ such that*

$$E|D(\phi)|^2 \leq \kappa^2 \sup_{-\tau \leq \theta \leq 0} E|\phi(\theta)|^2, \qquad \phi \in L^2_{\mathcal{F}}([-\tau,0]; R^d). \tag{6.2}$$

Let $q > (1-\kappa)^{-2}$. Assume furthermore that there is a $\lambda > 0$ such that

$$E\big[2(\phi(0) - D(\phi))^T f(\phi, t) + |g(\phi, t)|^2\big] \leq -\lambda E|\phi(0) - D(\phi)|^2 \tag{6.3}$$

for all $t \geq t_0$ and those $\phi \in L^2_{\mathcal{F}_t}([-\tau,0]; R^d)$ satisfying

$$E|\phi(\theta)|^2 < qE|\phi(0) - D(\phi)|^2, \qquad -\tau \leq \theta \leq 0.$$

Then for all $\xi \in L^2_{\mathcal{F}_{t_0}}([-\tau, 0]; R^d)$,

$$E|x(t;\xi)|^2 \leq q(1+\kappa)^2 e^{-\bar{\gamma}(t-t_0)} \sup_{-\tau \leq \theta \leq 0} E|\xi(\theta)|^2 \quad \text{on } t \geq t_0 \quad (6.4)$$

where

$$\bar{\gamma} = \min\left\{\lambda, \frac{1}{\tau}\log\left[\frac{q}{(1+\kappa\sqrt{q})^2}\right]\right\} > 0. \quad (6.5)$$

In other words, the trivial solution of equation (6.1) is exponentially stable in mean square.

In order to prove this theorem, let us present two useful lemmas.

Lemma 6.2 *Let (6.2) hold for some $\kappa \in (0,1)$. Then*

$$E|\phi(0) - D(\phi)|^2 \leq (1+\kappa)^2 \sup_{-\tau \leq \theta \leq 0} E|\phi(\theta)|^2$$

for all $\phi \in L^2_{\mathcal{F}}([-\tau, 0]; R^d)$.

Proof. Compute that

$$\begin{aligned} E|\phi(0) - D(\phi)|^2 &\leq E|\phi(0)|^2 + 2E(|\phi(0)||D(\phi)|) + E|D(\phi)|^2 \\ &\leq (1+\kappa)E|\phi(0)|^2 + (1+\kappa^{-1})E|D(\phi)|^2 \\ &\leq [1+\kappa+\kappa(1+\kappa)] \sup_{-\tau \leq \theta \leq 0} E|\phi(\theta)|^2 \\ &= (1+\kappa)^2 \sup_{-\tau \leq \theta \leq 0} E|\phi(\theta)|^2 \end{aligned}$$

as required.

Lemma 6.3 *Let (6.2) hold for some $\kappa \in (0,1)$. Let $\rho \geq t_0$ and $0 < \gamma < \tau^{-1}\log(1/\kappa^2)$. Let $x(t)$ be a solution of equation (6.1). If*

$$e^{\gamma(t-t_0)}E|x(t) - D(x_t)|^2 \leq (1+\kappa)^2 \sup_{-\tau \leq \theta \leq 0} E|x(t_0+\theta)|^2 \quad (6.6)$$

for all $t_0 \leq t \leq \rho$, then

$$e^{\gamma(t-t_0)}E|x(t)|^2 \leq \frac{(1+\kappa)^2}{(1-\kappa e^{\gamma\tau/2})^2} \sup_{-\tau \leq \theta \leq 0} E|x(t_0+\theta)|^2$$

for all $t_0 - \tau \leq t \leq \rho$.

Proof. Let $\kappa^2 e^{\gamma\tau} < \varepsilon < 1$. For $t_0 \leq t \leq \rho$, note that

$$\begin{aligned} E|x(t) - D(x_t)|^2 &\geq E|x(t)|^2 - 2E(|x(t)||D(x_t)|) + E|D(x_t)|^2 \\ &\geq (1-\varepsilon)E|x(t)|^2 - (\varepsilon^{-1} - 1)E|D(x_t)|^2. \end{aligned}$$

Hence

$$E|x(t)|^2 \leq \frac{1}{1-\varepsilon}E|x(t)-D(x_t)|^2 + \frac{\kappa^2}{\varepsilon}\sup_{-\tau\leq\theta\leq 0}E|x(t+\theta)|^2.$$

By condition (6.6), we then derive that for all $t_0 \leq t \leq \rho$,

$$e^{\gamma(t-t_0)}E|x(t)|^2 \leq \frac{1}{1-\varepsilon}\sup_{t_0\leq t\leq\rho}\left[e^{\gamma(t-t_0)}E|x(t)-D(x_t)|^2\right]$$

$$+ \frac{\kappa^2}{\varepsilon}\sup_{t_0\leq t\leq\rho}\left[e^{\gamma(t-t_0)}\sup_{-\tau\leq\theta\leq 0}E|x(t+\theta)|^2\right]$$

$$\leq \frac{(1+\kappa)^2}{1-\varepsilon}\sup_{-\tau\leq\theta\leq 0}E|x(t_0+\theta)|^2$$

$$+ \frac{\kappa^2 e^{\gamma\tau}}{\varepsilon}\sup_{t_0-\tau\leq t\leq\rho}\left[e^{\gamma(t-t_0)}E|x(t)|^2\right].$$

However, this holds for all $t_0 - \tau \leq t \leq t_0$ as well. Therefore

$$\sup_{t_0-\tau\leq t\leq\rho}\left[e^{\gamma(t-t_0)}E|x(t)|^2\right]$$

$$\leq \frac{(1+\kappa)^2}{1-\varepsilon}\sup_{-\tau\leq\theta\leq 0}E|x(t_0+\theta)|^2 + \frac{\kappa^2 e^{\gamma\tau}}{\varepsilon}\sup_{t_0-\tau\leq t\leq\rho}\left[e^{\gamma(t-t_0)}E|x(t)|^2\right].$$

Since $1 > \kappa^2 e^{\gamma\tau}/\varepsilon$, we obtain that

$$\sup_{t_0-\tau\leq t\leq\rho}\left[e^{\gamma(t-t_0)}E|x(t)|^2\right] \leq \frac{\varepsilon(1+\kappa)^2}{(1-\varepsilon)(\varepsilon-\kappa^2 e^{\gamma\tau})}\sup_{-\tau\leq\theta\leq 0}E|x(t_0+\theta)|^2.$$

Finally, the required assertion follows by taking $\varepsilon = \kappa e^{\gamma\tau/2}$.

We can now begin to prove theorem 6.1.

Proof of Theorem 6.1. First, note that $q/(1+\kappa\sqrt{q})^2 > 1$ since $q > (1-\kappa)^{-2}$, and hence $\bar{\gamma} > 0$. Now fix any $\xi \in L^2_{\mathcal{F}_{t_0}}([-\tau,0];R^d)$ and simply write $x(t;\xi) = x(t)$. Without any loss of generality we may assume that $\sup_{-\tau\leq\theta\leq 0}E|\xi(\theta)|^2 > 0$. Let $\gamma \in (0,\bar{\gamma})$ arbitrarily. It is easy to show that

$$0 < \gamma < \min\left\{\lambda, \frac{1}{\tau}\log\left(\frac{1}{\kappa^2}\right)\right\} \quad \text{and} \quad q > \frac{e^{\gamma\tau}}{(1-\kappa e^{\gamma\tau/2})^2}. \quad (6.7)$$

We now claim that

$$e^{\gamma(t-t_0)}E|x(t)-D(x_t)|^2 \leq (1+\kappa)^2\sup_{-\tau\leq\theta\leq 0}E|\xi(\theta)|^2 \quad \text{for all } t \geq t_0. \quad (6.8)$$

If so, an application of Lemma 6.3 to (6.8) yields that

$$e^{\gamma(t-t_0)} E|x(t)|^2 \leq \frac{(1+\kappa)^2}{(1-\kappa e^{\gamma\tau/2})^2} \sup_{-\tau \leq \theta \leq 0} E|\xi(\theta)|^2$$

$$\leq q(1+\kappa)^2 \sup_{-\tau \leq \theta \leq 0} E|\xi(\theta)|^2$$

for all $t \geq t_0$, where (6.7) has been used, and then the desired result (6.4) follows by letting $\gamma \to \bar{\gamma}$. The remainder of the proof is to show (6.8) by contradiction. Suppose (6.8) is not true. Then in view of Lemma 6.2, there is a $\rho \geq 0$ such that

$$e^{\gamma(t-t_0)} E|x(t) - D(x_t)|^2 \leq e^{\gamma(\rho-t_0)} E|x(\rho) - D(x_\rho)|^2$$
$$= (1+\kappa)^2 \sup_{-\tau \leq \theta \leq 0} E|\xi(\theta)|^2 \qquad (6.9)$$

for all $t_0 \leq t \leq \rho$ and, moreover, there is a sequence of $\{t_k\}_{k \geq 1}$ such that $t_k \downarrow \rho$ and

$$e^{\gamma(t_k-t_0)} E|x(t_k) - D(x_{t_k})|^2 > e^{\gamma(\rho-t_0)} E|x(\rho) - D(x_\rho)|^2. \qquad (6.10)$$

Applying Lemma 6.3, we derive from (6.9) that

$$e^{\gamma(t-t_0)} E|x(t)|^2 \leq \frac{(1+\kappa)^2}{(1-\kappa e^{\gamma\tau/2})^2} \sup_{-\tau \leq \theta \leq 0} E|\xi(\theta)|^2$$

$$= \frac{e^{\gamma(\rho-t_0)}}{(1-\kappa e^{\gamma\tau/2})^2} E|x(\rho) - D(x_\rho)|^2$$

for all $-\tau \leq t \leq \rho$. Particularly,

$$E|x(\rho+\theta)|^2 \leq \frac{e^{\gamma\tau}}{(1-\kappa e^{\gamma\tau/2})^2} E|x(\rho) - D(x_\rho)|^2$$
$$< qE|x(\rho) - D(x_\rho)|^2 \qquad (6.11)$$

for all $-\tau \leq \theta \leq 0$, where (6.7) has been used once again. By assumption (6.3), we then have

$$E\Big(2(x(\rho) - D(x_\rho))^T f(x_\rho, \rho) + |g(x_\rho, \rho)|^2\Big) \leq -\lambda E|x(\rho) - D(x_\rho)|^2.$$

Recalling $\gamma < \lambda$, we see by the continuity of the solution and the functionals D, f and g (this is the standing hypothesis in this section) that for all sufficiently small $h > 0$,

$$E\Big(2(x(t) - D(x_t))^T f(x_t, t) + |g(x_t, t)|^2\Big) \leq -\gamma E|x(t) - D(x_t)|^2$$

if $\rho \leq t \leq \rho + h$. Now by the Itô formula, for all sufficiently small $h > 0$, we have that

$$e^{\gamma(\rho+h-t_0)} E|x(\rho+h) - D(x_{\rho+h})|^2 - e^{\gamma(\rho-t_0)} E|x(\rho) - D(x_\rho)|^2$$
$$= \int_\rho^{\rho+h} e^{\gamma(t-t_0)} \Big[\gamma E|x(t) - D(x_t)|^2$$
$$+ E\Big(2(x(t) - D(x_t))^T f(x_t, t) + |g(t, x_t)|^2\Big)\Big] dt$$
$$\leq 0,$$

but this contradicts with (6.10), so (6.8) must hold. The proof is now complete.

We now turn to discuss the almost sure exponential stability. We need to prepare another lemma which is very useful in the study of the almost sure exponential stability of neutral stochastic functional differential equations.

Lemma 6.4 *Assume that there exists a constant $\kappa \in (0,1)$ such that*

$$|D(\varphi)| \leq \kappa \sup_{-\tau \leq \theta \leq 0} |\varphi(\theta)|, \qquad \varphi \in C([-\tau,0]; R^d). \tag{6.12}$$

Let $z : [t_0 - \tau, \infty) \to R^d$ be a continuous function and define $z_t = \{z(t+\theta) : -\tau \leq \theta \leq 0\}$ for $t \geq t_0$. Let $0 < \gamma < \tau^{-1} \log(1/\kappa^2)$ and $H > 0$. If

$$|z(t) - D(z_t)|^2 \leq He^{-\gamma(t-t_0)} \quad \text{for all } t \geq t_0,$$

then

$$\limsup_{t \to \infty} \frac{1}{t} \log |z(t)| \leq -\frac{\gamma}{2}.$$

Proof. Choose any $\varepsilon \in (\kappa^2 e^{\gamma\tau}, 1)$. In the same way as in the proof of Lemma 6.3, we can show that for any $T > t_0$,

$$\sup_{t_0 \leq t \leq T} \left[e^{\gamma(t-t_0)} |z(t)|^2 \right] \leq \frac{H}{1-\varepsilon} + \frac{\kappa^2 e^{\gamma\tau}}{\varepsilon} \sup_{t_0-\tau \leq t \leq T} \left[e^{\gamma(t-t_0)} |z(t)|^2 \right].$$

It then follows

$$\left(1 - \frac{\kappa^2 e^{\gamma\tau}}{\varepsilon}\right) \sup_{t_0 \leq t \leq T} \left[e^{\gamma(t-t_0)} |z(t)|^2 \right] \leq \frac{H}{1-\varepsilon} + \frac{\kappa^2 e^{\gamma\tau}}{\varepsilon} \sup_{t_0-\tau \leq t \leq t_0} |z(t)|^2.$$

This implies immediately that

$$\limsup_{t \to \infty} \frac{1}{t} \log |z(t)| \leq -\frac{\gamma}{2}$$

as required.

Theorem 6.5 *Let (6.2) hold for some $\kappa \in (0,1)$. Assume that there exists a positive constant $K > 0$ such that*

$$E\big(|f(\phi,t)|^2 + |g(\phi,t)|^2\big) \leq K \sup_{-\tau \leq \theta \leq 0} E|\phi(\theta)|^2 \tag{6.13}$$

for all $t \geq t_0$ and $\phi \in L^2_{\mathcal{F}}([-\tau,0]; R^d)$. Assume also that the trivial solution of equation (6.1) is exponentially stable in mean square, that is there exists a pair of positive constants γ and M such that

$$E|x(t;\xi)|^2 \leq Me^{-\gamma(t-t_0)} \sup_{-\tau \leq \theta \leq 0} E|\xi(\theta)|^2 \quad \text{on } t \geq t_0 \tag{6.14}$$

for all $\xi \in L^2_{\mathcal{F}_{t_0}}([-\tau,0]; R^d)$. Then

$$\limsup_{t\to\infty} \frac{1}{t} \log |x(t;\xi)| \leq -\frac{\bar{\gamma}}{2} \quad a.s. \qquad (6.15)$$

where $\bar{\gamma} = \min\{\gamma, \tau^{-1}\log(1/\kappa^2)\}$, that is, the trivial solution of equation (6.1) is also almost surely exponentially stable. In particular, if (6.2), (6.3) and (6.12) hold, then the trivial solution of equation (6.1) is almost surely exponentially stable.

Proof. Fix any initial data ξ and write the solution $x(t;\xi) = x(t)$ simply. By the well-known Doob martingale inequality, the Hölder inequality and the assumptions, we can derive that for any integer $k \geq 1$,

$$E\left(\sup_{0\leq\theta\leq\tau} |x(t_0+k\tau+\theta) - D(x_{t_0+k\tau+\theta})|^2\right)$$

$$\leq 3E|x(t_0+k\tau) - D(x_{t_0+k\tau})|^2$$

$$+ 3K(\tau+4) \int_{t_0+k\tau}^{t_0+(k+1)\tau} \left(\sup_{-\tau\leq\theta\leq 0} E|x(s+\theta)|^2\right) ds$$

$$\leq 6E|x(t_0+k\tau)|^2 + 6\kappa^2 \sup_{-\tau\leq\theta\leq 0} E|x(t_0+k\tau+\theta)|^2$$

$$+ 3KM(\tau+4)\left(\sup_{-\tau\leq\theta\leq 0} E|\xi(\theta)|^2\right) \int_{t_0+k\tau}^{t_0+(k+1)\tau} e^{-\bar{\gamma}(s-\tau-t_0)} ds$$

$$\leq 3M\left[2(1+\kappa^2) + K\tau(\tau+4)\right]e^{-\bar{\gamma}(k\tau-\tau)}\left(\sup_{-\tau\leq\theta\leq 0} E|\xi(\theta)|^2\right)$$

$$= Ce^{-\bar{\gamma}k\tau}, \qquad (6.16)$$

where $C = 3Me^{\bar{\gamma}\tau}\left[2(1+\kappa^2) + K\tau(\tau+4)\right]\sup_{-\tau\leq\theta\leq 0} E|\xi(\theta)|^2$. Let $\varepsilon \in (0, \bar{\gamma})$ be arbitrary. It then follows from (6.16) that

$$P\left(\omega: \sup_{0\leq\theta\leq\tau} |x(t_0+k\tau+\theta) - D(x_{t_0+k\tau+\theta})|^2 > e^{-(\bar{\gamma}-\varepsilon)k\tau}\right) \leq Ce^{-\varepsilon k\tau}.$$

In view of the well-known Borel–Cantelli lemma, we see that for almost all $\omega \in \Omega$,

$$\sup_{0\leq\theta\leq\tau} |x(t_0+k\tau+\theta) - D(x_{t_0+k\tau+\theta})|^2 \leq e^{-(\bar{\gamma}-\varepsilon)k\tau} \qquad (6.17)$$

holds for all but finitely many k. Hence for all $\omega \in \Omega$ excluding a P-null set, there exists a $k_0(\omega)$ for which (6.17) holds whenever $k \geq k_0$. In other words, for almost all $\omega \in \Omega$,

$$|x(t) - D(x_t)|^2 \leq e^{-(\bar{\gamma}-\varepsilon)(t-\tau-t_0)} \quad \text{if } t \geq t_0 + k_0\tau.$$

However, $|x(t) - D(x_t)|^2$ is finite on $[t_0, t_0+k_0\tau]$. Therefore, for almost all $\omega \in \Omega$, there exists a finite number $H = H(\omega)$ such that

$$|x(t) - D(x_t)|^2 \leq He^{-(\bar{\gamma}-\varepsilon)(t-t_0)} \quad \text{for all } t \geq t_0.$$

Since $C[-\tau; 0]; R^d) \subset L^2_{\mathcal{F}}([-\tau, 0]; R^d)$, we see that condition (6.2) implies condition (6.12). An application of Lemma 6.4 now yields

$$\limsup_{t \to \infty} \frac{1}{t} \log |x(t)| \leq -\frac{\bar{\gamma} - \varepsilon}{2} \quad a.s.$$

and the desired result (6.15) follows by letting $\varepsilon \to 0$. The proof is complete.

Let us now apply the above results to deal with special stochastic equations of neutral type.

(i) *Stochastically Perturbed Equations of Neutral Type*

Consider the neutral stochastic equation of the form

$$d[x(t) - D(x_t)] = [f_1(x(t), t) + f_2(x_t, t)]dt + g(x_t, t)dB(t) \tag{6.18}$$

on $t \geq t_0$ with initial data $x_0 = \xi \in L^2_{\mathcal{F}_{t_0}}([-\tau, 0]; R^d)$, where D, g are the same as before, $f_1 : R^d \times R_+ \to R^d$ and $f_2 : C([-\tau, 0]; R^d) \times R_+ \to R^d$ are sufficiently smooth and, moreover, $f_1(0, t) = f_2(0, t) \equiv 0$. This equation can be regarded as the stochastically perturbed system of the neutral ordinary functional differential equation

$$\frac{d}{dt}[x(t) - D(x_t)] = f_1(x(t), t).$$

Corollary 6.6 *Let (6.2) hold. Assume that there are two positive constants λ_1 and λ_2 such that*

$$E\Big(2(\phi(0) - D(\phi))^T [f_1(\phi(0), t) + f_2(\phi, t)] + |g(\phi, t)|^2\Big)$$
$$\leq -\lambda_1 E|\phi(0)|^2 + \lambda_2 \sup_{-\tau \leq \theta \leq 0} E|\phi(\theta)|^2 \tag{6.19}$$

for all $t \geq t_0$ and $\phi \in L^2_{\mathcal{F}}([-\tau, 0]; R^d)$. If

$$0 < \kappa < \frac{1}{2} \quad \text{and} \quad \lambda_1 > \frac{\lambda_2}{(1 - 2\kappa)^2}, \tag{6.20}$$

then the trivial solution of equation (6.18) is exponentially stable in mean square. If, in addition, there is a constant $K > 0$ such that

$$E\big(|f_1(\phi(0), t) + f_2(\phi, t)|^2 + |g(\phi, t)|^2\big) \leq K \sup_{-\tau \leq \theta \leq 0} E|\phi(\theta)|^2 \tag{6.21}$$

for all $t \geq t_0$ and $\phi \in L^2_{\mathcal{F}}([-\tau, 0]; R^d)$, then the trivial solution of equation (6.18) is also almost surely exponentially stable.

Proof. By condition (6.20), we can choose q such that

$$\frac{1}{\kappa^2} > q > \frac{1}{(1 - \kappa)^2} \quad \text{and} \quad \lambda_1 > \frac{\lambda_2 q}{(1 - \kappa\sqrt{q})^2}. \tag{6.22}$$

By defining $f(\varphi, t) = f_1(\varphi(0), t) + f_2(\varphi, t)$ for $t \geq t_0$ and $\varphi \in C([-\tau, 0]; R^d)$, equation (6.18) can be written as equation (6.1), so all that we need to do is verify condition (6.3). To do so, let $t \geq t_0$ and $\phi \in L^2_{\mathcal{F}_t}([-\tau, 0]; R^d)$, satisfying

$$E|\phi(\theta)|^2 < qE|\phi(0) - D(\phi)|^2, \qquad -\tau \leq \theta \leq 0. \tag{6.23}$$

Note that for any $\varepsilon > 0$,

$$E|\phi(0) - D(\phi)|^2 \leq (1+\varepsilon)E|\phi(0)|^2 + (1+\varepsilon^{-1})E|D(\phi)|^2.$$

Hence, using (6.2) and (6.23),

$$-E|\phi(0)|^2 \leq -\frac{1}{1+\varepsilon}E|\phi(0) - D(\phi)|^2 + \frac{1}{\varepsilon}E|D(\phi)|^2$$

$$\leq -\frac{1}{1+\varepsilon}E|\phi(0) - D(\phi)|^2 + \frac{\kappa^2}{\varepsilon}\sup_{-\tau \leq \theta \leq 0} E|\phi(\theta)|^2$$

$$\leq -\left(\frac{1}{1+\varepsilon} - \frac{\kappa^2 q}{\varepsilon}\right)E|\phi(0) - D(\phi)|^2. \tag{6.24}$$

It therefore follows from (6.19), (6.23) and (6.24) that

$$E\left(2(\phi(0) - D(\phi))^T[f_1(t, \phi(0)) + f_2(\phi, t)] + |g(t, \phi)|^2\right)$$

$$\leq -\lambda_1\left(\frac{1}{1+\varepsilon} - \frac{\kappa^2 q}{\varepsilon}\right)E|\phi(0) - D(\phi)|^2 + \lambda_2 q E|\phi(0) - D(\phi)|^2$$

$$= -\left[\lambda_1\left(\frac{1}{1+\varepsilon} - \frac{\kappa^2 q}{\varepsilon}\right) - \lambda_2 q\right]E|\phi(0) - D(\phi)|^2. \tag{6.25}$$

In particular, choose $\varepsilon = \kappa\sqrt{q}/(1 - \kappa\sqrt{q})$ and hence

$$\left[\lambda_1\left(\frac{1}{1+\varepsilon} - \frac{\kappa^2 q}{\varepsilon}\right) - \lambda_2 q\right] = \lambda_1(1 - \kappa\sqrt{q})^2 - \lambda_2 q > 0,$$

where (6.22) has been used. In other words, condition (6.3) is satisfied and hence the conclusions follow from Theorems 6.1 and 6.5. The proof is complete.

To state another result, let us introduce a new notation $\mathcal{W}([-\tau, 0]; R_+)$ which is the family of all Borel measurable bounded nonnegative functions $\eta(\theta)$ defined on $-\tau \leq \theta \leq 0$ such that $\int_{-\tau}^{0} \eta(\theta)d\theta = 1$. The functions in $\mathcal{W}([-\tau, 0]; R_+)$ are sometimes called *weighting functions*.

Corollary 6.7 *Assume that there is a positive constant κ and a function $\eta_1 \in \mathcal{W}([-\tau, 0]; R_+)$ such that*

$$|D(\varphi)|^2 \leq \kappa^2 \int_{-\tau}^{0} \eta_1(\theta)|\varphi(\theta)|^2 d\theta \qquad \text{for all } \varphi \in C([-\tau, 0]; R^d). \tag{6.26}$$

Assume also that there exists a function $\eta_2(.) \in \mathcal{W}([-\tau, 0]; R_+)$ and two positive constants λ_1 and λ_2 such that

$$2(\varphi(0) - D(\varphi))^T[f_1(t, \varphi(0)) + f_2(t, \varphi)] + |g(t, \varphi)|^2$$
$$\leq -\lambda_1|\varphi(0)|^2 + \lambda_2 \int_{-\tau}^{0} \eta_2(\theta)|\varphi(\theta)|^2 d\theta \quad (6.27)$$

for all $t \geq t_0$ and $\varphi \in C([-\tau, 0]; R^d)$. If $(6.20)^*$ is satisfied, then the trivial solution of equation (6.18) is exponentially stable in mean square. If, in addition, (6.21) is satisfied as well, then the trivial solution of equation (6.18) is also almost surely exponentially stable.

Proof. The conclusions follow from Corollary 6.6 provided we can verify that (6.26) and (6.27) imply (6.2) and (6.19), respectively. If (6.26) holds, then for any $\phi \in L^2_\mathcal{F}([-\tau, 0]; R^d)$,

$$E|D(\phi)|^2 \leq \kappa^2 \int_{-\tau}^{0} \eta_1(\theta) E|\phi(\theta)|^2 d\theta$$
$$\leq \kappa^2 \sup_{-\tau \leq \theta \leq 0} E|\phi(\theta)|^2 \int_{-\tau}^{0} \eta_1(\theta) d\theta = \kappa^2 \sup_{-\tau \leq \theta \leq 0} E|\phi(\theta)|^2,$$

that is (6.2) holds. Similarly,

$$E \int_{-\tau}^{0} \eta_2(\theta)|\phi(\theta)|^2 d\theta \leq \sup_{-\tau \leq \theta \leq 0} E|\phi(\theta)|^2$$

and hence (6.27) implies (6.19). The proof is complete.

(ii) *Neutral Stochastic Differential Delay Equations*

Consider the neutral stochastic differential delay equations

$$d[x(t) - \tilde{D}(x(t-\tau))]$$
$$= F(x(t), x(t-\tau), t)dt + G(x(t), x(t-\tau), t)dB(t) \quad (6.28)$$

on $t \geq t_0$, where $\tilde{D} : R^d \to R^d$, $F : R^d \times R^d \times R_+ \to R^d$ and $G : R^d \times R^d \times R_+ \to R^{d \times m}$. As before, assume that $\tilde{D}, F$ and G are smooth enough so that equation (6.28) has a unique global solution for any given initial data $x_0 = \xi \in L^2_{\mathcal{F}_{t_0}}([-\tau, 0]; R^d)$. The solution is still denoted by $x(t; \xi)$. Moreover, assume that $\tilde{D}(0) = 0$, $F(0,0,t) \equiv 0$ and $G(0,0,t) \equiv 0$. We first employ Corollary 6.6 to establish one useful result.

---

* Mao (1995b) showed using other techniques that (6.20) can be replaced by the much weaker conditions $\kappa \in (0,1)$ and $\lambda_1 > \lambda_2$.

Corollary 6.8 *Assume that there is a positive constant κ such that*
$$|\tilde{D}(x)| \leq \kappa |x| \quad \text{for all } x \in R^d.$$

Assume also that there are two positive constants λ_1, λ_2 such that
$$2(x - \tilde{D}(y))^T F(x,y,t) + |G(x,y,t)|^2 \leq -\lambda_1 |x|^2 + \lambda_2 |y|^2$$

for all $(x, y, t) \in R^d \times R^d \times [t_0, \infty)$. If (6.20) holds, then the trivial solution of equation (6.28) is exponentially stable in mean square. In addition, if there is a $K > 0$ such that
$$|F(x,y,t)|^2 + |G(x,y,t)|^2 \leq K(|x|^2 + |y|^2) \tag{6.29}$$

for all $(x, y, t) \in R^d \times R^d \times [t_0, \infty)$, then the trivial solution of equation (6.28) is also almost surely exponentially stable.

This corollary follows from Corollary 6.6 directly since equation (6.28) can be written as equation (6.18) by defining
$$D(\varphi) = \tilde{D}(\varphi(-\tau)), \quad f_1(x,t) = F(x,0,t),$$
$$f_2(\varphi,t) = -F(\varphi(0),0,t) + F(\varphi(0),\varphi(-\tau),t), \quad g(\varphi,t) = G(\varphi(0),\varphi(-\tau),t)$$

for $t \geq 0$, $x \in R^d$ and $\varphi \in C([-\tau,0]; R^d)$. Of course, we can apply Theorems 6.1 and 6.5 to obtain a more general result. For this purpose, let us recall the notation $L^2(\Omega; R^d)$ which denotes the family of all R^d-valued $\mathcal{F}$-measurable random variables X such that $E|X|^2 < \infty$.

Corollary 6.9 *Let (6.2) hold with $\kappa \in (0,1)$. Let $q > (1-\kappa)^{-2}$. Assume that there is a constant $\lambda > 0$ such that*
$$E\left[2(X - \tilde{D}(Y))^T F(X,Y,t) + |G(X,Y,t)|^2\right] \leq -\lambda E|X - \tilde{D}(Y)|^2 \tag{6.30}$$

for all $t \geq t_0$ and those $X, Y \in L^2(\Omega; R^d)$ satisfying $E|Y|^2 < qE|X - \tilde{D}(Y)|^2$. Then the trivial solution of equation (6.28) is exponentially stable in mean square. Furthermore, if (6.29) is satisfied, then the trivial solution of equation (6.28) is also almost surely exponentially stable.

This corollary follows from Theorems 6.1 and 6.5 directly since equation (6.28) can be written as equation (6.1) by defining
$$D(\varphi) = \tilde{D}(\varphi(-\tau)), \quad f(\varphi,t) = F(\varphi(0),\varphi(-\tau),t)$$
$$\text{and} \quad g(\varphi,t) = G(\varphi(0),\varphi(-\tau),t)$$

for $t \geq 0$ and $\varphi \in C([-\tau,0]; R^d)$.

(iii) *Linear Neutral Stochastic Functional Differential Equations*

As one more application, let us consider the linear neutral stochastic functional differential equation

$$d[x(t) - D(x_t)] = [-Ax(t) + G_0(x_t)]dt + \sum_{i=1}^{m} G_i(x_t)dB_i(t) \qquad (6.31)$$

on $t \geq t_0$ with initial data $x_0 = \xi \in L^2_{\mathcal{F}_{t_0}}([-\tau, 0]; R^d)$. Here A is a $d \times d$ constant matrix and

$$D(\varphi) = \int_{-\tau}^{0} d\gamma(\theta)\varphi(\theta), \qquad G_i(\varphi) = \int_{-\tau}^{0} d\beta_i(\theta)\varphi(\theta)$$

for $\varphi \in C([-\tau, 0]; R^d)$, $0 \leq i \leq m$, where

$$\gamma(\theta) = (\gamma^{kl}(\theta))_{d \times d} \quad \text{and} \quad \beta_i(\theta) = (\beta_i^{kl}(\theta))_{d \times d}$$

with all the elements $\gamma^{kl}(\theta)$ and $\beta_i^{kl}(\theta)$ being functions of bounded variation on $-\tau \leq \theta \leq 0$. Let $V_{\gamma^{kl}}(\theta)$ denote the total variations of γ^{kl} on the interval $[-\tau, \theta]$ and let $V_\gamma(\theta) = ||V_{\gamma^{kl}}(\theta)||$. We can define $V_{\beta_i}(\theta)$ similarly. In particular, let

$$\hat{\gamma} = V_\gamma(0) \quad \text{and} \quad \hat{\beta}_i = V_{\beta_i}(0), \quad 0 \leq i \leq m.$$

Let us now impose the first assumption:

$$0 < \hat{\gamma} < \frac{1}{2}. \qquad (6.32)$$

Then for any $\phi \in L^2_\mathcal{F}([-\tau, 0]; R^d)$,

$$E|D(\phi)|^2 \leq \hat{\gamma} E \int_{-\tau}^{0} dV_\gamma(\theta) |\phi(\theta)|^2 \leq \hat{\gamma}^2 \sup_{-\tau \leq \theta \leq 0} E|\phi(\theta)|^2. \qquad (6.33)$$

In other words, (6.2) is satisfied with $\kappa = \hat{\gamma}$. Moreover,

$$2E\big[|\phi(0)||D(\phi)|\big] \leq \frac{\hat{\gamma}}{1 - 2\hat{\gamma}} E|\phi(0)|^2 + \frac{1 - 2\hat{\gamma}}{\hat{\gamma}} E|D(\phi)|^2$$

$$\leq \frac{\hat{\gamma}}{1 - 2\hat{\gamma}} E|\phi(0)|^2 + \hat{\gamma}(1 - 2\hat{\gamma}) \sup_{-\tau \leq \theta \leq 0} E|\phi(\theta)|^2. \qquad (6.34)$$

Similarly, we can show that

$$2E\big[|\phi(0)||G_0(\phi)|\big] \leq \frac{\hat{\beta}_0}{1 - 2\hat{\gamma}} E|\phi(0)|^2 + \hat{\beta}_0(1 - 2\hat{\gamma}) \sup_{-\tau \leq \theta \leq 0} E|\phi(\theta)|^2, \qquad (6.35)$$

$$2E[|D(\phi)||G_0(\phi)|] \le 2\hat{\gamma}\hat{\beta}_0 \sup_{-\tau \le \theta \le 0} E|\phi(\theta)|^2, \qquad (6.36)$$

and

$$\sum_{i=1}^{m} E|G_i(\phi)|^2 \le \left[\sum_{i=1}^{m} \hat{\beta}_i^2\right] \sup_{-\tau \le \theta \le 0} E|\phi(\theta)|^2. \qquad (6.37)$$

Using (6.34)–(6.37), we then see that

$$E\left(2(\phi(0) - D(\phi))^T[-A\phi(0) + G_0(\phi)] + \sum_{i=1}^{m}|G_i(\phi)|^2\right)$$

$$\le -\left[\lambda_{\min}(A + A^T) - \frac{\hat{\gamma}||A|| + \hat{\beta}_0}{1 - 2\hat{\gamma}}\right]E|\phi(0)|^2$$

$$+ \left[(\hat{\gamma}||A|| + \hat{\beta}_0)(1 - 2\hat{\gamma}) + 2\hat{\gamma}\hat{\beta}_0 + \sum_{i=1}^{m}\hat{\beta}_i^2\right] \sup_{-\tau \le \theta \le 0} E|\phi(\theta)|^2. \qquad (6.38)$$

Applying Corollary 6.6 we conclude the following result.

Corollary 6.10 *Let (6.32) hold. If*

$$\lambda_{\min}(A + A^T) > \frac{2(\hat{\gamma}||A|| + \hat{\beta}_0)}{1 - 2\hat{\gamma}} + \frac{1}{(1 - 2\hat{\gamma})^2}\left[2\hat{\gamma}\hat{\beta}_0 + \sum_{i=1}^{m}\hat{\beta}_i^2\right], \qquad (6.39)$$

then the trivial solution of equation (6.31) is exponentially stable in mean square and is also almost surely exponentially stable.

(iv) *Examples*

Let us discuss a couple of examples to close this section.

Example 6.11 Consider the one-dimensional neutral stochastic differential delay equation

$$d[x(t) - \kappa x(t - \tau)] = -ax(t)dt + bx(t - \tau)dB(t) \qquad (6.40)$$

on $t \ge t_0$, where $B(t)$ is a one-dimensional Brownian motion, $a > 0$, $b > 0$ and $\kappa \in (0, \frac{1}{2})$. Let $\varepsilon > 0$. For $x, y \in R$, compute

$$2(x - \kappa y)(-ax) + b^2 y^2 = -2ax^2 + 2\kappa axy + b^2 y^2$$

$$\le -(2a - \kappa a\varepsilon)x^2 + \left(\frac{\kappa a}{\varepsilon} + b^2\right)y^2.$$

By Corollary 6.8, the trivial solution of equation (6.40) is exponentially stable both in mean square and almost surely provided we can find an $\varepsilon > 0$ for

$$(2a - \kappa a\varepsilon) > \frac{1}{(1 - 2\kappa)^2}\left(\frac{\kappa a}{\varepsilon} + b^2\right),$$

that is
$$2a > \frac{1}{(1-2\kappa)^2}\left(\frac{\kappa a}{\varepsilon} + b^2\right) + \kappa a\varepsilon.$$

Therefore, the stability condition becomes
$$2a > \min_{\varepsilon>0}\left[\frac{1}{(1-2\kappa)^2}\left(\frac{\kappa a}{\varepsilon} + b^2\right) + \kappa a\varepsilon\right]. \tag{6.41}$$

It is easy to show that the right-hand side of (6.41) reaches its minimum
$$\frac{2\kappa a}{1-2\kappa} + \frac{b^2}{(1-2\kappa)^2}$$

when $\varepsilon = (1-2\kappa)^{-1}$. Hence (6.41) becomes
$$2a > \frac{2\kappa a}{1-2\kappa} + \frac{b^2}{(1-2\kappa)^2}.$$

We therefore obtain the stability condition
$$2a(1-2\kappa)(1-3\kappa) > b^2 \tag{6.42}$$

for equation (6.40).

Example 6.12 Consider the d-dimensional neutral stochastic functional differential equation
$$d[x(t) - \tilde{D}(\Theta(x_t))] = f(x(t),t)dt + g(\Theta(x_t),t)dB(t) \tag{6.43}$$

on $t \geq t_0$. Here $\tilde{D}: R^d \to R^d$, $f: R^d \times R_+ \to R^d$, $g: R^d \times R_+ \to R^{d\times m}$, and Θ is a linear operator from $C([-\tau,0]; R^d)$ to R^d defined by
$$\Theta(\varphi) = \frac{1}{\tau}\int_{-\tau}^{0} \varphi(\theta)d\theta.$$

Assume that there are four positive constants $\lambda, \kappa, \kappa_1, \kappa_2$ with $\kappa \in (0, \frac{1}{2})$ such that
$$|\tilde{D}(x)| \leq \kappa|x|, \qquad -2x^T f(x,t) \leq -\lambda|x|^2,$$
$$|f(x,t)| \leq \kappa_1|x|, \qquad |g(x,t)|^2 \leq \kappa_2|x|^2.$$

Then, for any $\phi \in L^2_{\mathcal{F}}([-\tau,0]; R^d)$, we have that
$$E|\tilde{D}(\Theta(\phi))|^2 \leq \kappa^2 E|\Theta(\phi)|^2 \leq \frac{\kappa^2}{\tau^2}E\left|\int_{-\tau}^{0}\phi(\theta)d\theta\right|^2$$
$$\leq \frac{\kappa^2}{\tau}E\int_{-\tau}^{0}|\phi(\theta)|^2 d\theta \leq \kappa^2 \sup_{-\tau\leq\theta\leq 0} E|\phi(\theta)|^2.$$

In other words, condition (6.2) is satisfied (with $D(\cdot) = \tilde{D}(\Theta(\cdot))$). Similarly, we can show that
$$E|g(\phi,t)|^2 \leq \kappa_2 \sup_{-\tau \leq \theta \leq 0} E|\phi(\theta)|^2.$$

Moreover, we compute that

$$E\Big(2(\phi(0) - \tilde{D}(\Theta(\phi)))^T f(\phi(0),t) + |g(\phi,t)|^2\Big)$$
$$\leq -\lambda E|\phi(0)|^2 + 2E\big[|\tilde{D}(\Theta(\phi))||f(\phi(0),t)|\big] + E|g(\phi,t)|^2$$
$$\leq -\lambda E|\phi(0)|^2 + \frac{\kappa_1(1-2\kappa)}{\kappa} E|\tilde{D}(\Theta(\phi))|^2$$
$$\quad + \frac{\kappa}{\kappa_1(1-2\kappa)} E|f(\phi(0),t)|^2 + E|g(\phi,t)|^2$$
$$\leq \Big(\lambda - \frac{\kappa\kappa_1}{1-2\kappa}\Big) E|\phi(0)|^2 + \big[\kappa\kappa_1(1-2\kappa) + \kappa_2\big] \sup_{-\tau \leq \theta \leq 0} E|\phi(\theta)|^2.$$

In view of Corollary 6.6, we see that the condition for the mean square and the almost sure exponential stability is
$$\lambda - \frac{\kappa\kappa_1}{1-2\kappa} > \frac{\kappa\kappa_1(1-2\kappa) + \kappa_2}{(1-2\kappa)^2},$$

i.e.
$$\lambda > \frac{2\kappa\kappa_1(1-2\kappa) + \kappa_2}{(1-2\kappa)^2}. \tag{6.44}$$

7

Backward Stochastic Differential Equations

7.1 INTRODUCTION

In this chapter we shall study a new type of stochastic equations, namely the backward stochastic differential equations of the form

$$x(t) + \int_t^T f(x(s), y(s), s)ds + \int_t^T [g(x(s), s) + y(s)]dB(s) = X \qquad (1.1)$$

on $0 \leq t \leq T$. The equation for the adjoint process in optimal stochastic control (see e.g. Bensoussan (1982), Bismut (1973), Haussmann (1986)) is a linear version of the equation. In the field of control, we usually regard $y(t)$ as an adapted control and $x(t)$ the state of the system. The aim is to choose an adapted control $y(t)$ which drives the state $x(t)$ of the system to the given target X at time $t = T$. This is the so-called *reachability problem*. In the field of backward stochastic differential equations, we are looking for a pair of adapted processes $\{x(t), y(t)\}$ solving the equation. Such a pair is called an adapted solution of the equation. It is the freedom of choosing the process $y(t)$ that makes it possible to find an adapted solution.

Pardoux & Peng (1990) established some results on the existence and uniqueness of the adapted solution under the condition that $f(x, y, t)$ and $g(x, t)$ are uniformly Lipschitz continuous in (x, y) and in x, respectively. Mao (1995a) obtained some results in this direction under non-Lipschitz conditions. More importantly, Pardoux & Peng (1992) gave the probabilistic representation for the given solution of a certain system of quasilinear parabolic partial differential equation in terms of the solutions of the backward stochastic differential equations. In other words, they obtained a generalization of the well-known Feynman–Kac formula. In view of the powerfulness of the Feynman–Kac formula in the study of partial differential equations e.g. K.P.P. equation (cf. Freidlin (1985)), we may expect that this generalized Feynman–Kac formula will play an important role in the study of quasilinear parabolic partial differential equations. Hence from both viewpoints of the control theory and the study of partial differential equations, we see clearly the importance of the study of backward stochastic differential equations.

7.2 MARTINGALE REPRESENTATION THEOREM

In this section we shall introduce the useful martingale representation theorem that will play an important role in this chapter.

Unlike the other chapters, let us stress that in this chapter we are only given a complete probability space $(\Omega, \mathcal{F}, P)$ and an m-dimensional Brownian motion $B(t)$ on it (without a filtration). We then let $\{\mathcal{F}_t^B\}_{t \geq 0}$ be the natural filtration generated by the Brownian motion, that is $\mathcal{F}_t^B = \sigma\{B(s) : 0 \leq s \leq t\}$. Let $\{\mathcal{F}_t\}_{t \geq 0}$ be the augmentation under P of this natural filtration. Then $\{\mathcal{F}_t\}_{t \geq 0}$ is a filtration on $(\Omega, \mathcal{F}, P)$ satisfying the usual conditions and, moreover, $B(t)$ is a Brownian motion with respect to it (see Section 1.4).

Let $T > 0$. It was shown in Section 1.5 that for any $f \in \mathcal{M}^2([0,T]; R^{d \times m})$, the Itô integral

$$\int_0^t f(s) dB(s)$$

is a continuous square-integrable martingale with respect to $\{\mathcal{F}_t\}$ on $t \in [0, T]$. In this section we shall show the converse—any continuous square-integrable martingale with respect to $\{\mathcal{F}_t\}$ can be represented as an Itô integral. This result, known as the *martingale representation theorem*, is very useful in many applications and is described as follows.

Theorem 2.1 *Let $\{M_t\}_{0 \leq t \leq T}$ be a continuous R^d-valued square-integrable martingale with respect to $\{\mathcal{F}_t\}$. Then there is a unique stochastic process $f \in \mathcal{M}^2([0,T]; R^{d \times m})$ such that*

$$M_t = M_0 + \int_0^t f(s) dB(s) \quad on \ t \in [0, T]. \tag{2.1}$$

By uniqueness we mean that if there is any other process $g \in \mathcal{M}^2([0,T]; R^{d \times m})$ such that

$$M_t = M_0 + \int_0^t g(s) dB(s) \quad on \ t \in [0, T],$$

then

$$E \int_0^T |f(s) - g(s)|^2 ds = 0. \tag{2.2}$$

Clearly, we need only to show the theorem in the case of $d = 1$. To do so, we need to present several lemmas. Let $C_0^\infty(R^{m \times n}; R)$ denote the family of infinitely many times differentiable functions from $R^{m \times n}$ to R with compact support. Let $L_{\mathcal{F}_t}^2(\Omega; R)$ denote the family of all real-valued $\mathcal{F}_t$-measurable random variables ξ such that $E|\xi|^2 < \infty$. Let $L^2([0, T]; R^{1 \times m})$ denote the family of all Borel measurable functions h from $[0, T]$ to $R^{1 \times m}$ such that $\int_0^T |h(t)|^2 dt < \infty$. Note that the functions in $L^2([0, T]; R^{1 \times m})$ are deterministic and $L^2([0, T]; R^{1 \times m})$ is a subset of $\mathcal{M}^2([0, T]; R^{1 \times m})$.

Lemma 2.2 *The set of random variables*

$$\{\varphi(B(t_1), \cdots, B(t_n)) : t_i \in [0, T], \varphi \in C_0^\infty(R^{m \times n}; R), n = 1, 2, \cdots\}$$

is dense in $L_{\mathcal{F}_T}^2(\Omega; R)$.

Proof. Let $\{t_i\}_{i \geq 1}$ be a dense subset of $[0, T]$. For each integer $n \geq 1$, let $\mathcal{G}_n$ be the σ-algebra generated by $B(t_1), \cdots, B(t_n)$, i.e. $\mathcal{G}_n = \sigma\{B(t_1), \cdots, B(t_n)\}$. Obviously

$$\mathcal{G}_n \subset \mathcal{G}_{n+1} \quad \text{and} \quad \mathcal{F}_T = \sigma\left(\bigcup_{n=1}^\infty \mathcal{G}_n\right).$$

Let $g \in L_{\mathcal{F}_T}^2(\Omega; R)$ be arbitrary. By the Doob martingale convergence theorem (i.e. Theorem 1.3.5), we have that

$$E(g|\mathcal{G}_n) \to E(g|\mathcal{F}_T) = g \quad \text{as } n \to \infty$$

almost surely and in L^2 as well. On the other hand, by Lemma 1.2.1, for each n, there is a Borel measurable function $g_n : R^{m \times n} \to R$ such that

$$E(g|\mathcal{G}_n) = g_n(B(t_1), \cdots, B(t_n)).$$

However, such $g_n(B(t_1), \cdots, B(t_n))$ can be approximated in $L_{\mathcal{F}_T}^2(\Omega; R)$ by functions $\varphi_{n,k}(B(t_1), \cdots, B(t_n))$, where $\varphi_{n,k} \in C_0^\infty(R^{m \times n}; R)$, and hence the assertion follows.

Lemma 2.3 *The linear span of the random variables of the form*

$$\exp\left(\int_0^T h(t) dB(t) - \frac{1}{2} \int_0^T |h(t)|^2 dt\right), \quad h \in L^2([0, T]; R^{1 \times m}) \quad (2.3)$$

is dense in $L_{\mathcal{F}_T}^2(\Omega; R)$.

Proof. The assertion holds provided we can show that if $g \in L_{\mathcal{F}_T}^2(\Omega; R)$ is orthogonal (in $L_{\mathcal{F}_T}^2(\Omega; R)$) to all random variables of form (2.3), then $g = 0$. Let g be any such random variable. Then for all $\lambda = (\lambda_{ij})_{n \times m} \in R^{n \times m}$ and all $t_1, \cdots, t_n \in [0, T]$, we have

$$G(\lambda) := E\left\{g \exp\left(\text{trace}[\lambda(B(t_1), \cdots, B(t_n))]\right)\right\} = 0, \quad (2.4)$$

for
$$\exp\left(\int_0^T h(t)dB(t) - \frac{1}{2}\int_0^T |h(t)|^2 dt\right)$$
$$= \exp\left(trace[\lambda(B(t_1), \cdots, B(t_n))] - \frac{1}{2}\int_0^T |h(t)|^2 dt\right)$$

if set
$$h(t) = \sum_{i=1}^n (\lambda_i + \lambda_{i+1})I_{[t_i, t_{i-1})}(t),$$

where $t_0 = 0$, $\lambda_i = (\lambda_{i1}, \cdots, \lambda_{im})$ and $\lambda_{n+1} = 0$. The function $G(\lambda)$ is real analytic in $\lambda \in R^{n \times m}$ and hence has an analytic extension to the complex space $C^{n \times m}$ given by

$$G(z) = E\left\{g \exp\left(trace[z(B(t_1), \cdots, B(t_n))]\right)\right\}$$

for $z = (z_{ij})_{n \times m} \in C^{n \times m}$. Since $G = 0$ on $R^{n \times m}$ and G is analytic, we must have $G = 0$ on the whole $C^{n \times m}$. In particular,

$$G(iY) = E\left\{g \exp\left(i\, trace[Y(B(t_1), \cdots, B(t_n))]\right)\right\} = 0 \qquad (2.5)$$

for all $Y = (y_{ij})_{n \times m} \in R^{n \times m}$. Now, for any function $\varphi(X)$, $X = (x_{ij})_{m \times n}$, in $C_0^\infty(R^{m \times n}, R)$, let $\hat{\varphi}(Y)$ be the Fourier transform of $\varphi(X)$, namely

$$\hat{\varphi}(Y) = (2\pi)^{-\frac{nm}{2}} \int_{R^{m \times n}} \varphi(X) \exp[-i\, trace(YX)]dX.$$

Note from the inverse Fourier transform theorem that

$$\varphi(X) = (2\pi)^{-\frac{nm}{2}} \int_{R^{n \times m}} \hat{\varphi}(Y) \exp[i\, trace(YX)]dY.$$

We then compute
$$E[g\varphi(B(t_1), \cdots, B(t_n))]$$
$$= E\left[g(2\pi)^{-\frac{nm}{2}} \int_{R^{n \times m}} \hat{\varphi}(Y) \exp\left(i\, trace[Y(B(t_1), \cdots, B(t_n))]\right)dY\right]$$
$$= (2\pi)^{-\frac{nm}{2}} \int_{R^{n \times m}} \hat{\varphi}(Y) E\left\{g \exp\left(i\, trace[Y(B(t_1), \cdots, B(t_n))]\right)\right\}dY$$
$$= 0. \qquad (2.6)$$

This, together with Lemma 2.2, means that g is orthogonal to a dense subset of $L^2_{\mathcal{F}_T}(\Omega; R)$. We must therefore have that $g = 0$. The proof is therefore complete.

Lemma 2.4 *For any $\xi \in L^2_{\mathcal{F}_T}(\Omega; R)$, there exists a unique stochastic process $f \in \mathcal{M}^2([0,T]; R^{1\times m})$ such that*

$$\xi = E\xi + \int_0^T f(s)dB(s). \tag{2.7}$$

By uniqueness we mean that if there is any other process $g \in \mathcal{M}^2([0,T]; R^{1\times m})$ such that

$$\xi = E\xi + \int_0^T g(s)dB(s), \tag{2.8}$$

then

$$E\int_0^T |f(s) - g(s)|^2 ds = 0. \tag{2.9}$$

Proof. The uniqueness is rather obvious, for (2.7) and (2.8) give

$$\int_0^T [f(s) - g(s)]dB(s) = 0$$

which implies (2.9) by the property of the Itô integral. To show the existence, we first assume that ξ has the form of (2.3), that is

$$\xi = \exp\left(\int_0^T h(t)dB(t) - \frac{1}{2}\int_0^T |h(t)|^2 dt\right)$$

for some $h \in L^2([0,T]; R^{1\times m})$. Define

$$x(t) = \exp\left(\int_0^t h(s)dB(s) - \frac{1}{2}\int_0^t |h(s)|^2 ds\right), \quad 0 \le t \le T.$$

By Itô's formula,

$$dx(t) = x(t)\left[h(t)dB(t) - \frac{1}{2}|h(t)|^2 dt\right] + \frac{1}{2}x(t)|h(t)|^2 dt$$
$$= x(t)h(t)dB(t).$$

This yields that

$$x(t) = 1 + \int_0^t x(s)h(s)dB(s).$$

In particular,

$$\xi = x(T) = 1 + \int_0^T x(s)h(s)dB(s),$$

which gives $E\xi = 1$. Therefore the required assertion (2.7) holds in this case with $f(t) = x(t)h(t)$. By the linearity of (2.7), we see that (2.7) holds for any linear combination of the functions of form (2.3). Now, let $\xi \in L^2_{\mathcal{F}_T}(\Omega; R)$ be arbitrary. By Lemma 2.3, we can approximate ξ in $L^2_{\mathcal{F}_T}(\Omega; R)$ by $\{\xi_n\}$, where each ξ_n is a linear combination of the functions of form (2.3). So, for each n, we have a process $f_n \in \mathcal{M}^2([0,T]; R^{1\times m})$ such that

$$\xi_n = E\xi_n + \int_0^T f_n(s) dB(s). \tag{2.10}$$

Hence

$$E \int_0^T |f_n(s) - f_m(s)|^2 ds$$
$$= E \left| \int_0^T [f_n(s) - f_m(s)] dB(s) \right|^2$$
$$= E|\xi_n - E\xi_n - \xi_m + E\xi_m|^2$$
$$= E|\xi_n - \xi_m|^2 - |E\xi_n - E\xi_m|^2$$
$$\to 0 \quad \text{as } n, m \to \infty.$$

In other words, $\{f_n\}$ is a Cauchy sequence in $\mathcal{M}^2([0,T]; R^{1\times m})$ and hence converges to some $f \in \mathcal{M}^2([0,T]; R^{1\times m})$. We can now let $n \to \infty$ in (2.10) to obtain that

$$\xi = E\xi + \int_0^T f(s) dB(s)$$

as desired. The proof is complete.

We can now begin to prove the martingale representation theorem.

Proof Theorem 2.1. Without any loss of generality, we may assume that $d = 1$. Applying Lemma 2.4 to $\xi = M(T)$, we see that there exists a unique process (the uniqueness follows here) $f \in \mathcal{M}^2([0,T]; R^{1\times m})$ such that

$$M(T) = EM(T) + \int_0^T f(s) dB(s).$$

By the martingale property of $M(t)$ we have $EM(t) = EM(0)$. Since $M(0)$ is $\mathcal{F}_0$-measurable, it must be a constant almost surely and hence $EM(0) = M(0)$ a.s. Then

$$M(T) = M(0) + \int_0^T f(s) dB(s). \tag{2.11}$$

Now for any $0 \le t \le T$, by Theorem 1.5.21, we have that

$$M(t) = E(M(T)|\mathcal{F}_t) = M(0) + E\left(\int_0^T f(s) dB(s)|\mathcal{F}_t\right)$$
$$= M(0) + \int_0^t f(s) dB(s),$$

7.3 EQUATIONS WITH LIPSCHITZ COEFFICIENTS

Let $\mathcal{P}$ denotes the σ-algebra of $\mathcal{F}_t$-progressively measurable subsets of $[0, T] \times \Omega$. Let f be a mapping from $R^d \times R^{d \times m} \times [0, T] \times \Omega$ to R^d which is assumed to be $\mathcal{B}^d \otimes \mathcal{B}^{d \times m} \otimes \mathcal{P}$-measurable. Let g be a mapping from $R^d \times [0, T] \times \Omega$ to $R^{d \times m}$ which is assumed to be $\mathcal{B}_d \otimes \mathcal{P}$-measurable. Let X be a given $\mathcal{F}_T$-measurable R^d-valued random variable such that $E|X|^2 < \infty$, that is $X \in L^2_{\mathcal{F}_T}(\Omega; R^d)$.

In this section we shall discuss the following backward stochastic differential equation

$$x(t) + \int_t^T f(x(s), y(s), s)ds + \int_t^T [g(x(s), s) + y(s)]dB(s) = X \qquad (3.1)$$

on $t \in [0, T]$, where $x(\cdot)$ and $y(\cdot)$ are R^d-valued and $R^{d \times m}$-valued, respectively. If we write equation (3.1) as

$$x(T) - x(t) = \int_t^T f(x(s), y(s), s)ds + \int_t^T [g(x(s), s) + y(s)]dB(s),$$

we see clearly that $x(t)$ is an Itô process with the stochastic differential

$$dx(t) = f(x(t), y(t), t)dt + [g(x(t), t) + y(t)]dB(t). \qquad (3.2)$$

We may therefore interpret the backward equation (3.1) as the stochastic differential equation (3.2) with final value $x(T) = X$. It is this final value, instead of initial value, that makes the backward stochastic differential equations much different from the (forward) stochastic differential equations discussed in Chapter 2. Let us now give a precise definition of a solution to the backward stochastic differential equation.

Definition 3.1 *A pair of stochastic processes*

$$\{x(t), y(t)\}_{0 \le t \le T} \in \mathcal{M}^2([0, T]; R^d) \times \mathcal{M}^2([0, T]; R^{d \times m})$$

is called a solution of the backward stochastic differential equation (3.1) if it has the following properties:

(i) $f(x(\cdot), y(\cdot), \cdot) \in \mathcal{M}^2([0, T]; R^d)$ and $g(x(\cdot), \cdot) \in \mathcal{M}^2([0, T]; R^{d \times m})$;

(ii) *equation (3.1) holds for every $t \in [0, T]$ with probability 1.*

A solution $\{x(t), y(t)\}$ is said to be unique if for any other solution $\{\bar{x}(t), \bar{y}(t)\}$ we have

$$P\{x(t) = \bar{x}(t) \text{ for all } 0 \le t \le T\} = 1$$

and
$$E\int_0^T |y(s) - \bar{y}(s)|^2 ds = 0.$$

The following existence-and-uniqueness theorem is due to Pardoux & Peng (1990).

Theorem 3.2 *Assume that*
$$f(0,0,\cdot) \in \mathcal{M}^2([0,T]; R^d) \quad and \quad g(0,\cdot) \in \mathcal{M}^2([0,T]; R^{d\times m}). \quad (3.3)$$
Assume also that there exists a positive constant $K > 0$ such that
$$|f(x,y,t) - f(\bar{x},\bar{y},t)|^2 \leq K(|x-\bar{x}|^2 + |y-\bar{y}|^2) \quad a.s. \quad (3.4)$$
and
$$|g(x,t) - g(\bar{x},t)|^2 \leq K|x-\bar{x}|^2 \quad a.s. \quad (3.5)$$
for all $x, \bar{x} \in R^d$, $y, \bar{y} \in R^{d\times m}$ and $t \in [0,T]$. Then there exists a unique solution $\{x(t), y(t)\}$ to equation (3.1) in $\mathcal{M}^2([0,T]; R^d) \times \mathcal{M}^2([0,T]; R^{d\times m})$.

Let us present a number of lemmas in order to prove this theorem.

Lemma 3.3 *Let $f(\cdot) \in \mathcal{M}^2([0,T]; R^d)$ and $g(\cdot) \in \mathcal{M}^2([0,T]; R^{d\times m})$. Then there exists a unique pair $\{x(t), y(t)\}$ in $\mathcal{M}^2([0,T]; R^d) \times \mathcal{M}^2([0,T]; R^{d\times m})$ such that*
$$x(t) + \int_t^T f(s)ds + \int_t^T [g(s) + y(s)]dB(s) = X \quad (3.6)$$
for all $0 \leq t \leq T$.

Proof. Define
$$M(t) = E\left(X - \int_0^T f(s)ds \Big| \mathcal{F}_t\right), \quad 0 \leq t \leq T.$$

Then $M(t)$ is a square-integrable martingale. By Theorem 2.1, there is a unique process $\hat{y}(\cdot) \in \mathcal{M}^2([0,T]; R^{d\times m})$ such that
$$M(t) = M(0) + \int_0^t \hat{y}(s)dB(s), \quad 0 \leq t \leq T.$$

Define
$$x(t) = M(t) + \int_0^t f(s)ds \quad \text{and} \quad y(t) = \hat{y}(t) - g(t)$$
for $0 \leq t \leq T$. Clearly, $\{x(t), y(t)\} \in \mathcal{M}^2([0,T]; R^d) \times \mathcal{M}^2([0,T]; R^{d\times m})$. Moreover,
$$\int_t^T [g(s) + y(s)]dB(s) = \int_t^T \hat{y}(s)dB(s)$$
$$= \int_0^T \hat{y}(s)dB(s) - \int_0^t \hat{y}(s)dB(s) = M(T) - M(t).$$

Noting
$$M(T) = X - \int_0^T f(s)ds,$$
we obtain that
$$\int_t^T [g(s) + y(s)]dB(s) = X - \int_0^T f(s)ds - M(t) = X - x(t) - \int_t^T f(s)ds,$$
which is equation (3.6). To show the uniqueness, let $\{\bar{x}(t), \bar{y}(t)\}$ be another pair which solves equation (3.6). Then
$$x(t) - \bar{x}(t) = -\int_t^T [y(s) - \bar{y}(s)]dB(s), \quad 0 \le t \le T.$$
Hence, for every $t \in [0, T]$
$$x(t) - \bar{x}(t) = E(x(t) - \bar{x}(t) | \mathcal{F}_t)$$
$$= -E\left(\int_t^T [y(s) - \bar{y}(s)]dB(s) \Big| \mathcal{F}_t \right) = 0 \quad a.s.$$
Noting that $x(t)$ is continuous, we see easily that $x(t) = \bar{x}(t)$ for all $0 \le t \le T$ a.s. Now
$$0 = x(0) - \bar{x}(0) = -\int_0^T [y(s) - \bar{y}(s)]dB(s)$$
which yields immediately that
$$E \int_0^T |y(s) - \bar{y}(s)|^2 ds = 0.$$
The uniqueness has also been proved.

Lemma 3.4 *Let $g(\cdot) \in \mathcal{M}^2([0,T]; R^{d\times m})$. Let f be a mapping from $R^{d\times m} \times [0,T] \times \Omega$ to R^d which is $\mathcal{B}^{d\times m} \otimes \mathcal{P}$-measurable. Assume that*
$$f(0, \cdot) \in \mathcal{M}^2([0,T]; R^d).$$
Assume also that there exists a positive constant $K > 0$ such that
$$|f(y, t) - f(\bar{y}, t)|^2 \le K|y - \bar{y}|^2 \quad a.s. \tag{3.7}$$
for all $y, \bar{y} \in R^{d\times m}$ and $t \in [0,T]$. Then the backward stochastic differential equation
$$x(t) + \int_t^T f(y(s), s)ds + \int_t^T [g(s) + y(s)]dB(s) = X \tag{3.8}$$
has a unique solution $\{x(t), y(t)\}$ in $\mathcal{M}^2([0,T]; R^d) \times \mathcal{M}^2([0,T]; R^{d\times m})$.

Proof. We first prove the uniqueness. Let us $\{x(t), y(t)\}$ and $\{\bar{x}(t), \bar{y}(t)\}$ be two solutions. Then, recalling (3.2), we easily see that

$$d[x(t) - \bar{x}(t)] = [f(y(t), t) - f(\bar{y}(t), t)]dt + [y(t) - \bar{y}(t)]dB(t).$$

By Itô's formula, for all $0 \le t \le T$, we have that

$$d|x(t) - \bar{x}(t)|^2 = 2[x(t) - \bar{x}(t)]^T[f(y(t), t) - f(\bar{y}(t), t)]dt$$
$$+ |y(t) - \bar{y}(t)|^2 dt + 2[x(t) - \bar{x}(t)]^T[y(t) - \bar{y}(t)]dB(t).$$

Hence

$$-|x(t) - \bar{x}(t)|^2 = 2\int_t^T [x(s) - \bar{x}(s)]^T[f(y(s), s) - f(\bar{y}(s), s)]ds$$
$$+ \int_t^T |y(s) - \bar{y}(s)|^2 ds + 2\int_t^T [x(s) - \bar{x}(s)]^T[y(s) - \bar{y}(s)]dB(s).$$

Taking expectation on both sides yields that

$$E|x(t) - \bar{x}(t)|^2 + E\int_t^T |y(s) - \bar{y}(s)|^2 ds$$
$$= -2E\int_t^T [x(s) - \bar{x}(s)]^T[f(y(s), s) - f(\bar{y}(s), s)]ds.$$

Making use of the elementary inequality $2ab \le a^2/\varepsilon + \varepsilon b^2$ ($\varepsilon > 0$) and the Lipschitz condition (3.7) we obtain that

$$E|x(t) - \bar{x}(t)|^2 + E\int_t^T |y(s) - \bar{y}(s)|^2 ds$$
$$\le \frac{1}{\varepsilon} E\int_t^T |x(s) - \bar{x}(s)|^2 ds + \varepsilon KE\int_t^T |y(s) - \bar{y}(s)|^2 ds.$$

Setting $\varepsilon = 1/2K$ yields

$$E|x(t) - \bar{x}(t)|^2 + E\int_t^T |y(s) - \bar{y}(s)|^2 ds$$
$$\le 2KE\int_t^T |x(s) - \bar{x}(s)|^2 ds + \frac{1}{2}E\int_t^T |y(s) - \bar{y}(s)|^2 ds. \tag{3.8}$$

In particular, this implies that

$$E|x(t) - \bar{x}(t)|^2 \le 2KE\int_t^T |x(s) - \bar{x}(s)|^2 ds.$$

The Gronwall inequality now gives that

$$E|x(t) - \bar{x}(t)|^2 = 0 \quad \text{for all } 0 \le t \le T,$$

which implies that $x(t) = \bar{x}(t)$ for all $0 \leq t \leq T$ a.s. Substituting this into (3.8) we also see that
$$E\int_0^T |y(s) - \bar{y}(s)|^2 ds = 0.$$

The uniqueness has been proved.

Let us now proceed to prove the existence. Set $y_0(t) \equiv 0$. By Lemma 3.3, there is a unique pair $\{x_1(t), y_1(t)\}$ in $\mathcal{M}^2([0,T]; R^d) \times \mathcal{M}^2([0,T]; R^{d \times m})$ such that
$$x_1(t) + \int_t^T f(y_0(s), s)ds + \int_t^T [g(s) + y_1(s)]dB(s) = X.$$

Making use of Lemma 3.3 recursively, we can define, for every $n = 1, 2, \cdots$, a pair $\{x_n(t), y_n(t)\}$ in $\mathcal{M}^2([0,T]; R^d) \times \mathcal{M}^2([0,T]; R^{d \times m})$ by
$$x_n(t) + \int_t^T f(y_{n-1}(s), s)ds + \int_t^T [g(s) + y_n(s)]dB(s) = X. \qquad (3.9)$$

In the same way as in the proof of the uniqueness above, we can show that
$$E|x_{n+1}(t) - x_n(t)|^2 + E\int_t^T |y_{n+1}(s) - y_n(s)|^2 ds$$
$$\leq 2KE\int_t^T |x_{n+1}(s) - x_n(s)|^2 ds + \frac{1}{2}E\int_t^T |y_n(s) - y_{n-1}(s)|^2 ds. \qquad (3.10)$$

For every $n \geq 1$, define
$$u_n(t) = E\int_t^T |x_n(s) - x_{n-1}(s)|^2 ds$$
and
$$v_n(t) = E\int_t^T |y_n(s) - y_{n-1}(s)|^2 ds.$$

It then follows from (3.10) that
$$-\frac{d}{dt}\left(u_{n+1}(t)e^{2Kt}\right) + e^{2Kt}v_{n+1}(t) \leq \frac{1}{2}e^{2Kt}v_n(t). \qquad (3.11)$$

Integrating both sides from t to T, we obtain that
$$u_{n+1}(t)e^{2Kt} + \int_t^T e^{2Ks}v_{n+1}(s)ds \leq \frac{1}{2}\int_t^T e^{2Ks}v_n(s)ds.$$

Hence
$$u_{n+1}(t) + \int_t^T e^{2K(s-t)}v_{n+1}(s)ds \leq \frac{1}{2}\int_t^T e^{2K(s-t)}v_n(s)ds. \qquad (3.12)$$

In particular, this implies that

$$\int_0^T e^{2Ks} v_{n+1}(s) ds \leq \frac{1}{2} \int_0^T e^{2Ks} v_n(s) ds$$

$$\leq \frac{1}{2^n} \int_0^T e^{2Ks} v_1(s) ds \leq \frac{1}{2^n} v_1(0) \int_0^T e^{2Ks} ds \leq \frac{C e^{2KT}}{K 2^{n+1}}, \quad (3.13)$$

where $C = v_1(0) = E \int_0^T |y_1(s)|^2 ds$. Substituting this into (3.11) implies that

$$u_{n+1}(0) \leq \frac{C e^{2KT}}{K 2^{n+1}}. \quad (3.14)$$

It then follows from (3.10) and (3.14) that

$$v_{n+1}(0) \leq 2K u_{n+1}(0) + \frac{1}{2} v_n(0) \leq \frac{1}{2^n} C e^{2KT} + \frac{1}{2} v_n(0),$$

which implies immediately that

$$v_{n+1}(0) \leq \frac{1}{2^n} [n C e^{2KT} + v_1(0)]. \quad (3.15)$$

We now see from (3.14) and (3.15) that $\{x_n(\cdot)\}$ and $\{y_n(\cdot)\}$ are Cauchy sequences in $\mathcal{M}^2([0,T]; R^d)$ and $\mathcal{M}^2([0,T]; R^{d \times m})$, and denote their limits by $x(\cdot)$ and $y(\cdot)$, respectively. Finally, letting $n \to \infty$ in (3.9) we obtain that

$$x(t) + \int_t^T f(y(s), s) ds + \int_t^T [g(s) + y(s)] dB(s) = X,$$

that is, $\{x(t), y(t)\}$ is a solution. The existence has also been proved and therefore the proof of the lemma is complete.

We can now begin to prove Theorem 3.2.

Proof of Theorem 3.2. We first prove the uniqueness. Assume that $\{x(t), y(t)\}$ and $\{\bar{x}(t), \bar{y}(t)\}$ are two solutions. In the same way as in the proof of Lemma 3.4 we can show that

$$E|x(t) - \bar{x}(t)|^2 + E \int_t^T |y(s) - \bar{y}(s)|^2 ds$$

$$= -2E \int_t^T [x(s) - \bar{x}(s)]^T [f(x(s), y(s), s) - f(\bar{x}(s), \bar{y}(s), s)] ds$$

$$- E \int_t^T |g(x(s), s) - g(\bar{x}(s), s)|^2 ds$$

$$- 2E \int_t^T \text{trace}\Big([g(x(s), s) - g(\bar{x}(s), s)]^T [y(s) - \bar{y}(s)] \Big) ds$$

$$\leq 4KE \int_t^T |x(s) - \bar{x}(s)|^2 + \frac{1}{4K} E \int_t^T |f(x(s), y(s), s) - f(\bar{x}(s), \bar{y}(s), s)|^2 ds$$

$$+ 4E \int_t^T |g(x(s), s) - g(\bar{x}(s), s)|^2 + \frac{1}{4} E \int_t^T |y(s) - \bar{y}(s)|^2 ds$$

$$\leq (8K+1) E \int_t^T |x(s) - \bar{x}(s)|^2 + \frac{1}{2} E \int_t^T |y(s) - \bar{y}(s)|^2 ds. \quad (3.16)$$

The uniqueness then follows by applying the Gronwall inequality as we did in the proof of Lemma 3.4.

Let us now show the existence. Set $y_0(t) \equiv 0$. With the help of Lemma 3.4, we can define recursively, for every $n = 1, 2, \cdots$, a pair $\{x_n(t), y_n(t)\}$ in $\mathcal{M}^2([0,T]; R^d) \times \mathcal{M}^2([0,T]; R^{d \times m})$ by

$$x_n(t) + \int_t^T f(x_{n-1}(s), y_n(s), s) ds$$
$$+ \int_t^T [g(x_{n-1}(s), s) + y_n(s)] dB(s) = X. \qquad (3.17)$$

In the same way as in the proof of (3.16), we can show that

$$E|x_{n+1}(t) - x_n(t)|^2 + E \int_t^T |y_{n+1}(s) - y_n(s)|^2 ds$$
$$\leq 4KE \int_t^T |x_{n+1}(s) - x_n(s)|^2 ds$$
$$+ (4K+1)E \int_t^T |x_n(s) - x_{n-1}(s)|^2 ds$$
$$+ \frac{1}{2} E \int_t^T |y_{n+1}(s) - y_n(s)|^2 ds.$$

Hence

$$E|x_{n+1}(t) - x_n(t)|^2 + \frac{1}{2} E \int_t^T |y_{n+1}(s) - y_n(s)|^2 ds$$
$$\leq (4K+1)E \int_t^T \left[|x_{n+1}(s) - x_n(s)|^2 + |x_n(s) - x_{n-1}(s)|^2\right] ds. \qquad (3.18)$$

Define

$$u_n(t) = E \int_t^T |x_n(s) - x_{n-1}(s)|^2 ds.$$

It then follows from (3.18) that

$$-\frac{d}{dt}\left(u_{n+1}(t) e^{(4K+1)t}\right) \leq (4K+1) e^{(4K+1)t} u_n(t).$$

Integrating both sides from t to T yields that

$$u_{n+1}(t) \leq (4K+1) \int_t^T e^{(4K+1)(s-t)} u_n(s) ds$$
$$\leq (4K+1) e^{(4K+1)T} \int_t^T u_n(s) ds.$$

Iterating this inequality, we obtain that

$$u_{n+1}(0) \leq \frac{[(4K+1)Te^{(4K+1)T}]^n}{n!} u_1(0).$$

This, together with (3.18), implies that $\{x_n(\cdot)\}$ and $\{y_n(\cdot)\}$ are Cauchy sequences in $\mathcal{M}^2([0,T];R^d)$ and $\mathcal{M}^2([0,T];R^{d\times m})$. Denote their limits by $x(\cdot)$ and $y(\cdot)$, respectively. Finally, we can let $n \to \infty$ in (3.17) to obtain that

$$x(t) + \int_t^T f(x(s), y(s), s)ds + \int_t^T [g(x(s), s) + y(s)]dB(s) = X.$$

That is, $\{x(t), y(t)\}$ is a solution. The proof is therefore complete.

7.4 EQUATIONS WITH NON-LIPSCHITZ COEFFICIENTS

In the previous section, we established the existence-and-uniqueness theorem of the solution for the backward stochastic differential equation under the uniform Lipschitz condition. On the other hand, it is somewhat too strong to require the uniform Lipschitz continuity in applications e.g. in dealing with quasilinear parabolic partial differential equations. It is therefore important to find some weaker conditions than the Lipschitz one under which the backward stochastic differential equation still has a unique solution. In the first instance, we would perhaps like to try the local Lipschitz condition plus the linear growth condition, as these conditions guarantee the existence and uniqueness of the solution for a (forward) stochastic differential equation. To be precise, let us state these conditions as follows:

For each $n = 1, 2, \cdots$, there exists a constant $K_n > 0$ such that

$$|f(x,y,t) - f(\bar{x}, \bar{y}, t)|^2 \leq K_n(|x - \bar{x}|^2 + |y - \bar{y}|^2) \quad a.s.$$
$$|g(x,t) - g(\bar{x}, t)|^2 \leq K_n |x - \bar{x}|^2 \quad a.s.$$

for all $0 \leq t \leq T$, $x, \bar{x} \in R^d$, $y, \bar{y} \in R^{d\times m}$ with $\max\{|x|, |\bar{x}|, |y|, |\bar{y}|\} < n$. Moreover, there exists a constant $K > 0$ such that

$$|f(x,y,t)|^2 \leq K(1 + |x|^2 + |y|^2) \quad a.s.$$
$$|g(x,t)|^2 \leq K(1 + |x|^2) \quad a.s.$$

for all $0 \leq t \leq T$, $x \in R^d$ and $y \in R^{d\times m}$.

Unfortunately, it is still open whether these conditions guarantee the existence and uniqueness of the solution to the backward stochastic differential equation (3.1). The difficulty here is that the techniques of stopping time and localization seem not to work for backward stochastic differential equations. Now the question is: Are there any weaker conditions than the Lipschitz continuity under

Sec.7.4] Equations with Non-Lipschitz Coefficients

which the backward stochastic differential equation has a unique solution? The answer is of course positive, and the main aim of this section is to show the following conditions will do:

For all $0 \leq t \leq T$, $x, \bar{x} \in R^d$ and $y, \bar{y} \in R^{d \times m}$, we have

$$|f(x, y, t) - f(\bar{x}, \bar{y}, t)|^2 \leq \kappa(|x - \bar{x}|^2) + K|y - \bar{y}|^2 \quad a.s. \qquad (4.1)$$

and

$$|g(x, t) - g(\bar{x}, t)|^2 \leq \kappa(|x - \bar{x}|^2) \quad a.s. \qquad (4.2)$$

where K is a positive constant and $\kappa(\cdot)$ is a concave increasing function from R_+ to R_+ such that $\kappa(0) = 0$, $\kappa(u) > 0$ for $u > 0$ and

$$\int_{0+} \frac{du}{\kappa(u)} = \infty. \qquad (4.3)$$

Let us make a few comments about these conditions before we state the main result. First of all, since κ is concave and $\kappa(0) = 0$, we can find a pair of positive constants a and b such that

$$\kappa(u) \leq a + bu \quad \text{for all } u \geq 0. \qquad (4.4)$$

We therefore see that under conditions (3.3) and (4.1)–(4.3),

$$f(x(\cdot), y(\cdot), \cdot) \in \mathcal{M}^2([0, T]; R^d) \quad \text{and} \quad g(x(\cdot), y(\cdot), \cdot) \in \mathcal{M}^2([0, T]; R^{d \times m})$$

whenever

$$x(\cdot) \in \mathcal{M}^2([0, T]; R^d) \quad \text{and} \quad y(\cdot) \in \mathcal{M}^2([0, T]; R^{d \times m}).$$

Secondly, let us give a few examples for the function $\kappa(\cdot)$ in order to see that conditions (4.1)–(4.4) are irrestrictive. Let $K > 0$ and let $\delta \in (0, 1)$ be sufficiently small. Define

$$\kappa_1(u) = Ku \quad \text{for } u \geq 0;$$

$$\kappa_2(u) = \begin{cases} u \log(u^{-1}) & \text{for } 0 \leq u \leq \delta, \\ \delta \log(\delta^{-1}) + \dot{\kappa}_2(\delta-)(u - \delta) & \text{for } u > \delta; \end{cases}$$

$$\kappa_3(u) = \begin{cases} u \log(u^{-1}) \log \log(u^{-1}) & \text{for } 0 \leq u \leq \delta, \\ \delta \log(\delta^{-1}) \log \log(\delta^{-1}) + \dot{\kappa}_3(\delta-)(u - \delta) & \text{for } u > \delta. \end{cases}$$

It is easy to verify that these are all concave non-decreasing functions satisfying

$$\int_{0+} \frac{du}{\kappa_i(u)} = \infty.$$

In particular, we see clearly that if let $\kappa(u) = Ku$, then conditions (4.1)–(4.3) reduce to the Lipschitz conditions (3.4) and (3.5). In other words, conditions (4.1)–(4.3) are much weaker than the Lipschitz conditions (3.4) and (3.5). Therefore, the following result is a generalization of Theorem 3.2.

Theorem 4.1 *Assume that conditions (3.3) and (4.1)–(4.3) are fulfilled. Then there exists a unique solution $\{x(\cdot), y(\cdot)\}$ to the backward stochastic differential equation (3.1) in $\mathcal{M}^2([0,T]; R^d) \times \mathcal{M}^2([0,T]; R^{d \times m})$.*

The proof of this theorem is rather technical and we shall devote the remainder of this section to it.

We need to prepare a number of lemmas. Let us first construct an approximate sequence using an iteration of the Picard type with the help of Lemma 3.4. Let $x_0(t) \equiv 0$, and let $\{x_n(t), y_n(t) : 0 \leq t \leq T\}_{n \geq 1}$ be a sequence in $\mathcal{M}^2([0,T]; R^d) \times \mathcal{M}^2([0,T]; R^{d \times m})$ defined recursively by

$$x_n(t) + \int_t^T f(x_{n-1}(s), y_n(s), s) ds$$
$$+ \int_t^T [g(x_{n-1}(s), s) + y_n(s)] dB(s) = X \quad (4.5)$$

on $0 \leq t \leq T$. This sequence is well defined since once $x_{n-1}(\cdot) \in \mathcal{M}^2([0,T]; R^d)$ is given, $f(x_{n-1}(t), y, t)$ is Lipschitz continuous in y and

$$f(x_{n-1}(\cdot), 0, \cdot) \in \mathcal{M}^2([0,T]; R^d) \quad \text{and} \quad g(x_{n-1}(\cdot), \cdot) \in \mathcal{M}^2([0,T]; R^{d \times m}),$$

hence Lemma 3.4 can be used to define $x_n(t)$ and $y_n(t)$.

Lemma 4.2 *Assume that conditions (3.3) and (4.1)–(4.3) are fulfilled. Then for all $0 \leq t \leq T$ and $n \geq 1$,*

$$E|x_n(t)|^2 \leq C_1 \quad \text{and} \quad E\int_0^T |y_n(s)|^2 ds \leq C_2, \quad (4.6)$$

where C_1 and C_2 are both positive constants independent of n.

Proof. Applying Itô's formula to $|x_n(t)|^2$ we deduce that

$$|X|^2 - |x_n(t)|^2 = 2\int_t^T (x_n(s),\ f(x_{n-1}(s), y_n(s), s)) ds$$
$$+ 2\int_t^T (x_n(s),\ [g(x_{n-1}(s), s) + y_n(s)] dB(s))$$
$$+ \int_t^T |g(x_{n-1}(s), s) + y_n(s)|^2 ds.$$

Sec.7.4]   Equations with Non-Lipschitz Coefficients

Thus

$$E|x_n(t)|^2 + E\int_t^T |y_n(s)|^2 ds$$

$$= E|X|^2 - 2E\int_t^T (x_n(s),\ f(x_{n-1}(s),y_n(s),s))\,ds$$

$$- E\int_t^T \left(|g(x_{n-1}(s),s)|^2 + 2\,trace[g^T(x_{n-1}(s),s)y_n(s)]\right)ds.$$

Therefore, using the elementary inequality $2|uv| \le u^2/\alpha + \alpha v^2$ for any $\alpha > 0$, we see that

$$E|x_n(t)|^2 + E\int_t^T |y_n(s)|^2 ds$$

$$\le E|X|^2 + \frac{1}{\alpha}E\int_t^T |x_n(s)|^2 ds + \alpha E\int_t^T |f(x_{n-1}(s),y_n(s),s)|^2 ds$$

$$+ \frac{1}{\alpha}E\int_t^T |g(x_{n-1}(s),s)|^2 ds + \alpha E\int_t^T |y_n(s)|^2 ds. \tag{4.7}$$

But (4.1) and (4.4) we derive that

$$|f(x_{n-1}(s),y_n(s),s)|^2$$
$$\le 2|f(0,0,s)|^2 + 2|f(x_{n-1}(s),y_n(s),s) - f(0,0,s)|^2$$
$$\le 2|f(0,0,s)|^2 + 2\kappa(|x_{n-1}(s)|^2) + 2K|y_n(s)|^2$$
$$\le 2|f(0,0,s)|^2 + 2a + 2b|x_{n-1}(s)|^2 + 2K|y_n(s)|^2.$$

Similarly, it follows from (4.2) and (4.4) that

$$|g(x_{n-1}(s),s)|^2 \le 2|g(0,0,s)|^2 + 2a + 2b|x_{n-1}(s)|^2.$$

Substituting these into (4.7) gives that

$$E|x_n(t)|^2 + E\int_t^T |y_n(s)|^2 ds \le C_3(\alpha) + \frac{1}{\alpha}\int_t^T E|x_n(s)|^2 ds$$

$$+ 2b(\alpha + \frac{1}{\alpha})\int_t^T E|x_{n-1}(s)|^2 ds + \alpha(2K+1)E\int_t^T |y_n(s)|^2 ds,$$

where

$$C_3(\alpha) = E|X|^2 + 2a\left(\alpha + \frac{1}{\alpha}\right) + 2\alpha E\int_0^T |f(0,0,s)|^2 ds$$

$$+ \frac{2}{\alpha}E\int_0^T |g(0,0,s)|^2 ds.$$

In particular, choosing $\alpha = 1/(4K+2)$ and setting $C_4 = C_3(1/(4K+2))$, we get that

$$E|x_n(t)|^2 + \frac{1}{2} E \int_t^T |y_n(s)|^2 ds$$

$$\leq C_4 + (4K+2) \int_t^T E|x_n(s)|^2 ds$$

$$+ 2b \left[\frac{1}{4K+2} + 4K+2 \right] \int_t^T E|x_{n-1}(s)|^2 ds$$

$$\leq C_4 + C_5 \int_t^T \max\{E|x_{n-1}(s)|^2, E|x_n(s)|^2\} ds, \qquad (4.8)$$

where $C_5 = 4K + 2 + 2b[(4K+2)^{-1} + 4K + 2]$. Now let k be any positive integer. If $1 \leq n \leq k$, (4.8) implies (recalling $x_0(t) \equiv 0$) that

$$E|x_n(t)|^2 \leq C_4 + C_5 \int_t^T \left(\max_{1 \leq i \leq k} E|x_i(s)|^2 \right) ds.$$

Therefore

$$\max_{1 \leq n \leq k} E|x_n(t)|^2 \leq C_4 + C_5 \int_t^T \left(\max_{1 \leq n \leq k} E|x_n(s)|^2 \right) ds.$$

An application of the well-known Gronwall inequality implies

$$\max_{1 \leq n \leq k} E|x_n(t)|^2 \leq C_4 e^{C_5(T-t)} \leq C_4 e^{C_5 T}.$$

Since k is arbitrary, the first inequality of (4.6) follows by setting $C_1 = C_4 e^{C_5 T}$. Finally, it follows from (4.8) that

$$E \int_0^T |y_n(s)|^2 ds \leq 2(C_4 + C_5 C_1 T) := C_2.$$

The proof is complete.

Lemma 4.3 *Under conditions (3.3) and (4.1)–(4.3), there exists a constant $C_6 > 0$ such that*

$$E|x_{n+k}(t) - x_n(t)|^2 \leq C_6 \int_t^T \kappa \left(E|x_{n+k-1}(s) - x_{n-1}(s)|^2 \right) ds$$

for all $0 \leq t \leq T$ and $n, k \geq 1$.

Proof. Applying Itô's formula to $|x_{n+k}(t) - x_n(t)|^2$ we have that

$$- E|x_{n+k}(t) - x_n(t)|^2$$
$$= 2E \int_t^T (x_{n+k}(s) - x_n(s),$$
$$\quad f(x_{n+k-1}(s), y_{n+k}(s), s) - f(x_{n-1}(s), y_n(s), s))ds$$
$$+ E \int_t^T |g(x_{n+k-1}(s), s) + y_{n+k}(s) - g(x_{n-1}(s), s) - y_n(s)|^2 ds.$$

In the same way as in the proof of Lemma 4.2 we can then show that

$$E|x_{n+k}(t) - x_n(t)|^2 + \frac{1}{2} E \int_t^T |y_{n+k}(s) - y_n(s)|^2 ds$$
$$\leq (4K + 2) \int_t^T E|x_{n+k}(s) - x_n(s)|^2 ds$$
$$+ 2 \left[4K + 2 + \frac{1}{4K+2} \right] \int_t^T \kappa \left(E|x_{n+k-1}(s) - x_{n-1}(s)|^2 \right) ds. \quad (4.9)$$

Now fix $t \in [0, T]$ arbitrarily. If $t \leq r \leq T$, then

$$E|x_{n+k}(r) - x_n(r)|^2 \leq (4K + 2) \int_r^T E|x_{n+k}(s) - x_n(s)|^2 ds$$
$$+ 2 \left[4K + 2 + \frac{1}{4K+2} \right] \int_t^T \kappa \left(E|x_{n+k-1}(s) - x_{n-1}(s)|^2 \right) ds.$$

In view of the Gronwall inequality we see that

$$E|x_{n+k}(t) - x_n(t)|^2 \leq 2 \left[4K + 2 + \frac{1}{4K+2} \right] e^{(4K+2)(T-t)}$$
$$\times \int_t^T \kappa(E|x_{n+k-1}(s) - x_{n-1}(s)|^2) ds.$$

Hence the required assertion follows by setting

$$C_6 = 2 \left[4K + 2 + \frac{1}{4K+2} \right] e^{(4K+2)T}.$$

The proof is complete.

Lemma 4.4 *Under conditions (3.3) and (4.1)–(4.3), there exists a constant $C_7 > 0$ such that*
$$E|x_{n+k}(t) - x_n(t)|^2 \leq C_7 (T - t)$$
for all $0 \leq t \leq T$ and $n, k \geq 1$.

Proof. By Lemmas 4.2 and 4.3, we have that

$$E|x_{n+k}(t) - x_n(t)|^2 \leq C_6 \int_t^T \kappa(4C_1)ds = C_6\kappa(4C_1)(T-t)$$

and the assertion follows by letting $C_7 = C_6\kappa(4C_1)$. The proof is complete.

We now begin to present a key lemma. Set $\bar{\kappa}(u) = C_6\kappa(u)$. Choose $T_1 \in [0,T)$ for

$$\bar{\kappa}(C_7(T-t)) \leq C_7 \quad \text{for all } T_1 \leq t \leq T. \tag{4.10}$$

Fix $k \geq 1$ arbitrarily and define two sequences of functions $\{\varphi_n(t) : 0 \leq t \leq T\}_{n\geq 1}$ and $\{\tilde{\varphi}_{n,k}(t) : 0 \leq t \leq T\}_{n\geq 1}$ as follows:

$$\varphi_1(t) = C_7(T-t),$$

$$\varphi_{n+1}(t) = \int_t^T \bar{\kappa}(\varphi_n(s))ds, \quad n = 1, 2, \cdots,$$

$$\tilde{\varphi}_{n,k}(t) = E|x_{n+k}(t) - x_n(t)|^2, \quad n = 1, 2, \cdots.$$

Lemma 4.5 *Assume that conditions (3.3) and (4.1)–(4.3) are satisfied. Then for each $k \geq 1$ and all $n \geq 1$, we have that*

$$0 \leq \tilde{\varphi}_{n,k}(t) \leq \varphi_n(t) \leq \varphi_{n-1}(t) \leq \cdots \leq \varphi_1(t) \tag{4.11}$$

whenever $t \in [T_1, T]$.

Proof. Let $t \in [T_1, T]$. First of all, by Lemma 4.4,

$$\tilde{\varphi}_{1,k}(t) = E|x_{1+k}(t) - x_1(t)|^2 \leq C_7(T-t) = \varphi_1(t),$$

that is, (4.11) holds for $n = 1$. Next, by Lemma 4.3,

$$\tilde{\varphi}_{2,k}(t) = E|x_{2+k}(t) - x_2(t)|^2 \leq C_6 \int_t^T \kappa(E|x_{1+k}(s) - x_1(s)|^2)ds$$

$$= \int_t^T \bar{\kappa}(\tilde{\varphi}_{1,k}(s))ds \leq \int_t^T \bar{\kappa}(\varphi_1(s))ds = \varphi_2(t).$$

But by (4.10), we also have

$$\varphi_2(t) = \int_t^T \bar{\kappa}(C_7(T-s))ds \leq \int_t^T C_7 ds = C_7(T-t) = \varphi_1(t).$$

In other words, we have already shown that

$$\tilde{\varphi}_{2,k}(t) \leq \varphi_2(t) \leq \varphi_1(t) \quad \text{if } t \in [T_1, T],$$

i.e. (4.11) holds also for $n = 2$. We now assume that (4.11) holds for some $n \geq 2$. Then by Lemma 4.3 again,

$$\tilde{\varphi}_{n+1,k}(t) \leq \int_t^T \bar{\kappa}(\tilde{\varphi}_{n,k}(s))ds \leq \int_t^T \bar{\kappa}(\varphi_n(s))ds = \varphi_{n+1}(t)$$

$$\leq \int_t^T \bar{\kappa}(\varphi_{n-1}(s))ds = \varphi_n(t),$$

that is, (4.11) holds for $n + 1$ as well. By induction, (4.11) must therefore hold for all $n \geq 1$. The proof is complete.

At last we can begin to prove the main result Theorem 4.1

Proof of Theorem 4.1. Existence: We first prove the existence of a solution. This will be done by four steps. In the following proof please bear in mind that the constants C_1–C_7 as well as T_1 have already been defined above.

Step 1. We claim that

$$\sup_{T_1 \leq t \leq T} E|x_n(t) - x_i(t)|^2 \to 0 \quad \text{as } n, i \to \infty. \tag{4.12}$$

In fact, note that for each $n \geq 1$, $\varphi_n(t)$ is continuous and decreasing on $[T_1, T]$ and, by Lemma 4.5, for each t, $\varphi_n(t)$ is non-increasing monotonically as $n \to \infty$. Therefore we can define function $\varphi(t)$ on $[T_1, T]$ by $\varphi_n(t) \downarrow \varphi(t)$. It is easy to verify that $\varphi(t)$ is continuous and non-increasing on $[T_1, T]$. By the definition of $\varphi_n(t)$ and $\varphi(t)$, we see that

$$\varphi(t) = \lim_{n \to \infty} \varphi_{n+1}(t) = \lim_{n \to \infty} \int_t^T \bar{\kappa}(\varphi_n(s))ds = \int_t^T \bar{\kappa}(\varphi(s))ds, \quad t \in [T_1, T].$$

Hence for any $\varepsilon > 0$,

$$\varphi(t) \leq \varepsilon + \int_t^T \bar{\kappa}(\varphi(s))ds, \quad t \in [T_1, T].$$

Applying the Bihari inequality (i.e. Theorem 1.8.2), we obtain that

$$\varphi(t) \leq G^{-1}(G(\varepsilon) + T - t) \leq G^{-1}(G(\varepsilon) + T - T_1), \quad t \in [T_1, T], \tag{4.13}$$

where

$$G(r) = \int_1^r \frac{du}{\bar{\kappa}(u)} \quad \text{on } r > 0$$

and $G^{-1}(\cdot)$ is the inverse function of G. By condition (4.3) and the definition of $\bar{\kappa}(\cdot)$, we have

$$\int_{0+} \frac{du}{\bar{\kappa}(u)} = \infty,$$

which implies

$$\lim_{\varepsilon \to 0} G(\varepsilon) = -\infty \quad \text{and then} \quad \lim_{\varepsilon \to 0} G^{-1}(G(\varepsilon) + T - T_1) = 0.$$

Therefore, by letting $\varepsilon \to 0$ in (4.13), we obtain that

$$\varphi(t) = 0 \quad \text{for all } t \in [T_1, T].$$

In particular, we see that $\varphi_n(T_1) \downarrow \varphi(T_1) = 0$ as $n \to \infty$. So for any $\varepsilon > 0$, we can find an integer $N \geq 1$ such that $\varphi_n(T_1) < \varepsilon$ whenever $n \geq N$. Now for any $k \geq 1$ and $n \geq N$, by Lemma 4.5, we have that

$$\sup_{T_1 \leq t \leq T} E|x_{n+k}(t) - x_n(t)|^2 = \sup_{T_1 \leq t \leq T} \tilde{\varphi}_{n,k}(t)$$

$$\leq \sup_{T_1 \leq t \leq T} \varphi_n(t) = \varphi_n(T_1) < \varepsilon,$$

and (4.12) must therefore hold.

Step 2. Define

$$T_2 = \inf\{s \in [0,T] : \sup_{s \leq t \leq T} E|x_n(t) - x_i(t)|^2 \to 0 \text{ as } n, i \to \infty\}.$$

We see immediately from Step 1 that $0 \leq T_2 \leq T_1 < T$. In this step we shall show that

$$\sup_{T_2 \leq t \leq T} E|x_n(t) - x_i(t)|^2 \to 0 \quad \text{as } n, i \to \infty. \tag{4.14}$$

Let $\varepsilon > 0$ be arbitrary. Choose $\delta \in (0, T - T_2)$ for

$$C_6 \kappa(4C_1)\delta < \frac{\varepsilon}{2}. \tag{4.15}$$

Since $\kappa(0) = 0$, we can find a constant $\theta \in (0, \varepsilon)$ such that

$$T C_6 \kappa(\theta) < \frac{\varepsilon}{2}. \tag{4.16}$$

By the definition of T_2 we observe that for a sufficiently large N,

$$E|x_n(t) - x_i(t)|^2 < \theta \quad \text{for } t \in [T_2 + \delta, T] \text{ if } n, i \geq N. \tag{4.17}$$

Now let $n, i \geq N+1$. By Lemmas 4.3 and 4.2 as well as inequalities (4.15)–(4.17), we can derive that if $T_2 \leq t \leq T_2 + \delta$,

$$E|x_n(t) - x_i(t)|^2 \leq C_6 \int_{T_2}^{T_2+\delta} \kappa(E|x_{n-1}(s) - x_{i-1}(s)|^2) ds$$

$$+ C_6 \int_{T_2+\delta}^{T} \kappa(E|x_{n-1}(s) - x_{i-1}(s)|^2) ds \leq C_6 \kappa(4C_1)\delta + T C_6 \kappa(\theta) < \varepsilon.$$

This, together with (4.17) and $\theta < \varepsilon$, yields

$$\sup_{T_2 \leq t \leq T} E|x_n(t) - x_i(t)|^2 < \varepsilon \quad \text{whenever } n, i \geq N+1.$$

That is, (4.14) holds.

Step 3. In this step, we shall show that $T_2 = 0$. Assume otherwise that $T_2 > 0$. By step 2, we can choose a sequence of numbers $\{a_i\}_{i \geq 1}$ such that $a_i \downarrow 0$ as $i \to \infty$ and

$$\sup_{T_2 \leq t \leq T} E|x_n(t) - x_i(t)|^2 \leq a_i \quad \text{whenever } n > i \geq 1. \tag{4.18}$$

If $0 \leq t \leq T_2$ and $n > i \geq 2$, by Lemmas 4.3 and 4.2 together with (4.18), we derive that

$$E|x_n(t) - x_i(t)|^2 \leq C_6 \int_t^T \kappa(E|x_{n-1}(s) - x_{i-1}(s)|^2) ds$$

$$\leq TC_6\kappa(a_{i-1}) + C_6 \int_t^{T_2} \kappa(E|x_{n-1}(s) - x_{i-1}(s)|^2) ds$$

$$\leq TC_6\kappa(a_{i-1}) + C_6\kappa(4C_1)(T_2 - t). \tag{4.19}$$

We shall now show an assertion which is similar to Lemma 4.5. In order to state the assertion, we need introduce some new notations. Choose a positive number $\delta \in (0, T_2)$ and a positive integer $j \geq 1$ for

$$TC_6\kappa(a_j) + C_6\kappa(4C_1)\delta \leq 4C_1. \tag{4.20}$$

Define a sequence of functions $\{\phi_k(t)\}_{k \geq 1}$ on $T_2 - \delta \leq t \leq T_2$ by:

$$\phi_1(t) = TC_6\kappa(a_j) + C_6\kappa(4C_1)(T_2 - t),$$

$$\phi_{k+1}(t) = TC_6\kappa(a_{j+k}) + C_6 \int_t^{T_2} \kappa(\phi_k(s)) ds, \quad k \geq 1.$$

Fix $l \geq 1$ arbitrarily and define a sequence of functions $\{\tilde{\phi}_{k,l}(t)\}_{k \geq 1}$ by

$$\tilde{\phi}_{k,l}(t) = E|x_{l+j+k}(t) - x_{j+k}(t)|^2, \quad T_2 - \delta \leq t \leq T_2.$$

We claim that

$$\tilde{\phi}_{k,l}(t) \leq \phi_k(t) \leq \phi_{k-1}(t) \leq \cdots \leq \phi_1(t), \quad T_2 - \delta \leq t \leq T_2. \tag{4.21}$$

In fact, it follows from (4.19) that

$$\tilde{\phi}_{1,l}(t) = E|x_{l+j+1}(t) - x_{j+1}(t)|^2$$
$$\leq TC_6\kappa(a_j) + C_6\kappa(4C_1)(T_2 - t) = \phi_1(t),$$

this is (4.21) holds for $k = 1$. Then, by (4.19) and (4.20), we derive that

$$\tilde{\phi}_{2,l}(t) = E|x_{l+j+2}(t) - x_{j+2}(t)|^2$$
$$\leq TC_6\kappa(a_{j+1}) + C_6 \int_t^{T_2} \kappa(E|x_{l+j+1}(s) - x_{j+1}(s)|^2)ds$$
$$= TC_6\kappa(a_{j+1}) + C_6 \int_t^{T_2} \kappa(\tilde{\phi}_{1,l}(s))ds$$
$$\leq TC_6\kappa(a_{j+1}) + C_6 \int_t^{T_2} \kappa(\phi_1(s))ds = \phi_2(t)$$
$$\leq TC_6\kappa(a_j) + C_6 \int_t^{T_2} \kappa[C_6\kappa(a_j) + C_6\kappa(4C_1)(T_2 - t)]ds$$
$$\leq TC_6\kappa(a_j) + C_6\kappa(4C_1)(T_2 - t) = \phi_1(t),$$

In other words, we have already shown that

$$\tilde{\phi}_{2,l}(t) \leq \phi_2(t) \leq \phi_1(t) \quad \text{on } T_2 - \delta \leq t \leq T_2,$$

this is (4.21) holds for $k = 2$. Now assume that (4.21) holds for some $k \geq 2$. Then, by (4.19),

$$\tilde{\phi}_{k+1,l}(t) = E|x_{l+j+k+1}(t) - x_{j+k+1}(t)|^2$$
$$\leq TC_6\kappa(a_{j+k}) + C_6 \int_t^{T_2} \kappa(E|x_{l+j+k}(s) - x_{j+k}(s)|^2)ds$$
$$= TC_6\kappa(a_{j+k}) + C_6 \int_t^{T_2} \kappa(\tilde{\phi}_{k,l}(s))ds$$
$$\leq TC_6\kappa(a_{j+k}) + C_6 \int_t^{T_2} \kappa(\phi_k(s))ds = \phi_{k+1}(t)$$
$$\leq TC_6\kappa(a_{j+k-1}) + C_6 \int_t^{T_2} \kappa(\phi_{k-1}(s))ds = \phi_k(t),$$

that is, (4.21) holds for $k+1$ as well. So, by induction, (4.21) holds for all $k \geq 1$. Note that for each $k \geq 1$, $\phi_k(t)$ is continuous and decreasing on $[T_2 - \delta, T_2]$ and, moreover, for each t, $\phi_k(t)$ is non-increasing monotonically as $k \to \infty$. Therefore we can define function $\phi(t)$ on $[T_2 - \delta, T_2]$ by $\phi_k(t) \downarrow \phi(t)$. It is easy to verify that $\phi(t)$ is continuous and non-increasing on $[T_2 - \delta, T_2]$. By the definition of $\phi_n(t)$ and $\phi(t)$ we also have that

$$\phi(t) = \lim_{k \to \infty} \phi_{k+1}(t) = \lim_{k \to \infty} \left[TC_6\kappa(a_{j+k}) + C_6 \int_t^{T_2} \kappa(\phi_k(s))ds \right]$$
$$= C_6 \int_t^{T_2} \kappa(\phi(s))ds \quad \text{on } T_2 - \delta \leq t \leq T_2.$$

In the same way as in Step 1, we can then apply the Bihari inequality to show that
$$\phi(t) = 0 \quad \text{on } T_2 - \delta \leq t \leq T_2.$$

In particular, we see that $\phi_k(T_2 - \delta) \downarrow = \phi(T_2 - \delta)0$ as $k \to \infty$. Hence, for any $\varepsilon > 0$, we can find an integer $k_0 \geq 1$ such that $\phi_k(T_2 - \delta) < \varepsilon$ whenever $k \geq k_0$. It then follows from (4.21) that

$$\sup_{T_2-\delta \leq t \leq T_2} E|x_{l+j+k}(t) - x_{j+k}(t)|^2 \leq \phi_k(T_2 - \delta) < \varepsilon \tag{4.22}$$

whenever $k \geq k_0$. Since $l \geq 1$ is arbitrary and k_0 is independent of l, (4.22) means that

$$\sup_{T_2-\delta \leq t \leq T_2} E|x_n(t) - x_i(t)|^2 \to 0 \quad \text{as } n, i \to \infty.$$

This, together with (4.14), yields

$$\sup_{T_2-\delta \leq t \leq T} E|x_n(t) - x_i(t)|^2 \to 0 \quad \text{as } n, i \to \infty.$$

But this is in contradiction with the definition of T_2. So we must have $T_2 = 0$. In other words, we have already shown that

$$\sup_{0 \leq t \leq T} E|x_n(t) - x_i(t)|^2 \to 0 \quad \text{as } n, i \to \infty. \tag{4.23}$$

Step 4. Applying (4.23) to (4.9) we see that $\{x_n(\cdot)\}$ is a Cauchy sequence in $\mathcal{M}^2([0,T]; R^d)$ and $\{y_n(\cdot)\}$ is a Cauchy sequence in $\mathcal{M}^2([0,T]; R^{d \times m})$. Define their limits by $x(\cdot)$ and $y(\cdot)$, respectively. Letting $n \to \infty$ in (4.5) we finally obtain

$$x(t) + \int_t^T f(x(s), y(s), s)ds + \int_t^T [g(x(s), x) + y(s)]dB(s) = X$$

on $0 \leq t \leq T$. The existence of the solution has been proved.

Uniqueness: To show the uniqueness, let $\{x(\cdot), y(\cdot)\}$ and $\{\bar{x}(\cdot), \bar{y}(\cdot)\}$ be two solutions of equation (3.1). Then, in the same way as in the proof of Lemma 4.2, we can show that

$$E|x(t) - \bar{x}(t)|^2 + \frac{1}{2} E \int_t^T |y(s) - \bar{y}(s)|^2 ds$$

$$\leq 2\left[4K + 2 + \frac{1}{4K+2}\right]$$

$$\times \int_t^T [E|x(s) - \bar{x}(s)|^2 + \kappa(E|x(s) - \bar{x}(s)|^2)]ds \tag{4.24}$$

for $0 \le t \le T$. Since $\kappa(\cdot)$ is a concave function with $\kappa(0) = 0$, we have
$$\kappa(u) \ge \kappa(1)u \quad \text{for } 0 \le u \le 1.$$
So
$$\int_{0+} \frac{du}{u + \kappa(u)} \ge \frac{\kappa(1)}{\kappa(1)+1} \int_{0+} \frac{du}{\kappa(u)} = \infty.$$
Therefore we can apply the Bihari inequality to (4.24) to obtain that
$$E|x(t) - \bar{x}(t)|^2 = 0 \quad \text{for all } 0 \le t \le T.$$
This implies immediately that $x(t) = \bar{x}(t)$ for all $0 \le t \le T$ almost surely. It then follows from (4.23) that
$$E \int_0^T |y(s) - \bar{y}(s)|^2 ds = 0.$$
The uniqueness has also been proved and the proof of the theorem is therefore complete.

7.5 REGULARITIES

In the previous sections, we observed clearly that the second moment of of the solution to the backward stochastic differential equation (3.1) is finite. In this section, we shall discuss the higher order moments of the solution. For this purpose, we impose the following hypotheses: Let $p \ge 2$. Assume that
$$f(0,0,\cdot) \in M^p([0,T]; R^d) \quad \text{and} \quad g(0,\cdot) \in M^p([0,T]; R^{d \times m}). \quad (5.1)$$
Assume also that there exists a positive constant $K > 0$ such that
$$|f(x,y,t) - f(0,0,t)|^2 \le K(1 + |x|^2 + |y|^2) \quad \text{a.s.} \quad (5.2)$$
and
$$|g(x,t) - g(0,t)|^2 \le K(1 + |x|^2) \quad \text{a.s} \quad (5.3)$$
for all $x \in R^d$, $y \in R^{d \times m}$ and $t \in [0,T]$. Obviously, (3.4) and (3.5) imply (5.2) and (5.3), respectively. It is also not difficult to see that (4.1) and (4.2) imply (5.2) and (5.3), respectively. For example, making use of (4.4), we derive from (4.1) that
$$|f(x,y,t) - f(0,0,t)|^2 \le a + b|x|^2 + K|y|^2 \le (a \vee b \vee K)(1 + |x|^2 + |y|^2).$$
In other words, hypotheses (5.2) and (5.3) follow usually from the conditions imposed for the existence and uniqueness of the solution. Moreover, (5.1) reduces to (3.3) if $p = 2$ but we naturally require (5.1) when discuss the pth moment.

Theorem 5.1 *Let $p \geq 2$ and $X \in L^p_{\mathcal{F}_T}(\Omega; R^d)$. Let (5.1)–(5.3) hold. Then the solution of equation (3.1) has the properties that*

$$E|x(t)|^p \leq (E|X|^p + \bar{C}_1)e^{2pT(4K+1)} \quad \text{for all } 0 \leq t \leq T \quad (5.4)$$

and

$$E \int_0^T |x(t)|^{p-2}|y(t)|^2 dt \leq \frac{4}{p}(E|X|^p + \bar{C}_1)\left[1 + e^{2pT(4K+1)}\right], \quad (5.5)$$

where

$$\bar{C}_1 = \frac{2}{p}E \int_0^T \left[\frac{p}{8K}(|f(0,0,s)|^2 + K) + 4p(|g(0,s)|^2 + K)\right]^{\frac{p}{2}} ds < \infty.$$

Proof. By Itô's formula, we have that

$$|X|^p - |x(t)|^p = p\int_t^T |x(s)|^{p-2}x^T(s)f(x(s), y(s), s)ds$$

$$+ \frac{p}{2}\int_t^T |x(s)|^{p-2}|g(x(s), s) + y(s)|^2 ds$$

$$+ \frac{p(p-2)}{2}\int_t^T |x(s)|^{p-4}|x^T(s)[g(x(s), s) + y(s)]|^2 ds$$

$$+ p\int_t^T |x(s)|^{p-2}x^T(s)[g(x(s), s) + y(s)]dB(s).$$

This implies that

$$|x(t)|^p + \frac{p}{2}\int_t^T |x(s)|^{p-2}|y(s)|^2 ds$$

$$\leq |X|^p - p\int_t^T |x(s)|^{p-2}x^T(s)f(x(s), y(s), s)ds$$

$$- p\int_t^T |x(s)|^{p-2}\text{trace}[g^T(x(s), s)y(s)]ds$$

$$- p\int_t^T |x(s)|^{p-2}x^T(s)[g(x(s), s) + y(s)]dB(s). \quad (5.6)$$

Taking expectation on both sides we obtain that

$$E|x(t)|^p + \frac{p}{2}E\int_t^T |x(s)|^{p-2}|y(s)|^2 ds$$

$$\leq E|X|^p - pE\int_t^T |x(s)|^{p-2}x^T(s)f(x(s), y(s), s)ds$$

$$- pE\int_t^T |x(s)|^{p-2}\text{trace}[g^T(x(s), s)y(s)]ds. \quad (5.7)$$

By condition (5.2), we have that
$$|f(x(s),y(s),s)|^2 \leq 2|f(0,0,s)|^2 + 2|f(x(s),y(s),s) - f(0,0,s)|^2$$
$$\leq 2|f(0,0,s)|^2 + 2K(1+|x(s)|^2+|y(s)|^2).$$

We then estimate that
$$-px^T(s)f(x(s),y(s),s)$$
$$\leq 4pK|x(s)|^2 + \frac{p}{16K}|f(x(s),y(s),s)|^2$$
$$\leq \frac{p}{8K}(|f(0,0,s)|^2 + K) + p(4K+1)|x(s)|^2 + \frac{p}{8}|y(s)|^2. \qquad (5.8)$$

Similarly, we can use condition (5.3) to show that
$$-p\,\text{trace}[g^T(x(s),s)y(s)]$$
$$\leq 4p(|g(0,s)|^2 + K) + 4pK|x(s)|^2 + \frac{p}{8}|y(s)|^2. \qquad (5.9)$$

Substituting (5.8) and (5.9) into (5.7) yields that
$$E|x(t)|^p + \frac{p}{2}E\int_t^T |x(s)|^{p-2}|y(s)|^2 ds$$
$$\leq E|X|^p + p(8K+1)E\int_t^T |x(s)|^p ds + \frac{p}{4}E\int_t^T |x(s)|^{p-2}|y(s)|^2 ds$$
$$+ E\int_t^T |x(s)|^{p-2}\left[\frac{p}{8K}(|f(0,0,s)|^2 + K) + 4p(|g(0,s)|^2 + K)\right] ds. \qquad (5.10)$$

On the other hand, using the elementary inequality $u^\alpha v^{1-\alpha} \leq \alpha u + (1-\alpha)v$ for $\alpha \in [0,1]$ and $u,v \geq 0$, we derive that
$$E\int_t^T |x(s)|^{p-2}\left[\frac{p}{8K}(|f(0,0,s)|^2 + K) + 4p(|g(0,s)|^2 + K)\right] ds$$
$$\leq \frac{p-2}{p}E\int_t^T |x(s)|^p ds$$
$$+ \frac{2}{p}E\int_t^T \left[\frac{p}{8K}(|f(0,0,s)|^2 + K) + 4p(|g(0,s)|^2 + K)\right]^{\frac{p}{2}} ds$$
$$\leq E\int_t^T |x(s)|^p ds + \bar{C}_1, \qquad (5.11)$$

where $\bar{C}_1$ has been defined in the statement of the theorem and, by condition (5.1), $\bar{C}_1 < \infty$. Substituting (5.11) into (5.10) we get that
$$E|x(t)|^p + \frac{p}{2}E\int_t^T |x(s)|^{p-2}|y(s)|^2 ds$$
$$\leq E|X|^p + \bar{C}_1 + 2p(4K+1)\int_t^T E|x(s)|^p ds$$
$$+ \frac{p}{4}E\int_t^T |x(s)|^{p-2}|y(s)|^2 ds. \qquad (5.12)$$

In particular, this gives that

$$E|x(t)|^p \leq E|X|^p + \bar{C}_1 + 2p(4K+1) \int_t^T E|x(s)|^p ds.$$

The Gronwall inequality now implies that

$$E|x(t)|^p \leq (E|X|^p + \bar{C}_1)e^{2p(4K+1)(T-t)} \quad \text{for all } 0 \leq t \leq T, \tag{5.13}$$

and the required assertion (5.4) follows. Finally, we derive from (5.12) and (5.13) that

$$E \int_0^T |x(s)|^{p-2}|y(s)|^2 ds$$
$$\leq \frac{4}{p}(E|X|^p + \bar{C}_1) + 8(4K+1) \int_0^T E|x(s)|^p ds$$
$$\leq (E|X|^p + \bar{C}_1)\left[\frac{4}{p} + 8(4K+1) \int_0^T e^{2p(4K+1)(T-s)} ds\right]$$
$$\leq \frac{4}{p}(E|X|^p + \bar{C}_1)\left[1 + e^{2pT(4K+1)}\right],$$

which is the required assertion (5.5). The proof is complete.

Theorem 5.2 *Let $p \geq 2$ and $X \in L^p_{\mathcal{F}_T}(\Omega; R^d)$. Let (5.1)–(5.3) hold. Then the solution of equation (3.1) has the properties that*

$$E\left(\sup_{0 \leq t \leq T} |x(t)|^p\right) < \infty \tag{5.14}$$

and

$$E\left(\int_0^T |y(t)|^2 dt\right)^{\frac{p}{2}} < \infty. \tag{5.15}$$

Proof. Substituting (5.8) and (5.9) into (5.6) yields that

$$|x(t)|^p + \frac{p}{2}\int_t^T |x(s)|^{p-2}|y(s)|^2 ds$$
$$\leq |X|^p + p(8K+1) \int_t^T |x(s)|^p ds + \frac{p}{4} \int_t^T |x(s)|^{p-2}|y(s)|^2 ds$$
$$+ \int_t^T |x(s)|^{p-2}\left[\frac{p}{8K}(|f(0,0,s)|^2 + K) + 4p(|g(0,s)|^2 + K)\right] ds$$
$$- p \int_t^T |x(s)|^{p-2}x^T(s)[g(x(s),s) + y(s)]dB(s).$$

Hence

$$E\left(\sup_{0\le t\le T}|x(t)|^p\right)$$
$$\le E|X|^p + p(8K+1)\int_0^T E|x(s)|^p ds$$
$$+ E\int_0^T |x(s)|^{p-2}\left[\frac{p}{8K}(|f(0,0,s)|^2+K)+4p(|g(0,s)|^2+K)\right]ds$$
$$+ E\left(\sup_{0\le t\le T}\left[-p\int_t^T |x(s)|^{p-2}x^T(s)[g(x(s),s)+y(s)]dB(s)\right]\right). \quad (5.16)$$

But, letting $t=0$ in (5.11), we see that

$$E\int_0^T |x(s)|^{p-2}\left[\frac{p}{8K}(|f(0,0,s)|^2+K)+4p(|g(0,s)|^2+K)\right]ds$$
$$\le \int_0^T E|x(s)|^p ds + \bar{C}_1.$$

Substituting this into (5.16) we obtain that

$$E\left(\sup_{0\le t\le T}|x(t)|^p\right)$$
$$\le \bar{C}_1 + E|X|^p + [p(8K+1)+1]\int_0^T E|x(s)|^p ds$$
$$+ E\left(\sup_{0\le t\le T}\left[-p\int_t^T |x(s)|^{p-2}x^T(s)[g(x(s),s)+y(s)]dB(s)\right]\right). \quad (5.17)$$

On the other hand, by the Burkholder–Davis–Gundy inequality etc., we can derive that

$$E\left(\sup_{0\le t\le T}\left[-p\int_t^T |x(s)|^{p-2}x^T(s)[g(x(s),s)+y(s)]dB(s)\right]\right)$$
$$- E\left(\sup_{0\le t\le T}\left[-p\int_0^T |x(s)|^{p-2}x^T(s)[g(x(s),s)+y(s)]dB(s)\right.\right.$$
$$\left.\left.+ p\int_0^t |x(s)|^{p-2}x^T(s)[g(x(s),s)+y(s)]dB(s)\right]\right)$$
$$= pE\left(\sup_{0\le t\le T}\int_0^t |x(s)|^{p-2}x^T(s)[g(x(s),s)+y(s)]dB(s)\right)$$
$$\le 4\sqrt{2}pE\left(\int_0^T |x(s)|^{2p-2}|g(x(s),s)+y(s)|^2 ds\right)^{\frac{1}{2}}$$
$$\le 4\sqrt{2}pE\left(\left[\sup_{0\le s\le T}|x(s)|^p\right]\int_0^T |x(s)|^{p-2}|g(x(s),s)+y(s)|^2 ds\right)^{\frac{1}{2}}$$

$$\leq \frac{1}{2} E\left(\sup_{0\leq s\leq T} |x(s)|^p\right) + 16p^2 E\int_0^T |x(s)|^{p-2}|g(x(s),s)+y(s)|^2 ds$$

$$\leq \frac{1}{2} E\left(\sup_{0\leq s\leq T} |x(s)|^p\right) + 32p^2 E\int_0^T |x(s)|^{p-2}|y(s)|^2 ds$$

$$+ 32p^2 E\int_0^T |x(s)|^{p-2}|g(x(s),s)|^2 ds. \tag{5.18}$$

However, in the same way as in the proof of Theorem 5.1, we can estimate that

$$E\int_0^T |x(s)|^{p-2}|g(x(s),s)|^2 ds$$

$$\leq E\int_0^T |x(s)|^{p-2}\Big[2|g(0,s)|^2 + 2K(1+|x(s)|^2)\Big] ds$$

$$\leq \left(2K + \frac{p-2}{p}\right) E\int_0^T |x(s)|^p ds + \frac{2}{p} E\int_0^T [2|g(0,s)|^2 + 2K]^{\frac{p}{2}} ds$$

$$\leq (2K+1)\int_0^T E|x(s)|^p ds + \bar{C}_1.$$

Substituting this into (5.18) gives that

$$E\left(\sup_{0\leq t\leq T}\left[-p\int_t^T |x(s)|^{p-2} x^T(s)[g(x(s),s)+y(s)]dB(s)\right]\right)$$

$$\leq \frac{1}{2} E\left(\sup_{0\leq s\leq T} |x(s)|^p\right) + 32p^2 E\int_0^T |x(s)|^{p-2}|y(s)|^2 ds$$

$$+ 32p^2(2K+1)\int_0^T E|x(s)|^p ds + 32p^2 \bar{C}_1. \tag{5.19}$$

Substituting (5.19) into (5.17) we obtain that

$$E\left(\sup_{0\leq t\leq T} |x(t)|^p\right)$$

$$\leq 2\bar{C}_1(1+32p^2) + 2E|X|^p$$

$$+ 2\big[32p^2(2K+1) + p(8K+1) + 1\big]\int_0^T E|x(s)|^p ds$$

$$+ 64p^2 E\int_0^T |x(s)|^{p-2}|y(s)|^2 ds$$

$$:= \bar{C}_2. \tag{5.20}$$

But by Theorem 5.1, $\bar{C}_2 < \infty$ and hence the required assertion (5.14) follows.

We now begin to show (5.15). Clearly, (5.15) holds if we can show that

$$E\left(\int_u^v |y(s)|^2 ds\right)^{\frac{p}{2}} < \infty \tag{5.21}$$

for any $0 \leq u < v \leq T$ satisfying
$$\bar{C}_p > 3^{p-1}[4K(v-u)]^{\frac{p}{2}}, \tag{5.22}$$
where $c_p = 1$ or $(2p)^{-p/2}$ according to $p = 2$ or $p > 2$, respectively. Fix such u and v arbitrarily. For $u \leq t \leq v$, we have
$$\int_u^t y(s)dB(s) = x(t) - x(u) - \int_u^t f(x(s), y(s), s)ds - \int_u^t g(x(s), s)dB(s).$$
Hence
$$E\left(\sup_{u \leq t \leq v} \left|\int_u^t y(s)dB(s)\right|^p\right) \leq 3^{p-1} E\left(\sup_{u \leq t \leq v} |x(t) - x(u)|^p\right)$$
$$+ 3^{p-1} E\left(\int_u^v |f(x(s), y(s), s)|ds\right)^p$$
$$+ 3^{p-1} E\left(\sup_{u \leq t \leq v} \left|\int_u^t g(x(s), s)dB(s)\right|^p\right). \tag{5.23}$$

Note from (5.20) that
$$E\left(\sup_{u \leq t \leq v} |x(t) - x(u)|^p\right) \leq 2^p E\left(\sup_{u \leq t \leq v} |x(t)|^p\right) \leq 2^p \bar{C}_2 < \infty. \tag{5.24}$$
Also, by the Burkholder–Davis–Gundy inequality etc., we derive that
$$E\left(\sup_{u \leq t \leq v} \left|\int_u^t g(x(s), s)dB(s)\right|^p\right)$$
$$\leq C_p E\left(\int_u^v |g(x(s), s)|^2 ds\right)^{\frac{p}{2}}$$
$$\leq C_p (v-u)^{\frac{p-2}{2}} E \int_u^v \left[2|g(0,s)|^2 + 2K(1+|x(s)|^2)\right]^{\frac{p}{2}} ds$$
$$\leq C_p 6^{\frac{p}{2}} (v-u)^{\frac{p-2}{2}} E \int_u^v \left[|g(0,s)|^p + K^{\frac{p}{2}} + K^{\frac{p}{2}}|x(s)|^p\right] ds$$
$$< \infty, \tag{5.25}$$
where C_p is the constant given by the Burkholder–Davis–Gundy inequality, that is $C_p = 4$ or $[p^{p+1}/2(p-1)^{p-1}]^{p/2}$ according to $p = 2$ or $p > 2$, respectively. Moreover, we have that
$$E\left(\int_u^v |f(x(s), y(s), s)|ds\right)^p$$
$$\leq E\left((v-u)\int_u^v |f(x(s), y(s), s)|^2 ds\right)^{\frac{p}{2}}$$
$$\leq (v-u)^{\frac{p}{2}} E\left(\int_u^v \left[2|f(0,0,s)|^2 + 2K(1+|x(s)|^2 + |y(s)|^2)\right] ds\right)^{\frac{p}{2}}$$
$$\leq \bar{C}_3 + [4K(v-u)]^{\frac{p}{2}} E\left(\int_u^v |y(s)|^2 ds\right)^{\frac{p}{2}}, \tag{5.26}$$

where

$$\bar{C}_3 = [2(v-u)]^{\frac{p}{2}} E\left(\int_u^v \left[2|f(0,0,s)|^2 + 2K(1+|x(s)|^2)\right]ds\right)^{\frac{p}{2}}$$

$$\leq 2^p(v-u)^{p-1} E\int_u^v \left[|f(0,0,s)|^2 + K + K|x(s)|^2\right]^{\frac{p}{2}} ds$$

$$\leq 6^p(v-u)^{p-1} E\int_u^v \left[|f(0,0,s)|^p + K^{\frac{p}{2}} + K^{\frac{p}{2}}|x(s)|^p\right]ds$$

$$< \infty.$$

On the other hand, by the Burkholder–Davis–Gundy inequality, we have that

$$c_p E\left(\int_u^v |y(s)|^2 ds\right)^{\frac{p}{2}} \leq E\left(\sup_{u\leq t\leq v}\left|\int_u^t y(s)dB(s)\right|^p\right), \quad (5.27)$$

where c_p has been defined above. Combining (5.23)–(5.27) yields that

$$\left(c_p - 3^{p-1}[4K(v-u)]^{\frac{p}{2}}\right) E\left(\int_u^v |y(s)|^2 ds\right)^{\frac{p}{2}} < \infty.$$

Finally, making use of (5.22), we obtain (5.21) and therefore the proof is complete.

7.6 BSDE AND QUASILINEAR PDE

In Section 2.8, we established the Feynman–Kac formula which gives the *explicit* representation for the solution to a *linear* partial differential equation in terms of the solutions of the corresponding stochastic differential equations. However, we also pointed out there that for a solution of a *quasilinear* partial differential equation, the stochastic representation is *not* explicit. For example, let us recall the quasilinear parabolic partial differential equation (2.8.27), namely

$$\begin{cases} \frac{\partial}{\partial t}u(x,t) + \mathcal{L}u(x,t) + c(x,u)u(x,t) = 0 & \text{in } R^d \times [0,T), \\ u(x,T) = \phi(x) & \text{in } R^d. \end{cases} \quad (6.1)$$

Here

$$\mathcal{L} = \frac{1}{2}\sum_{i,j=1}^d a_{ij}(x,t)\frac{\partial^2}{\partial x_i \partial x_j} + \sum_{i=1}^d f_i(x,t)\frac{\partial}{\partial x_i},$$

$a_{ij}(x,t)$, $f_i(x,t)$ and $c(x,u)$ are all the same as defined in Section 2.8. Moreover, set

$$f(x,t) = (f_1,\cdots,f_d)^T \quad \text{and} \quad a(x,t) = (a_{ij}(x,t))_{d\times d}.$$

Let $g(x,t) = (g_{ij}(x,t))_{d\times d}$ be the square root of $a(x,t)$, i.e.

$$g(x,t)g^T(x,t) = a(x,t).$$

Let $B(t)$ be a d-dimensional Brownian motion. For every $(x,t) \in R^d \times [0,T)$, solve the stochastic differential equation

$$\begin{cases} d\xi_{x,t}(s) = f(\xi_{x,t}(s),s)ds + g(\xi_{x,t}(s),s)dB(s), & t \le s \le T, \\ \xi_{x,t}(t) = x. \end{cases} \quad (6.2)$$

The Feynman–Kac formula tells us that the solution of equation (6.1) can be expressed as

$$u(x,t) = E\left[\phi(\xi_{x,t}(T))\exp\left(\int_t^T c(\xi_{x,t}(s), u(\xi_{x,t}(s),s))ds\right)\right]. \quad (6.3)$$

Of course, this is no longer an explicit representation. However, in this section, we shall establish an explicit representation in terms of the solutions of the corresponding backward stochastic differetial equations.

In general, let us consider the quasilinear parabolic partial differential equation

$$\begin{cases} \frac{\partial}{\partial t}u(x,t) + \mathcal{L}u(x,t) + F(x,u,\nabla u g, t) = 0 & \text{in } R^d \times [0,T), \\ u(x,T) = \phi(x) & \text{in } R^d. \end{cases} \quad (6.4)$$

Here F is a real-valued function defined on $R^d \times R \times R^d \times [0,T]$ and ∇u is the gradient of u, i.e.

$$\nabla u = \left(\frac{\partial u}{\partial x_1}, \ldots, \frac{\partial u}{\partial x_d}\right).$$

For every $(x,t) \in R^d \times [0,T)$, let $\xi_{x,t}(s)$, $t \le s \le T$ be the solution of stochastic differential equation (6.2). Consider the corresponding backward stochastic differential equation

$$X_{x,t}(s) = \phi(\xi_{x,t}(T)) + \int_s^T F(\xi_{x,t}(r), X_{x,t}(r), Y_{x,t}(r), r)dr$$
$$- \int_s^T Y_{x,t}(r)dB(r) \quad \text{on } t \le s \le T, \quad (6.5)$$

where $X_{x,t}(\cdot)$ takes values in R but $Y_{x,t}(\cdot)$ in $R^{1\times d}$. The following theorem is called the generalized Feynman–Kac formula and is due to Pardoux & Peng (1992).

Theorem 6.1 *Assume that all the functions a, f, g, F, ϕ are sufficiently smooth so that equation (6.4) has a unique $C^{2,1}$-solution and equations (6.2) and (6.5) have their own unique solutions as well. Then the unique solution of equation (6.4) can be represented as*

$$u(x,t) = EX_{x,t}(t). \quad (6.6)$$

Proof. Applying the Itô formula to $u(\xi_{x,t}(r), r)$ we have that

$$du(\xi_{x,t}(r), r) = \left[\frac{\partial}{\partial t}u(\xi_{x,t}(r), r) + \mathcal{L}u(\xi_{x,t}(r), r)\right]dr$$
$$+ \nabla u(\xi_{x,t}(r), r)g(\xi_{x,t}(r), r)dB(r).$$

Since $u(x, t)$ is the solution of equation (6.4), we see that

$$du(\xi_{x,t}(r), r) = -F\big(\xi_{x,t}(r), u(\xi_{x,t}(r), r), \nabla u(\xi_{x,t}(r), r)g(\xi_{x,t}(r), r), r\big) dr$$
$$+ \nabla u(\xi_{x,t}(r), r)g(\xi_{x,t}(r), r)dB(r).$$

Integrating both sides from $r = s$ to $r = T$ implies that

$$u(\xi_{x,t}(T), T) - u(\xi_{x,t}(s), s)$$
$$= -\int_s^T F\big(\xi_{x,t}(r), u(\xi_{x,t}(r), r), \nabla u(\xi_{x,t}(r), r)g(\xi_{x,t}(r), r), r\big) dr$$
$$+ \int_s^T \nabla u(\xi_{x,t}(r), r)g(\xi_{x,t}(r), r)dB(r).$$

Noting that $u(x, T) = \phi(x)$, we obtain that

$$u(\xi_{x,t}(s), s) = \phi(\xi_{x,t}(T))$$
$$= \int_s^T F\big(\xi_{x,t}(r), u(\xi_{x,t}(r), r), \nabla u(\xi_{x,t}(r), r)g(\xi_{x,t}(r), r), r\big) dr$$
$$- \int_s^T \nabla u(\xi_{x,t}(r), r)g(\xi_{x,t}(r), r)dB(r). \qquad (6.7)$$

Comparing equation (6.7) with equation (6.5), we see, by the uniqueness, that $\{u(\xi_{x,t}(s), s), \nabla u(\xi_{x,t}(s), s)g(\xi_{x,t}(s), s)\}_{t \leq s \leq T}$ must coincide with the unique solution $\{X_{x,t}(s), Y_{x,t}(s)\}_{t \leq s \leq T}$ of equation (6.5). In particular, we have that

$$u(x, t) = u(\xi_{x,t}(t), t) = X_{x,t}(t) \quad \text{a.s.}$$

Taking expectation on both sides and bearing in mind that $u(x, t)$ is deterministic, we obtain the required assertion (6.6). The proof is complete.

To show that Theorem 6.1 is a generalization of the Feynman–Kac formula, let $F(x, u, y, t) = c(x, t)u$. In this case, equation (6.4) reduces to the linear equation

$$\begin{cases} \frac{\partial}{\partial t}u(x, t) + \mathcal{L}u(x, t) + c(x, t)u(x, t) = 0 & \text{in } R^d \times [0, T), \\ u(x, T) = \phi(x) & \text{in } R^d. \end{cases} \qquad (6.8)$$

Moreover, the backward stochastic differential equation (6.5) becomes

$$X_{x,t}(s) = \phi(\xi_{x,t}(T)) + \int_s^T c(\xi_{x,t}(r), r)X_{x,t}(r)dr - \int_s^T Y_{x,t}(r)dB(r). \qquad (6.9)$$

Note that equation (6.9) has the explicit solution

$$X_{x,t}(s) = \phi(\xi_{x,t}(T))exp\left[\int_s^T c(\xi_{x,t}(r),r)dr\right]$$
$$- \int_s^T exp\left[\int_s^r c(\xi_{x,t}(v),v)dv\right]Y_{x,t}(r)dB(r)$$

Substituting this into (6.6) gives

$$u(x,t) = EX_{x,t}(t) = E\left(\phi(\xi_{x,t}(T))exp\left[\int_t^T c(\xi_{x,t}(r),r)dr\right]\right),$$

but this is the same as the classical Feynman–Kac formula obtained in Section 2.8.

8

Stochastic Oscillators

8.1 INTRODUCTION

A stochastic oscillator is described mathematically as the solution of an appropriate ordinary differential equation, which is driven by an external disturbance of white noise. Accordingly, such solutions are stochastic processes.

In this chapter, we shall mainly discuss the stochastic oscillators described by the d-dimensional second order stochastic differential equations of the form

$$\ddot{x}(t) + f(x(t), \dot{x}(t), t) = g(x(t), \dot{x}(t), t)\dot{B}(t) \quad \text{on } t \geq 0, \tag{1.1}$$

where $\dot{B}(t)$ is an m-dimensional white noise, $x(t)$ takes values in R^d, $f: R^d \times R^d \times R_+ \to R^d$ and $g: R^d \times R^d \times R_+ \to R^{d \times m}$. Introducing the new variable $y(t) = \dot{x}(t)$, we can write equation (1.1) as the $2d$-dimensional Itô equation

$$\begin{cases} dx(t) = y(t)dt, \\ dy(t) = -f(x(t), y(t), t)dt + g(x(t), y(t), t)dB(t). \end{cases} \tag{1.2}$$

Clearly, this is a special Itô stochastic differential equation in the sense that $x(t)$ and $y(t)$ have the simple relation $dx(t) = y(t)dt$ which does not involve stochastic differential. It is this special form that makes the equation have a number of important properties. In this chapter we shall discuss some of these properties e.g. the oscillations, the statistical distribution of their zeros and the energy bounds.

The materials in this chapter are mainly based on Markus & Weerasinghe (1988) and Mao & Markus (1991), except the Cameron–Martin–Girsanov transformation theorem which is classical.

8.2 THE CAMERON–MARTIN–GIRSANOV THEOREM

In this section we shall present the well-known Cameron–Martin–Girsanov transformation theorem of Brownian motions, which will play an important role in this chapter.

As usual, let $(\Omega, \mathcal{F}, \{\mathcal{F}_t\}_{t\geq 0}, P)$ be a complete probability space with a filtration $\{\mathcal{F}_t\}$ satisfying the usual conditions. If $\tilde{P}$ is a measure on $(\Omega, \mathcal{F})$ given by
$$\tilde{P}(A) = \int_A f(\omega) dP(\omega), \quad A \in \mathcal{F},$$
we then write $d\tilde{P}(\omega) = f(\omega) dP(\omega)$.

Theorem 2.1 (The Cameron–Martin–Girsanov Theorem) *Assume that $\{B(t)\}_{0\leq t\leq T}$ is an m-dimensional Brownian motion defined on the probability space $(\Omega, \mathcal{F}, \{\mathcal{F}_t\}_{t\geq 0}, P)$. Let $\phi = (\phi_1, \cdots, \phi_m)^T \in \mathcal{L}^2([0,T]; R^m)$. Define*

$$\zeta_0^T = -\frac{1}{2}\int_0^T |\phi(u)|^2 du + \int_0^T \phi^T(u) dB(u), \tag{2.1}$$

$$\tilde{B}(t) = B(t) - \int_0^t \phi(u) du, \tag{2.2}$$

$$d\tilde{P}(\omega) = \exp\bigl[\zeta_0^T(\phi)\bigr] dP(\omega). \tag{2.3}$$

If

$$\tilde{P}(\Omega) = 1, \tag{2.4}$$

then $\{\tilde{B}(t)\}_{0\leq t\leq T}$ is a new m-dimensional Brownian motion defined on the probability space $(\Omega, \mathcal{F}, \tilde{P})$ with respect to the same filtration $\{\mathcal{F}_t\}$. Moreover, a sufficient condition for (2.4) to hold is that there are two positive constants μ and C such that

$$E\Bigl[e^{\mu|\phi(t)|^2}\Bigr] \leq C \quad \text{for all } 0 \leq t \leq T. \tag{2.5}$$

The proof of the Cameron–Martin–Girsanov theorem is omitted here since it can be found in many texts e.g. Friedman (1975). Instead, we shall explain how this theorem can be used to transform a complicated Itô equation into a relatively easier one without loss of certain properties.

Let $\tilde{P}(\Omega) = 1$. If $g \in \mathcal{L}^2([0,T]; R^{d\times m})$, then, by definition,
$$P\Bigl\{\omega : \int_0^T |g(t,\omega)|^2 dt < \infty\Bigr\} = 1.$$

Since $\tilde{P}$ is absolutely continuous with respect to P,
$$\tilde{P}\Bigl\{\omega : \int_0^T |g(t,\omega)|^2 dt < \infty\Bigr\} = 1.$$

Hence the stochastic integral of g with respect to $\tilde{B}$

$$\int_0^t g(s)d\tilde{B}(s) \quad \text{on } 0 \le t \le T$$

is well defined. Let $\{g_k\}$ be a sequence of simple processes in $\mathcal{L}^2([0,T]; R^{d\times m})$ such that

$$\int_0^T |g_k(s) - g(s)|^2 ds \to 0 \quad P\text{-a.s.}$$

Then

$$\int_0^T |g_k(s) - g(s)|^2 ds \to 0 \quad \tilde{P}\text{-a.s.}$$

Consequently

$$\int_0^t g_k(s)dB(s) \to \int_0^t g(s)dB(s) \quad \text{in probability } P,$$

and

$$\int_0^t g_k(s)d\tilde{B}(s) \to \int_0^t g(s)d\tilde{B}(s) \quad \text{in probability } \tilde{P}.$$

Therefore, for a subsequence $\{k'\}$,

$$\int_0^t g_{k'}(s)dB(s) \to \int_0^t g(s)dB(s) \quad P\text{-a.s.} \tag{2.6}$$

and

$$\int_0^t g_{k'}(s)d\tilde{B}(s) \to \int_0^t g(s)d\tilde{B}(s) \quad \tilde{P}\text{-a.s.} \tag{2.7}$$

From (2.2) we easily see that

$$\int_0^t g_{k'}(s)dB(s) = \int_0^t g_{k'}(s)d\tilde{B}(s) + \int_0^t g_{k'}(s)\phi(s)ds. \tag{2.8}$$

It is also clear that

$$\int_0^t g_{k'}(s)\phi(s)ds \to \int_0^t g(s)\phi(s)ds.$$

Hence, taking $k' \to \infty$ in (2.8) and using (2.6) and (2.7), we get

$$\int_0^t g(s)dB(s) = \int_0^t g(s)d\tilde{B}(s) + \int_0^t g(s)\phi(s)ds \tag{2.9}$$

almost surely in P (or in $\tilde{P}$). This equality along with Theorem 2.1 yields the following result, which is known as the Girsanov theorem.

Theorem 2.2 (The Girsanov Theorem) *Let $\{B(t)\}_{0 \leq t \leq T}$ be an m-dimensional Brownian motion defined on the probability space $(\Omega, \mathcal{F}, \{\mathcal{F}_t\}_{t \geq 0}, P)$. Let $x(t)$ be a d-dimensional Itô process on $[0, T]$, namely*

$$x(t) = x(0) + \int_0^t f(s)ds + \int_0^t g(s)dB(s) \qquad (2.10)$$

with $f \in \mathcal{L}^2([0,T]; R^d)$ and $g \in \mathcal{L}^2([0,T]; R^{d \times m})$. Let $\phi = (\phi_1, \cdots, \phi_m) \in \mathcal{L}^2([0,T]; R^m)$. Let $\tilde{B}(t)$ and $\tilde{P}$ be the same as defined in Theorem 2.1. If $\tilde{P}(\Omega) = 1$, then $x(t)$ is still an Itô process on the probability space $(\Omega, \mathcal{F}, \tilde{P})$ with respect to the Brownian motion $\tilde{B}(t)$. More precisely, we have

$$x(t) = x(0) + \int_0^t [f(s) + g(s)\phi(s)]ds + \int_0^t g(s)d\tilde{B}(s). \qquad (2.11)$$

If $x(t)$ is the solution to the stochastic differential equation

$$dx(t) = f(x(t), t)dt + g(x(t), t)dB(t), \quad 0 \leq t \leq T \qquad (2.12)$$

on the probability space $(\Omega, \mathcal{F}, P)$, then the Girsanov theorem tells us that $x(t)$ is the solution to the following new equation

$$dx(t) = [f(x(t), t) + g(x(t), t)\phi(t)]dt + g(x(t), t)d\tilde{B}(t), \quad 0 \leq t \leq T \qquad (2.13)$$

on the probability space $(\Omega, \mathcal{F}, \tilde{P})$. In many situations, we can choose $\phi(t)$ to make the term $f(x,t) + g(x,t)\phi(t)$ relatively simpler so that equation (2.13) can be dealt with much more easily than the original equation (2.12). For example, if $d = m$ and $g(x,t)$ is invertible, and both $f(x,t)$ and $g^{-1}(x,t)$ are bounded, we can let

$$\phi(t) = -g^{-1}(x(t), t)f(x(t), t).$$

In this case, (2.5) is fulfilled due to the boundedness of $\phi(t)$ and so $\tilde{P}(\Omega) = 1$. Moreover, equation (2.13) reduces to

$$dx(t) = g(x(t), t)d\tilde{B}(t), \quad 0 \leq t \leq T$$

and this is clearly much simpler than equation (2.12).

8.3 NONLINEAR STOCHASTIC OSCILLATORS

Let $B(t)$ be a 1-dimensional Brownian motion. Consider the scalar nonlinear oscillator

$$\ddot{x}(t) + \kappa(x(t), \dot{x}(t), t) = h\dot{B}(t) \quad \text{on } t \geq 0, \qquad (3.1)$$

where h is a positive constant and κ is a real function on $R^2 \times R_+$. By introducing the new variable $y(t) = \dot{x}(t)$, the corresponding Itô equation is

$$d\begin{bmatrix} x(t) \\ y(t) \end{bmatrix} = \begin{bmatrix} y(t) \\ -\kappa(x(t), y(t), t) \end{bmatrix} dt + \begin{bmatrix} 0 \\ h \end{bmatrix} dB(t). \qquad (3.1)'$$

We shall assume that $\kappa(x, y, t)$ is locally Lipschitz continuous in (x, y) and is bounded on the whole domain $R^2 \times R_+$. By the theory of stochastic differential equations, we know that for any given initial value $(x_0, y_0) \in R^2$, equation (3.1)' has a unique solution on $t \geq 0$. The boundedness of function $\kappa(x, y, t)$ is of course unnecessary, for example, the linear growth condition would do, but for simplicity of further treatments we would rather impose this condition of boundedness.

In this section we shall demonstrate that to study certain properties of equation (3.1)', for example, the property that the solution initiating at $(x_0, y_0) \neq 0$ will almost surely miss the origin for all $t \geq 0$, it is enough to study the relatively simpler equation

$$d\begin{bmatrix} x(t) \\ y(t) \end{bmatrix} = \begin{bmatrix} y(t) \\ 0 \end{bmatrix} dt + \begin{bmatrix} 0 \\ h \end{bmatrix} d\tilde{B}(t) \qquad (3.2)$$

with respect to a new Brownian motion $\tilde{B}(t)$ of course.

Lemma 3.1 *Let $T > 0$ and $(x(t), y(t))$ be a solution to the equation*

$$d\begin{bmatrix} x(t) \\ y(t) \end{bmatrix} = \begin{bmatrix} y(t) \\ -\kappa(x(t), y(t), t) \end{bmatrix} dt + \begin{bmatrix} 0 \\ h \end{bmatrix} dB(t) \quad on \ 0 \leq t \leq T. \qquad (3.3)$$

Let $\phi(t) = h^{-1}\kappa(x(t), y(t), t)$ for $0 \leq t \leq T$ and set

$$\zeta_0^T = -\frac{1}{2} \int_0^T |\phi(u)|^2 du + \int_0^T \phi(u) dB(u),$$

$$\tilde{B}(t) = B(t) - \int_0^t \phi(u) du,$$

$$d\tilde{P}(\omega) = \exp[\zeta_0^T(\phi)] dP(\omega).$$

Then $\tilde{P}(\Omega) = 1$, $\{\tilde{B}(t)\}_{0 \leq t \leq T}$ is a one-dimensional Brownian motion on the probability space $(\Omega, \mathcal{F}, \{\mathcal{F}_t\}, \tilde{P})$, and all P-null sets are also of $\tilde{P}$-null. Moreover, $(x(t), y(t))$ is a solution to the following equation

$$d\begin{bmatrix} x(t) \\ y(t) \end{bmatrix} = \begin{bmatrix} y(t) \\ 0 \end{bmatrix} dt + \begin{bmatrix} 0 \\ h \end{bmatrix} d\tilde{B}(t) \quad on \ 0 \leq t \leq T. \qquad (3.4)$$

Proof. The boundedness of $\kappa(x, y, t)$ guarantees that condition (2.5) is fulfilled. Hence, by the Cameron–Martin-Girsanov theorem, $\tilde{P}(\Omega) = 1$ and $\{\tilde{B}(t)\}_{0 \leq t \leq T}$

is a one-dimensional Brownian motion on $(\Omega, \mathcal{F}, \tilde{P})$. Since $\tilde{P}$ is absolutely continuous with respect to P, any P-null set is of $\tilde{P}$-null. Moreover, by the Girsanov theorem, $(x(t), y(t))$ is a solution of equation (3.4) on the probability space $(\Omega, \mathcal{F}, \tilde{P})$.

Lemma 3.2 *Let $T > 0$. Let $(\Omega, \mathcal{F}, \{\mathcal{F}_t\}, \tilde{P})$ be a complete probability space and $\{\tilde{B}(t)\}_{0 \leq t \leq T}$ be a one-dimensional Brownian motion defined on the space. Let $(x(t), y(t))$ be a solution of the equation*

$$d\begin{bmatrix} x(t) \\ y(t) \end{bmatrix} = \begin{bmatrix} y(t) \\ 0 \end{bmatrix} dt + \begin{bmatrix} 0 \\ h \end{bmatrix} d\tilde{B}(t) \quad \text{on } 0 \leq t \leq T.$$

Let $\phi(t) = -h^{-1}\kappa(x(t), y(t), t)$ for $0 \leq t \leq T$ and set

$$\zeta_0^T = -\frac{1}{2}\int_0^T |\phi(u)|^2 du + \int_0^T \phi(u) d\tilde{B}(u),$$

$$\hat{B}(t) = \tilde{B}(t) - \int_0^t \phi(u) du,$$

$$d\hat{P}(\omega) = \exp[\zeta_0^T(\phi)] d\tilde{P}(\omega).$$

Then $\hat{P}(\Omega) = 1$, $\{\hat{B}(t)\}_{0 \leq t \leq T}$ is a one-dimensional Brownian motion on the probability space $(\Omega, \mathcal{F}, \{\mathcal{F}_t\}, \hat{P})$, and all $\tilde{P}$-null sets are of $\hat{P}$-null as well. Moreover, $(x(t), y(t))$ is a solution to the following equation

$$d\begin{bmatrix} x(t) \\ y(t) \end{bmatrix} = \begin{bmatrix} y(t) \\ -\kappa(x(t), y(t), t) \end{bmatrix} dt + \begin{bmatrix} 0 \\ h \end{bmatrix} d\hat{B}(t) \quad \text{on } 0 \leq t \leq T. \tag{3.5}$$

The proof is the same as that of Lemma 3.1. Using these two lemmas, we can obtain the following useful results.

Theorem 3.3 *Let $(x(t), y(t)), \tilde{B}(t)$ etc. be the same as defined in Lemmas 3.1 and 3.2. Let $\mathcal{G}_T$ be the σ-algebra generated by the solution $(x(t), y(t))$, that is $\mathcal{G}_T = \sigma\{(x(t), y(t)) : 0 \leq t \leq T\}$. Then for any set $A \in \mathcal{G}_T$, $P(A) = 0$ if and only if $\tilde{P}(A) = 0$. In consequence,*

$$P\{|x(t)|^2 + |y(t)|^2 = 0 \text{ for some } 0 \leq t \leq T\} = 0$$

if and only if

$$\tilde{P}\{|x(t)|^2 + |y(t)|^2 = 0 \text{ for some } 0 \leq t \leq T\} = 0.$$

Proof. Comparing equation (3.3) with (3.5), we see clearly that the probability distribution of $(x(t), y(t))$ under probability measure P is the same as that under probability measure $\hat{P}$. In other words,

$$P(A) = \hat{P}(A) \quad \text{for all } A \in \mathcal{G}_T.$$

On the other hand, by Lemmas 3.1 and 3.2,

$$P(A) = 0 \Rightarrow \tilde{P}(A) = 0 \Rightarrow \hat{P}(A) = 0.$$

Hence, for any set $A \in \mathcal{G}_T$, $P(A) = 0$ if and only if $\tilde{P}(A) = 0$.

Theorem 3.4 *Let $(x(t), y(t))$ be the unique solution, initiating at $(x_0, y_0) \neq 0$ in R^2, for the nonlinear stochastic oscillator*

$$\ddot{x}(t) + \kappa(x(t), \dot{x}(t), t) = h\dot{B}(t) \quad \text{on } t \geq 0$$

or the Itô equation

$$d\begin{bmatrix} x(t) \\ y(t) \end{bmatrix} = \begin{bmatrix} y(t) \\ -\kappa(x(t), y(t), t) \end{bmatrix} dt + \begin{bmatrix} 0 \\ h \end{bmatrix} dB(t).$$

Then

$$P\{|x(t)|^2 + |y(t)|^2 > 0 \text{ for all } 0 \leq t < \infty\} = 1. \tag{3.6}$$

Proof. Clearly, (3.6) follows if we can show that for every $T > 0$,

$$P\{|x(t)|^2 + |y(t)|^2 > 0 \text{ for all } 0 \leq t \leq T\} = 1$$

or

$$P\{|x(t)|^2 + |y(t)|^2 = 0 \text{ for some } 0 \leq t \leq T\} = 0.$$

But by Theorem 3.3, this is equivalent to

$$\tilde{P}\{|x(t)|^2 + |y(t)|^2 = 0 \text{ for some } 0 \leq t \leq T\} = 0, \tag{3.7}$$

where $\tilde{P}$ and the following $\tilde{B}(t)$ are the same as defined in Lemma 3.1. Note from Lemma 3.1 that $(x(t), y(t))$ is a solution to the equation

$$d\begin{bmatrix} x(t) \\ y(t) \end{bmatrix} = \begin{bmatrix} y(t) \\ 0 \end{bmatrix} dt + \begin{bmatrix} 0 \\ h \end{bmatrix} d\tilde{B}(t) \quad \text{on } 0 \leq t \leq T \tag{3.8}$$

on the probability space $(\Omega, \mathcal{F}, \tilde{P})$. We now define a C^∞-function

$$V(x, y) = \int_0^\infty p(t, x, y) dt, \tag{3.9}$$

on its domain $R^2 - (0, 0)$, where

$$p(t, x, y) = \frac{\sqrt{3}}{\pi t} \exp\left(\frac{-2}{h^2 t^3}\left[t^2 y^2 + 3txy + 3x^2\right]\right). \tag{3.10}$$

It is not difficult to verify that $V(x, y)$ has the following three properties:

(i) $V(x,y) > 0$,

(ii) $\lim_{|x|+|y|\to 0} V(x,y) = 0$,

(iii) $LV(x,y) \equiv 0$ on the whole domain $R^2 - (0,0)$, where L is the diffusion operator associated with equation (3.8), namely

$$L = y\frac{\partial}{\partial x} + \frac{h^2}{2}\frac{\partial}{\partial y^2}.$$

The routine calculations verifying these properties are left to the reader as an exercise. Define the stopping time

$$\rho = \inf\{t \geq 0 : |x(t)|^2 + |y(t)|^2 = 0\}.$$

Also, for each $k = 1, 2, \cdots$, define

$$\rho_k = \inf\{t \geq 0 : |x(t)|^2 + |y(t)|^2 \leq 1/k\}.$$

Obviously, $\rho_k \uparrow \rho$ as $k \to \infty$ $\tilde{P}$-a.s. From Itô's formula and property (iii) it follows directly that

$$\tilde{E}V(x(\rho_k \wedge T), y(\rho_k \wedge T)) = V(x_0, y_0) < \infty,$$

where $\tilde{E}$ denotes the expection with respect to the probability measure $\tilde{P}$. Letting $v_k = \min\{V(x,y) : |x|^2 + |y|^2 = 1/k\}$, we then have that

$$v_k \tilde{P}\{\rho_k \leq T\} \leq V(x_0, y_0).$$

But, it follows from property (ii) that $v_k \to \infty$ as $k \to \infty$. We hence let $k \to \infty$ to obtain that

$$\tilde{P}\{\rho \leq T\} = 0$$

which is (3.7). The proof is complete.

To close this section, let us mention that the motivation for seeking function $V(x,y)$ in the form of (3.9) and (3.10) can be found in Markus & Weerasinghe (1988).

8.4 LINEAR STOCHASTIC OSCILLATORS

Again let $B(t)$ be a 1-dimensional Brownian motion. Consider the scalar linear stochastic oscillator

$$\ddot{x}(t) + \kappa x(t) = h\dot{B}(t) \quad \text{on } t \geq 0, \tag{4.1}$$

where κ and h are positive constants. Of course this is a special case of the nonlinear stochastic oscillator

$$\ddot{x}(t) + \kappa(x(t), \dot{x}(t), t) = h\dot{B}(t) \quad \text{on } t \geq 0.$$

However, as discussed in the previous section, it is enough to study the linear stochastic oscillator in order to understand many properties of the nonlinear stochastic oscillator as long as these properties hold almost surely, for instance, the property that the solution almost surely misses the origin. On the other hand, certain probabilistic results hold only with some positive probability depending on the parameters κ and h, and are particular to the theory of linear stochastic oscillators.

By introducing the new variable $y(t) = \dot{x}(t)$, the corresponding Itô equation is
$$d\begin{bmatrix} x(t) \\ y(t) \end{bmatrix} = A \begin{bmatrix} x(t) \\ y(t) \end{bmatrix} dt + \begin{bmatrix} 0 \\ h \end{bmatrix} dB(t), \qquad (4.1)'$$
where
$$A = \begin{bmatrix} 0 & 1 \\ -\kappa & 0 \end{bmatrix}.$$

Given any initial value $(x(0), y(0)) = (x_0, y_0) \in R^2$, by the theory established in Chapter 3, we know that equation $(4.1)'$ has the unique solution
$$\begin{bmatrix} x(t) \\ y(t) \end{bmatrix} = e^{At} \begin{bmatrix} x_0 \\ y_0 \end{bmatrix} + \int_0^t e^{A(t-s)} \begin{bmatrix} 0 \\ h \end{bmatrix} dB(s). \qquad (4.2)$$

Noting that $A^2 = -\kappa I$ with I the 2×2 identity matrix, we have that
$$\begin{aligned} e^{At} &= \sum_{i=0}^{\infty} \left(\frac{A^{2i} t^{2i}}{(2i)!} + \frac{A^{2i+1} t^{2i+1}}{(2i+1)!} \right) \\ &= \sum_{i=0}^{\infty} \left(\frac{(-\kappa)^i t^{2i} I}{(2i)!} + \frac{(-\kappa)^i t^{2i+1} A}{(2i+1)!} \right) \\ &= \sum_{i=0}^{\infty} (-1)^i \begin{bmatrix} \frac{(\sqrt{\kappa}t)^{2i}}{(2i)!}, & \frac{(\sqrt{\kappa}t)^{2i+1}}{\sqrt{\kappa}(2i+1)!} \\ -\frac{\sqrt{\kappa}(\sqrt{\kappa}t)^{2i+1}}{(2i+1)!}, & \frac{(\sqrt{\kappa}t)^{2i}}{(2i)!} \end{bmatrix} \\ &= \begin{bmatrix} \cos(\sqrt{\kappa}t), & \frac{1}{\sqrt{\kappa}}\sin(\sqrt{\kappa}t) \\ -\sqrt{\kappa}\sin(\sqrt{\kappa}t), & \cos(\sqrt{\kappa}t) \end{bmatrix}. \end{aligned}$$

Substituting this into (4.2) yields
$$\begin{cases} x(t) = x_0 \cos(\sqrt{\kappa}t) + \dfrac{y_0}{\sqrt{\kappa}} \sin(\sqrt{\kappa}t) + \dfrac{h}{\sqrt{\kappa}} \displaystyle\int_0^t \sin(\sqrt{\kappa}(t-s))dB(s), \\[2mm] y(t) = -x_0 \sqrt{k} \sin(\sqrt{\kappa}t) + y_0 \cos(\sqrt{\kappa}t) + h \displaystyle\int_0^t \cos(\sqrt{\kappa}(t-s))dB(s). \end{cases} \qquad (4.3)$$

For simplicity of treatment we shall consider primarily the case $\kappa = 1$, $x_0 = 1$ and $y_0 = 0$, that is, the stochastic oscillator
$$\ddot{x}(t) + x(t) = h\dot{B}(t) \qquad (4.4)$$

or
$$d\begin{bmatrix} x(t) \\ y(t) \end{bmatrix} = \begin{bmatrix} y(t) \\ -x(t) \end{bmatrix} dt + \begin{bmatrix} 0 \\ h \end{bmatrix} dB(t) \qquad (4.4)'$$

with initial value $(1,0)$. In this case, (4.3) reduces to

$$\begin{cases} x(t) = \cos t + h \int_0^t \sin(t-s) dB(s), \\ y(t) = -\sin t + h \int_0^t \cos(t-s) dB(s). \end{cases} \qquad (4.5)$$

Of course, when $\kappa \neq 1$, $x_0 \neq 1$ but still $y_0 = 0$, we can apply appropriate changes of scale to reduce the study of equation (4.1) to the case of equation (4.4).

Theorem 4.1 *Consider the scalar stochastic process $x(t)$ satisfying the linear stochastic oscillator:*

$$\ddot{x}(t) + x(t) = h\dot{B}(t) \quad \text{on } t \geq 0$$

from $x(0) = 1$, $\dot{x}(0) = 0$, with parameter $h > 0$. Then, almost surely, $x(t)$ has infinitely many zeros, all simple, on each half line $[t_0, \infty)$ for every $t_0 \geq 0$.

Proof. It follows from (4.5) that

$$x(t) = \cos t - h \cos t \int_0^t \sin s\, dB(s) + h \sin t \int_0^t \cos s\, dB(s).$$

Consider $x(t)$ at the discrete instants $t = (2k + \frac{1}{2})\pi$, for $k = 1, 2, \cdots$, when

$$x\left((2k + \frac{1}{2})\pi\right) = h \int_0^{(2k+\frac{1}{2})\pi} \cos s\, dB(s).$$

Define a sequence of random variables $\{Y_k\}_{k \geq 1}$ by

$$Y_k = h \int_{(2(k-1)+\frac{1}{2})\pi}^{(2k+\frac{1}{2})\pi} \cos s\, dB(s).$$

Obviously

$$x\left((2k + \frac{1}{2})\pi\right) = \sum_{i=1}^k Y_i.$$

Also, $\{Y_k\}_{k \geq 1}$ is a sequence of independent random variables, because of the independence of the increments of the Brownian motion on disjoint intervals. Moreover, each Y_k is normally distributed with mean zero and variance

$$E(Y_k^2) = h^2 \int_{(2(k-1)+\frac{1}{2})\pi}^{(2k+\frac{1}{2})\pi} \cos^2 s\, ds = h^2 \pi.$$

Now the familiar theorems on the limits of sums of independent random variables (e.g. the law of the iterated logarithm) show that, almost surely, the terms of the sequence $\{x((2k + \frac{1}{2})\pi)\}$ have infinitely many switches of sign as $k \to \infty$. Since almost all sample path of the solution $x(t)$ is continuous on $[0, \infty)$, $x(t)$ must have, almost surely, infinitely many zeros on each half line $[t_0, \infty)$ for every $t_0 \geq 0$. The simplicity of the zeros of $x(t)$ has already been proved in Theorem 3.4. The proof is complete.

Let us now turn to discuss the first zero of the oscillation.

Theorem 4.2 *Consider the scalar stochastic process $x(t)$ satisfying the linear stochastic oscillator:*

$$\ddot{x}(t) + x(t) = h\dot{B}(t) \quad on\ t \geq 0$$

from $x(0) = 1$, $\dot{x}(0) = 0$, with parameter $h > 0$. Let τ_1 be the time of the first zero of $x(t)$ on $[0, \infty)$, i.e.

$$\tau_1 = \inf\{t \geq 0 : x(t) = 0\}.$$

Then

$$E\tau_1 \geq 2\operatorname{arc cot} h\ \operatorname{Erf}\left(\frac{1}{\sqrt{\operatorname{arc cot} h}}\right), \tag{4.6}$$

where $\operatorname{Erf}(\cdot)$ is the error function given by

$$\operatorname{Erf}(z) = \frac{1}{\sqrt{2\pi}} \int_0^z e^{-u^2/2} du \quad on\ z \geq 0.$$

In order to prove the theorem, we need to present a lemma.

Lemma 4.3 *For any $b > 0$ and $T > 0$,*

$$P\{B(t) > -b\ on\ 0 \leq t \leq T\} = 2\operatorname{Erf}\left(\frac{b}{\sqrt{T}}\right). \tag{4.7}$$

Proof. The proof uses the well-known reflection principle. Let τ be the first time of the Brownian motion reaching $-b$, i.e.

$$\tau = \inf\{t \geq 0 : B(t) = -b\}.$$

Obviously

$$P\{\tau < T\} = P\{\tau < T,\ B(T) < -b\} + P\{\tau < T,\ B(T) > -b\}.$$

If $\tau < T$ and $B(T) > -b$, then sometime before time T the Brownian path reached level $-b$ and then in the remaining time it travelled from $-b$ to a point

$-b+c$ (with $c > 0$). Because of the symmetry with respect to $-b$ of a Brownian motion starting at b, the probability of doing this is the same as the probability of travelling from $-b$ to the point $-b-c$ (please see Figure 8.4.1 below).

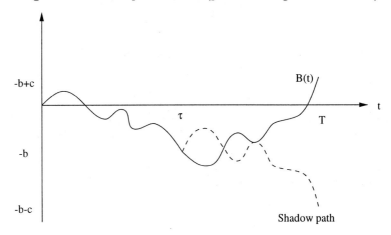

Figure 8.4.1: Reflection principle

This reflection principle shows that

$$P\{\tau < T, \ B(T) > -b\} = P\{\tau < T, \ B(T) < -b\}.$$

Hence

$$P\{\tau < T\} = 2P\{\tau < T, \ B(T) < -b\}$$

But, it is quite easy to see that

$$P\{\tau < T, \ B(T) < -b\} = P\{B(T) < -b\}.$$

We therefore obtain that

$$P\{\tau < T\} = 2P\{B(T) < -b\} = 2P\{B(T) > b\}$$
$$= \frac{2}{\sqrt{2\pi T}} \int_b^\infty e^{-u^2/2T} du = \frac{2}{\sqrt{2\pi}} \int_{b/\sqrt{T}}^\infty e^{-u^2/2} du.$$

That is

$$P\left\{\inf_{0 \leq t \leq T} B(t) \leq -b\right\} = 1 - 2\operatorname{Erf}\left(\frac{b}{\sqrt{T}}\right),$$

and the required assertion (4.7) follows.

We can now prove Theorem 4.2.

Proof of Theorem 4.2. Noting from the integration-by-parts formula that

$$\int_0^t \sin(t-s) dB(s) = \int_0^t B(s) \cos(t-s) ds,$$

we obtain from (4.5) that

$$x(t) = \cos t + h \int_0^t B(s) \cos(t-s) ds.$$

Set

$$\bar{\Omega} = \{\omega : B(t) > -1 \text{ on } 0 \le t \le \text{arc cot } h\}.$$

By Lemma 4.3,

$$P(\bar{\Omega}) = 2 \operatorname{Erf}\left(\frac{1}{\sqrt{\text{arc cot } h}}\right).$$

Since $\cos(t-s) > 0$ on $0 \le s \le t < \pi/2$, we find that for all $\omega \in \bar{\Omega}$,

$$x(t) > \cos t - h \int_0^t \cos(t-s) ds = \cos t - h \sin t > 0$$

if $0 \le t \le \text{arc cot } h$. Hence

$$\tau_1(\omega) \ge \text{arc cot } h \quad \text{for } \omega \in \bar{\Omega}.$$

Finally

$$E\tau_1 \ge E(\tau_1 I_{\bar{\Omega}}) \ge \text{arc cot } h \, P(\bar{\Omega}) = 2 \text{ arc cot } h \operatorname{Erf}\left(\frac{1}{\sqrt{\text{arc cot } h}}\right)$$

as desired. The proof is complete.

We now demonstrate that the first zero time τ_1 has finite moments, and we estimate the first two moments $E\tau_1$ and $E\tau_1^2$ from above.

Theorem 4.4 *Consider the scalar stochastic process $x(t)$ satisfying the linear stochastic oscillator:*

$$\ddot{x}(t) + x(t) = h\dot{B}(t) \quad \text{on } t \ge 0$$

from $x(0) = 1$, $\dot{x}(0) = 0$, with parameter $h > 0$. Let τ_1 be the time of the first zero of $x(t)$ on $[0, \infty)$. Then

$$P\{\tau_1 > T\} < \frac{4c(h)}{2^{T/\pi}} \quad \text{for each } T \ge \pi, \tag{4.8}$$

where the constant $c(h) = \frac{1}{2} - \operatorname{Erf}(h^{-1}\sqrt{2/\pi})$, so $\lim_{h \to 0} = 0$. In consequence,

$$P\{\tau_1 < \infty\} = 1,$$

and every moment of τ_1 is finite, with the first two moments having the upper bounds:

$$E\tau_1 < \pi(1 + 2c(h)) \le 2\pi \tag{4.9}$$

and
$$E\tau_1^2 < \pi^2(1 + 22c(h)) \leq 12\pi^2. \qquad (4.10)$$

Proof. Evaluate $x(t)$ at the discrete instants $t = k\pi$ for $k = 1, 2, \cdots$ to obtain
$$x(k\pi) = \cos(k\pi)[1 - h\bar{B}(k\pi)],$$
where
$$\bar{B}(k\pi) = \int_0^{k\pi} \sin s\, dB(s).$$
Hence
$$x(k\pi) > 0$$
if and only if
$$\bar{B}(k\pi) \begin{cases} > 1/h & \text{for } k = 1, 3, \cdots, \\ < 1/h & \text{for } k = 2, 4, \cdots. \end{cases}$$
Set
$$Y_k = \int_{(k-1)\pi}^{k\pi} \sin s\, dB(s).$$
Then Y_k's are independent normal random variables with mean zero and variance
$$E(Y_k^2) = \int_{(k-1)\pi}^{k\pi} \sin^2 s\, ds = \frac{\pi}{2}.$$
Moreover
$$\bar{B}(k\pi) = \sum_{i=1}^{k} Y_k.$$
Combining the above arguments, we see that
$$\{\tau_1 > \pi\} = \{x(t) > 0 \text{ for all } 0 \leq t \leq \pi\}$$
$$\subset \{x(\pi) > 0\} = \{\bar{B}(\pi) > 1/h\} = \{Y_1 > 1/h\}.$$
Thus
$$P\{\tau_1 > \pi\} \leq P\{Y_1 > 1/h\} = \frac{1}{2} - \text{Erf}(h^{-1}\sqrt{2/\pi}) \equiv c(h).$$
Furthermore, we have that
$$\{\tau_1 > 2\pi\} = \{x(t) > 0 \text{ for all } 0 \leq t \leq 2\pi\}$$
$$\subset \{x(\pi) > 0, x(2\pi) > 0\} = \{\bar{B}(\pi) > 1/h, \bar{B}(2\pi) < 1/h\}$$
$$= \{Y_1 > 1/h, Y_1 + Y_2 < 1/h\} \subset \{Y_1 > 1/h, Y_2 < 0\}.$$
Hence
$$P\{\tau_1 > 2\pi\} \leq P\{Y_1 > 1/h, Y_2 < 0\} = P\{Y_1 > 1/h\}P\{Y_2 < 0\} = \frac{c(h)}{2}.$$

Continuing this argument, we find that

$$P\{\tau_1 > k\pi\} \leq \frac{c(h)}{2^{k-1}} \quad \text{for } k = 1, 2, \cdots.$$

In order to estimate the probability that τ_1 exceeds an arbitrary positive number $T \geq \pi$, we let $[T/\pi]$ be the greatest integer not exceeding T/π, so $[T/\pi]\pi \leq T$. Then

$$P\{\tau_1 > T\} \leq P\{\tau_1 > [T/\pi]\pi\} \leq \frac{c(h)}{2^{[T/\pi]-1}} < \frac{4c(h)}{2^{T/\pi}},$$

which is the required assertion (4.8). From this upper estimate on the probability distribution for τ_1 we easily conclude that

$$\lim_{T \to \infty} P\{\tau_1 > T\} = 0 \quad \text{or} \quad P\{\tau_1 < \infty\} = 1.$$

Now re-write (4.8) as

$$P\{\tau_1 > T\} < 4c(h) \exp\left(-\frac{\log 2}{\pi} T\right),$$

we see that $P\{\tau_1 > T\}$ satisfies a bound of exponential decay, and we are assured that each moment of τ_1 is finite. We now establish the upper bounds for its first two moments. Note that for any non-negative random variable ξ, we have

$$E\xi \leq \sum_{k=1}^{\infty} kP\{k-1 < \xi \leq k\} = \sum_{k=1}^{\infty} \sum_{i=1}^{k} P\{k-1 < \xi \leq k\}$$

$$= \sum_{i=1}^{\infty} \sum_{k=i}^{\infty} P\{k-1 < \xi \leq k\} = \sum_{i=1}^{\infty} P\{i-1 < \xi\}$$

$$\leq 1 + \sum_{k=1}^{\infty} P\{\xi > k\}. \tag{4.11}$$

Let $\xi = \tau_1/\pi$ to obtain that

$$\frac{1}{\pi} E\tau_1 \leq 1 + \sum_{k=1}^{\infty} P\{\tau_1 > k\pi\} \leq 1 + c(h) \sum_{k=1}^{\infty} \frac{1}{2^{k-1}} = 1 + 2c(h).$$

Thus

$$E\tau_1 \leq \pi[1 + 2c(h)] \leq 2\pi.$$

For the second moment we estimate

$$E\tau_1^2 \leq \sum_{k=0}^{\infty} [(k+1)\pi]^2 P\{k\pi < \tau_1 \leq (k+1)\pi\}$$

$$\leq \pi^2 \left[1 + c(h) \sum_{k=1}^{\infty} \frac{(k+1)^2}{2^{k-1}}\right] = \pi^2 + 4\pi^2 c(h) \left[\sum_{k=0}^{\infty} \frac{(k+1)^2}{2^{k+1}} - \frac{1}{2}\right].$$

The sum of the infinite series is easily calculated:

$$\sum_{k=0}^{\infty} \frac{(k+1)^2}{2^{k+1}} = (zq'(z))'\big|_{z=1} = 6,$$

where

$$q(z) = \sum_{k=0}^{\infty} \frac{z^{k+1}}{2^{k+1}} = \frac{z}{2-z}.$$

Thus we have the required estimate

$$E\tau_1^2 \leq \pi^2[1 + 22c(h)] \leq 12\pi^2.$$

The proof is complete.

Let us now proceed to study the times of subsequent zeros of $x(t)$. For each $i = 2, 3, \cdots$, define by τ_i the time of the ith zero of $x(t)$, namely

$$\tau_i = \inf\{t > \tau_{i-1} : x(t) = 0\}.$$

Since the zeros of $x(t)$ are all simple almost surely, these stopping times are all well defined. Clearly, they are increasing, namely $\tau_1 < \tau_2 < \tau_3 < \cdots$.

Theorem 4.5 *Consider the scalar stochastic process $x(t)$ satisfying the linear stochastic oscillator:*

$$\ddot{x}(t) + x(t) = h\dot{B}(t) \quad \text{on } t \geq 0$$

from $x(0) = 1$, $\dot{x}(0) = 0$, with parameter $h > 0$. Let τ_i be the time of the ith zero of $x(t)$. Then

$$E\tau_i \leq 2i\pi \quad \text{for each } i = 1, 2, \cdots. \tag{4.12}$$

In order to prove this theorem, let us present a lemma.

Lemma 4.6 *Consider the scalar stochastic process $z(t)$ satisfying the linear stochastic oscillator:*

$$\ddot{z}(t) + z(t) = h\dot{B}(t) \quad \text{on } t \geq 0$$

from $z(0) = 0$, $\dot{z}(0) \neq 0$, with parameter $h > 0$. Let θ_1 be the time of the first zero of $z(t)$ on $t > 0$, that is

$$\theta_1 = \inf\{t > 0 : z(t) = 0\}.$$

Then

$$E\theta_1 \leq 2\pi. \tag{4.13}$$

Proof. It follows from (4.3) that

$$z(t) = \dot{z}(0) \sin t + h \int_0^t \sin(t-s) dB(s)$$

$$= \dot{z}(0) \sin t + h \sin t \int_0^t \cos s \, dB(s) - h \cos t \int_0^t \sin s \, dB(s).$$

Examine $z(t)$ at the discrete times $t = k\pi$ for $k = 1, 2, \cdots$, when

$$z(k\pi) = h(-1)^{k+1} \bar{B}(k\pi),$$

where

$$\bar{B}(k\pi) = \int_0^{k\pi} \sin s \, dB(s).$$

Hence

$$z(k\pi) > 0$$

if and only if

$$\bar{B}(k\pi) \begin{cases} > 0 & \text{for } k = 1, 3, \cdots, \\ < 0 & \text{for } k = 2, 4, \cdots. \end{cases}$$

These results do not depend on the value of $\dot{z}(0) \neq 0$. From this calculation we note that

$$z(\pi) > 0 \quad \text{if and only if} \quad \bar{B}(\pi) > 0$$

so

$$P\{z(t) > 0 \text{ on } 0 < t \leq \pi\} \leq P\{z(\pi) > 0\} = P\{\bar{B}(\pi) > 0\} = \frac{1}{2}.$$

We also compute that

$$P\{z(t) > 0 \text{ on } 0 < t \leq 2\pi\}$$
$$\leq P\{z(\pi) > 0, \ z(2\pi) > 0\}$$
$$= P\{\bar{B}(\pi) > 0, \ \bar{B}(2\pi) < 0\}$$
$$= P\{\bar{B}(\pi) > 0, \ \bar{B}(\pi) + [\bar{B}(2\pi) - \bar{B}(\pi)] < 0\}$$
$$\leq P\{\bar{B}(\pi) > 0, \ \bar{B}(2\pi) - \bar{B}(\pi) < 0\}$$
$$= P\{\bar{B}(\pi) > 0\} P\{\bar{B}(2\pi) - \bar{B}(\pi) < 0\}$$
$$= \frac{1}{2^2}.$$

Repeating this argument, we obtain that

$$P\{z(t) > 0 \text{ on } 0 < t \leq k\pi\} \leq \frac{1}{2^k}.$$

Therefore

$$P\{\theta_1 > k\pi\} \leq \frac{1}{2^k}.$$

We now let $\xi = \theta_1/\pi$ in (4.11) to obtain that

$$\frac{1}{\pi}E\theta_1 \leq 1 + \sum_{k=1}^{\infty} P\{\theta_1 > k\pi\} \leq 1 + \sum_{k=1}^{\infty} \frac{1}{2^k} = 2.$$

That is $E\theta_1 \leq 2\pi$ as desired.

We can now easily prove Theorem 4.5.

Proof. We already know from Theorem 4.4 that $E\tau_1 \leq 2\pi$. In order to estimate $E\tau_2$, we shift the time scale from $t \geq 0$ to $s = t - \tau_1$. Write $z(s) = x(s + \tau_1)$ and note that

$$\ddot{z}(s) + z(s) = h\dot{w}(s)$$

with $z(0) = 0$ and $\dot{z}(0) \neq 0$, where $w(s) = B(s + \tau_1) - B(\tau_1)$ which is a new Brownian motion on $s \geq 0$. Define

$$\theta_1 = \inf\{s > 0 : z(s) = 0\}.$$

Then $\tau_2 = \tau_1 + \theta_1$. But by Lemma 4.5, $E\theta_1 \leq 2\pi$. We therefore obtain that $E\tau_2 \leq 4\pi$. An elementary repetition of this argument yields the desired result

$$E\tau_i \leq 2i\pi \quad \text{for each } i = 1, 2, \cdots.$$

8.5 ENERGY BOUNDS

In this section we let $B(t)$ be an m-dimensional Brownian motion. Motivated by the classical theory of nonlinear oscillations, we consider the d-dimensional second order stochastic differential equation

$$\ddot{x}(t) + b(x(t), \dot{x}(t)) + \nabla G(x(t)) = \sigma(x(t), \dot{x}(t), t)\dot{B}(t) + h(t) \quad (5.1)$$

on $t \geq 0$, where $x = (x_1, \cdots, x_d)^T \in R^d$, $-\nabla G(x)$ is the restoring force, $-b(x, \dot{x})$ is a dissipative force, $h(t)$ is an external driving force, and $\sigma(x, \dot{x}, t)$ represents the intensity of a stochastic disturbance. In terms of mathematics, we have that $b : R^d \times R^d \to R^d$, $G \in C^1(R^d; R_+)$ and $\nabla G(x) = (G_{x_1}, \cdots, G_{x_d})^T$, $\sigma : R^d \times R^d \times R_+ \to R^{d \times m}$ and $h : R_+ \to R^d$. The corresponding 2d-dimensional Itô equation is

$$\begin{cases} dx(t) = y(t)dt, \\ dy(t) = [-b(x(t), y(t)) - \nabla G(x(t)) + h(t)]dt + \sigma(x(t), y(t), t)dB(t). \end{cases} \quad (5.1)'$$

Assume that all the functions b, G, σ and h are sufficiently smooth so that equation (5.1)' has a unique global solution $(x(t), y(t))$ on $t \geq 0$ for any given initial value $(x_0, y_0) \in R^d \times R^d$.

Define the energy of the system by

$$U(t) = \frac{1}{2}|y(t)|^2 + G(x(t)), \quad t \geq 0. \qquad (5.2)$$

Bearing in mind that G is a nonnegative function, we see that the energy is always nonnegative. Moreover, although the function $\frac{1}{2}|y|^2 + G(x)$ is only once differentiable in x (of course twice in y), we can still apply the Itô formula to $U(t)$ due to the relation $dx(t) = y(t)dt$ which does not involve the stochastic integral. The aim of this section is to establish the asymptotic bounds for the energy. To make the theory more understandable, we shall first discuss the unforced system and then return to the above general case.

(i) *Unforced Dynamics*

If there is no external driving force, i.e. $h(t) \equiv 0$, equation (5.1)' becomes

$$\begin{cases} dx(t) = y(t)dt, \\ dy(t) = [-b(x(t), y(t)) - \nabla G(x(t))]dt + \sigma(x(t), y(t), t)dB(t). \end{cases} \qquad (5.3)$$

We shall prove several theorems on the energy bounds to allow for polynomial-exponential growth of the diffusion term. We begin with the following theorem.

Theorem 5.1 *Let $p(t)$ be a real polynomial of degree q on $t \geq 0$ with the leading coefficient $K > 0$ and the other coefficients nonnegative. Assume that*

$$y^T b(x, y, t) \geq 0 \qquad (5.4)$$

and

$$|\sigma(x, y, t)|^2 \leq p(t) \qquad (5.5)$$

for all $x, y \in R^d$ and $t \geq 0$. Then the energy of system (5.3) has the property that

$$\limsup_{t \to \infty} \frac{U(t)}{t^{q+1} \log \log t} \leq \frac{Ke}{q+1} \quad a.s. \qquad (5.6)$$

To prove this theorem, let us present a useful inequality.

Lemma 5.2 *Let $T > 0$ and $z_0 \geq 0$. Let z, u and v be continuous nonnegative functions defined on $[0, T]$. If*

$$z(t) \leq z_0 + \int_0^t [u(s) + v(s)z(s)]ds \qquad (5.6)$$

for all $0 \leq t \leq T$, then

$$z(t) \leq \exp\left(\int_0^t v(s)ds\right)\left[z_0 + \int_0^t \exp\left(-\int_0^s v(r)dr\right)u(s)ds\right] \qquad (5.7)$$

for all $0 \leq t \leq T$ as well.

Proof. Set
$$\bar{z}(t) = z_0 + \int_0^t [u(s) + v(s)z(s)]ds.$$

It then follows from (5.6) that
$$z(t) \leq \bar{z}(t) \quad \text{on } 0 \leq t \leq T.$$

Moreover, we compute that
$$\frac{d}{dt}\left[\exp\left(-\int_0^t v(s)ds\right)\bar{z}(t)\right]$$
$$= \exp\left(-\int_0^t v(s)ds\right)\left[\frac{d\bar{z}(t)}{dt} - v(t)\bar{z}(t)\right]$$
$$= \exp\left(-\int_0^t v(s)ds\right)\left[u(t) + v(t)z(t) - v(t)\bar{z}(t)\right]$$
$$\leq \exp\left(-\int_0^t v(s)ds\right)u(t).$$

Integrating both sides from 0 to t yields that
$$\exp\left(-\int_0^t v(s)ds\right)\bar{z}(t) - z_0 \leq \int_0^t \exp\left(-\int_0^s v(r)dr\right)u(s)ds.$$

This implies that
$$\bar{z}(t) \leq \exp\left(\int_0^t v(s)ds\right)\left[z_0 + \int_0^t \exp\left(-\int_0^s v(r)dr\right)u(s)ds\right]$$

and the required inequality (5.7) follows.

Proof of Theorem 5.1. Applying Itô's formula along with hypotheses (5.4) and (5.5), we derive that
$$U(t) = U(0) + \int_0^t \left(-y^T(s)b(x(s),y(s)) + \frac{1}{2}|\sigma(x(s),y(s),s)|^2\right)ds$$
$$+ \int_0^t y^T(s)\sigma(x(s),y(s),s)dB(s)$$
$$\leq U(0) + \frac{1}{2}\int_0^t p(s)ds + \int_0^t y^T(s)\sigma(x(s),y(s),s)dB(s). \quad (5.8)$$

For any positive constants γ, δ and τ, by the exponential martingale inequality (i.e. Theorem 1.7.4), we have that
$$P\left\{\sup_{0 \leq t \leq \tau}\left[\int_0^t y^T \sigma dB(s) - \frac{\gamma}{2}\int_0^t |y^T \sigma|^2 ds\right] > \delta\right\} \leq e^{-\gamma\delta}. \quad (5.9)$$

Fix $\theta > 1$ and $\beta > 0$ arbitrarily. For each $k = 1, 2, \cdots$, we let
$$\gamma = \beta\theta^{-k(q+1)}, \quad \delta = \beta^{-1}\theta^{k(q+1)+1}\log k, \quad \tau = \theta^k$$
in (5.9) to obtain that
$$P\left\{\sup_{0 \le t \le \theta^k}\left[\int_0^t y^T\sigma dB(s) - \frac{\beta\theta^{-k(q+1)}}{2}\int_0^t |y^T\sigma|^2 ds\right]\right.$$
$$\left. > \beta^{-1}\theta^{k(q+1)+1}\log k\right\} \le \frac{1}{k^\theta}.$$

By the Borel–Cantelli lemma, we see that for almost all $\omega \in \Omega$,
$$\sup_{0 \le t \le \theta^k}\left[\int_0^t y^T\sigma dB(s) - \frac{\beta\theta^{-k(q+1)}}{2}\int_0^t |y^T\sigma|^2 ds\right] \le \beta^{-1}\theta^{k(q+1)+1}\log k$$
holds for all but finitely many k. Hence, for almost all $\omega \in \Omega$, there exists a random integer $k_0 = k_0(\omega)$ such that
$$\int_0^t y^T\sigma dB(s) \le \beta^{-1}\theta^{k(q+1)+1}\log k + \frac{\beta\theta^{-k(q+1)}}{2}\int_0^t |y^T\sigma|^2 ds$$
holds for all $0 \le t \le \theta^k$ whenever $k \ge k_0$. Noting that
$$\frac{1}{2}\int_0^t |y^T\sigma|^2 ds \le \int_0^t p(s)U(s)ds,$$
we obtain that
$$\int_0^t y^T\sigma dB(s) \le \beta^{-1}\theta^{k(q+1)+1}\log k + \beta\theta^{-k(q+1)}\int_0^t p(s)U(s)ds$$
for all $0 \le t \le \theta^k$, $k \ge k_0$ almost surely. Substituting this into (5.8) yields that
$$U(t) \le U(0) + \frac{1}{2}\int_0^t p(s)ds + \beta^{-1}\theta^{k(q+1)+1}\log k$$
$$+ \beta\theta^{-k(q+1)}\int_0^t p(s)U(s)ds$$
for all $0 \le t \le \theta^k$, $k \ge k_0$ almost surely. Applying Lemma 5.2, we then derive that
$$U(t) \le \exp\left(\beta\theta^{-k(q+1)}\int_0^t p(s)ds\right)$$
$$\times \left[U(0) + \beta^{-1}\theta^{k(q+1)+1}\log k + \frac{1}{2}\int_0^t \exp\left(-\beta\theta^{-k(q+1)}\int_0^s p(r)dr\right)p(s)ds\right]$$
$$\le \exp[\beta\theta^{-k(q+1)}\bar{p}(\theta^k)]\left(U(0) + \beta^{-1}\theta^{k(q+1)+1}\log k + \frac{1}{2\beta}\theta^{k(q+1)}\right) \quad (5.10)$$

for all $0 \leq t \leq \theta^k$, $k \geq k_0$ almost surely, where

$$\bar{p}(t) = \int_0^t p(s)ds$$

which is a polynomial of degree $q+1$ with the leading coefficient $K/(q+1)$ and so

$$\theta^{-k(q+1)}\bar{p}(\theta^k) \to \frac{K}{q+1} \quad \text{as } k \to \infty.$$

Note that for $\theta^{k-1} \leq t \leq \theta^k$,

$$\log(k-1) + \log\log\theta \leq \log\log t \leq \log k + \log\log\theta.$$

It follows from (5.10) that for almost all $\omega \in \Omega$, if $\theta^{k-1} \leq t \leq \theta^k$ and $k \geq k_0$,

$$\frac{U(t)}{t^{q+1}\log\log t} \leq \frac{1}{\theta^{(k-1)(q+1)}[\log(k-1) + \log\log\theta]}$$
$$\times \exp[\beta\theta^{-k(q+1)}\bar{p}(\theta^k)]\left(U(0) + \beta^{-1}\theta^{k(q+1)+1}\log k + \frac{1}{2\beta}\theta^{k(q+1)}\right).$$

Therefore

$$\limsup_{t\to\infty} \frac{U(t)}{t^{q+1}\log\log t}$$
$$\leq \limsup_{k\to\infty}\left\{\frac{1}{\theta^{(k-1)(q+1)}[\log(k-1) + \log\log\theta]}\right.$$
$$\left.\times \exp[\beta\theta^{-k(q+1)}\bar{p}(\theta^k)]\left(U(0) + \beta^{-1}\theta^{k(q+1)+1}\log k + \frac{1}{2\beta}\theta^{k(q+1)}\right)\right\}$$
$$= \beta^{-1}\theta^{q+2}\exp\left(\frac{\beta K}{q+1}\right) \quad \text{a.s.}$$

Since $\theta > 1$ is arbitrary, we must have that

$$\limsup_{t\to\infty} \frac{U(t)}{t^{q+1}\log\log t} \leq \beta^{-1}\exp\left(\frac{\beta K}{q+1}\right) \quad \text{a.s.}$$

Further note the trivial arithmetic statement

$$\min_{\beta>0}\left\{\beta^{-1}\exp\left(\frac{\beta K}{q+1}\right)\right\} = \beta^{-1}\exp\left(\frac{\beta K}{q+1}\right)\bigg|_{\beta=(q+1)/K} = \frac{Ke}{q+1}.$$

We can therefore choose $\beta = (q+1)/K$ to obtain

$$\limsup_{t\to\infty} \frac{U(t)}{t^{q+1}\log\log t} \leq \frac{Ke}{q+1} \quad \text{a.s.}$$

as required. The proof is complete.

Corollary 5.3 *Let $\bar{U}(t) = \sup_{0 \leq s \leq t} U(s)$ on $0 \leq t < \infty$. Then, under the hypotheses of Theorem 5.1,*

$$\limsup_{t \to \infty} \frac{\bar{U}(t)}{t^q \log \log t} = \limsup_{t \to \infty} \frac{U(t)}{t^q \log \log t} \quad a.s.$$

This corollary follows from Theorem 5.1 and the following lemma.

Lemma 5.4 *Let $g : R_+ \to R_+$ and $f : R_+ \to (0, \infty)$ be two functions with f nondecreasing and $f(t) \to \infty$ as $t \to \infty$. Let $\bar{g}(t) = \sup_{0 \leq s \leq t} g(s)$ on $0 \leq t < \infty$. If*

$$\limsup_{t \to \infty} \frac{g(t)}{f(t)} < \infty,$$

then

$$\limsup_{t \to \infty} \frac{\bar{g}(t)}{f(t)} = \limsup_{t \to \infty} \frac{g(t)}{f(t)}.$$

Proof. Let

$$\xi = \limsup_{t \to \infty} \frac{g(t)}{f(t)} < \infty.$$

Assign $\varepsilon > 0$ arbitrarily and then there exists a $T(\varepsilon) > 0$ such that

$$\frac{g(t)}{f(t)} \leq \xi + \varepsilon \quad \text{for all } t \geq T(\varepsilon).$$

That is,

$$g(t) \leq (\xi + \varepsilon) f(t) \quad \text{for all } t \geq T(\varepsilon).$$

Thus

$$\bar{g}(t) \leq \bar{g}(T(\varepsilon)) + \sup_{T(\varepsilon) \leq s \leq t} g(s) \leq \bar{g}(T(\varepsilon)) + f(t)(\xi + \varepsilon)$$

for all $t \geq T(\varepsilon)$. In consequence,

$$\limsup_{t \to \infty} \frac{\bar{g}(t)}{f(t)} \leq \xi + \varepsilon.$$

Since $\varepsilon > 0$ is arbitrary, we get

$$\limsup_{t \to \infty} \frac{\bar{g}(t)}{f(t)} \leq \limsup_{t \to \infty} \frac{g(t)}{f(t)}.$$

But $\bar{g}(t) \geq g(t)$ and so the equality must hold, and the proof is complete.

Theorem 5.5 *Let $p(t)$ be a real polynomial of degree q on $t \geq 0$ with the leading coefficient $K > 0$ and the other coefficients nonnegative. Let $\rho > 0$. Assume that*
$$y^T b(x,y,t) \geq 0 \tag{5.11}$$
and
$$|\sigma(x,y,t)|^2 \leq p(t) e^{\rho t} \tag{5.12}$$
for all $x, y \in R^d$ and $t \geq 0$. Then the energy of system (5.3) has the property that
$$\limsup_{t \to \infty} \frac{U(t)}{e^{\rho t} t^q \log t} \leq \frac{Ke}{\rho} \quad a.s. \tag{5.13}$$
Moreover (in the notation of Corollary 5.3),
$$\limsup_{t \to \infty} \frac{\bar{U}(t)}{e^{\rho t} t^q \log t} = \limsup_{t \to \infty} \frac{U(t)}{e^{\rho t} t^q \log t} \quad a.s. \tag{5.14}$$

Proof. We use the same notations as in the proof of Theorem 5.1. By Itô's formula and the hypotheses we can easily show that
$$U(t) \leq U(0) + \frac{1}{2} \int_0^t p(s) e^{\rho s} ds + \int_0^t y^T(s) \sigma(x(s), y(s), s) dB(s). \tag{5.15}$$

Fix $\theta > 1$ and $\beta, \xi > 0$ arbitrarily. For each $k = 1, 2, \cdots$, we let
$$\gamma = \beta (k\xi)^{-q} e^{-\rho k \xi}, \quad \delta = \frac{\theta}{\beta}(k\xi)^q e^{\rho k \xi} \log k, \quad \tau = k\xi.$$

in (5.9) to obtain that
$$P\left\{ \sup_{0 \leq t \leq k\xi} \left[\int_0^t y^T \sigma dB(s) - \frac{1}{2}\beta(k\xi)^{-q} e^{-\rho k\xi} \int_0^t |y^T \sigma|^2 ds \right] \right.$$
$$\left. > \frac{\theta}{\beta}(k\xi)^q e^{\rho k\xi} \log k \right\} \leq \frac{1}{k^\theta}.$$

By the Borel–Cantelli lemma, we then see that for almost all $\omega \in \Omega$, there exists a random integer $k_0 = k_0(\omega)$ such that
$$\int_0^t y^T \sigma dB(s) \leq \frac{\theta}{\beta}(k\xi)^q e^{\rho k\xi} \log k + \frac{1}{2}\beta(k\xi)^{-q} e^{-\rho k\xi} \int_0^t |y^T \sigma|^2 ds$$
$$\leq \frac{\theta}{\beta}(k\xi)^q e^{\rho k\xi} \log k + \beta(k\xi)^{-q} e^{-\rho k\xi} \int_0^t p(s) e^{\rho s} U(s) ds$$

for all $0 \leq t \leq k\xi$ and $k \geq k_0$. Substituting this into (5.15) yields that
$$U(t) \leq U(0) + \frac{\theta}{\beta}(k\xi)^q e^{\rho k\xi} \log k$$
$$+ \frac{1}{2} \int_0^t p(s) e^{\rho s} ds + \beta(k\xi)^{-q} e^{-\rho k\xi} \int_0^t p(s) e^{\rho s} U(s) ds$$

for all $0 \leq t \leq k\xi$, $k \geq k_0$ almost surely. By Lemma 5.2, we then derive that

$$U(t) \leq \exp\left(\beta(k\xi)^{-q} e^{-\rho k\xi} \int_0^t p(s) e^{\rho s} ds\right)$$

$$\times \left\{ U(0) + \frac{\theta}{\beta}(k\xi)^q e^{\rho k\xi} \log k \right.$$

$$\left. + \frac{1}{2} \int_0^t p(s) e^{\rho s} \exp\left(-\beta(k\xi)^{-q} e^{-\rho k\xi} \int_0^s p(r) e^{\rho r} dr\right) ds \right\}$$

$$\leq \exp\left[\frac{\beta}{\rho}(k\xi)^{-q} p(k\xi)\right]$$

$$\times \left\{ U(0) + \frac{\theta}{\beta}(k\xi)^q e^{\rho k\xi} \log k + \frac{1}{2\beta}(k\xi)^q e^{\rho k\xi} \right\}$$

for all $0 \leq t \leq k\xi$, $k \geq k_0$ almost surely. Therefore, for almost all $\omega \in \Omega$, if $(k-1)\xi \leq t \leq k\xi$ and $k \geq k_0$,

$$\frac{U(t)}{e^{\rho t} t^q \log t} \leq \left(e^{\rho(k-1)\xi}[(k-1)\xi]^q [\log(k-1) + \log \xi]\right)^{-1}$$

$$\times \exp\left[\frac{\beta}{\rho}(k\xi)^{-q} p(k\xi)\right]$$

$$\times \left\{ U(0) + \frac{\theta}{\beta}(k\xi)^q e^{\rho k\xi} \log k + \frac{1}{2\beta}(k\xi)^q e^{\rho k\xi} \right\}.$$

Noting that
$$(k\xi)^{-q} p(k\xi) \to K \quad \text{as } k \to \infty,$$

we see
$$\limsup_{t \to \infty} \frac{U(t)}{e^{\rho t} t^q \log t} \leq \frac{\theta}{\beta} e^{\rho \xi + \beta K/\rho} \quad a.s.$$

Letting $\theta \to 1$ and $\xi \to 0$ yields that

$$\limsup_{t \to \infty} \frac{U(t)}{e^{\rho t} t^q \log t} \leq \frac{1}{\beta} e^{\beta K/\rho} \quad a.s.$$

Since this holds for arbitrary $\beta > 0$, we can specially choose $\beta = \rho/K$ to obtain that
$$\limsup_{t \to \infty} \frac{U(t)}{e^{\rho t} t^q \log t} \leq \frac{Ke}{\rho} \quad a.s.$$

as desired. The proof is complete.

(ii) *Forced Dynamics*

We shall now turn to consider the energy bounds for the more general equation (5.1) under the external force, namely

$$\begin{cases} dx(t) = y(t)dt, \\ dy(t) = [-b(x(t), y(t)) - \nabla G(x(t)) + h(t)]dt + \sigma(x(t), y(t), t)dB(t). \end{cases} \quad (5.16)$$

Theorem 5.6 Let $p_i(t)$, $1 \leq i \leq 4$, be real polynomials of degrees q_i on $t \geq 0$ with the leading coefficients $K_i > 0$ and the other coefficients nonnegative. Assume that

$$y^T b(x,y,t) \geq -p_1(t) - p_2(t)\left(\frac{1}{2}|y|^2 + G(x)\right),$$

$$|\sigma(x,y,t)|^2 \leq p_3(t),$$

$$|h(t)| \leq p_4(t)$$

for all $x, y \in R^d$ and $t \geq 0$. Then the energy of system (5.16) has the property that

$$\limsup_{t \to \infty} \frac{U(t)}{\exp[t(p_2(t) + p_4(t))] \, t^{q+1} \log\log t}$$
$$\leq \begin{cases} K_3 e/(q_3 + 1) & \text{if } q_3 \geq q_1 \vee q_2, \\ 0 & \text{otherwise,} \end{cases} \quad (5.17)$$

almost surely, where $q = q_1 \vee q_2 \vee q_3$.

Proof. By Itô's formula,

$$U(t) = U(0) + \int_0^t \left(y^T(s)[-b(x(s),y(s)) + h(s)] + \frac{1}{2}|\sigma(x(s),y(s),s)|^2\right) ds$$
$$+ \int_0^t y^T(s)\sigma(x(s),y(s),s) dB(s).$$

Noting that

$$y^T(s)h(s) \leq |h(s)||y(s)| \leq \frac{1}{2}|h(s)|(1 + |y(s)|^2) \leq |h(s)|(1 + U(s))$$

and using the hypotheses we then see that

$$U(t) \leq U(0) + \int_0^t \left[\bar{p}(s) + [p_2(s) + p_4(s)]U(s)\right] ds$$
$$+ \int_0^t y^T(s)\sigma(x(s),y(s),s) dB(s), \quad (5.18)$$

where $\bar{p}(s) = p_1(s) + \frac{1}{2}p_3(s) + p_4(s)$ which is a polynomial of degree $q := q_1 \vee q_2 \vee q_3$. Assign $\theta > 1$ and $\beta > 0$ arbitrarily. In the same way as in the proof of Theorem 5.1, we can show that for almost all $\omega \in \Omega$, there is a random integer $k_0 = k_0(\omega)$ such that

$$\int_0^t y^T \sigma dB(s) \leq \beta^{-1}\theta^{k(q+1)+1} \log k + \beta\theta^{-k(q+1)} \int_0^t p_3(s)U(s)ds$$

for all $0 \leq t \leq \theta^k$, $k \geq k_0$. Substituting this into (5.18) yields that

$$U(t) \leq U(0) + \beta^{-1}\theta^{k(q+1)+1}\log k + \bar{p}(\theta^k)\theta^k$$
$$+ \int_0^t \left[\beta\theta^{-k(q+1)}p_3(s) + p_2(s) + p_4(s)\right]U(s)ds$$

for all $0 \leq t \leq \theta^k$, $k \geq k_0$ almost surely. By the Gronwall inequality, we then have that

$$U(t) \leq \left[U(0) + \beta^{-1}\theta^{k(q+1)+1}\log k + \bar{p}(\theta^k)\theta^k\right]$$
$$\times \exp\left(\int_0^t \left[\beta\theta^{-k(q+1)}p_3(s) + p_2(s) + p_4(s)\right]ds\right)$$
$$\leq \left[U(0) + \beta^{-1}\theta^{k(q+1)+1}\log k + \bar{p}(\theta^k)\theta^k\right]$$
$$\times \exp\left(\beta\theta^{-k(q+1)}\int_0^{\theta^k} p_3(s)ds + t[p_2(t) + p_4(t)]\right)$$

for all $0 \leq t \leq \theta^k$, $k \geq k_0$ almost surely. Therefore, for almost all $\omega \in \Omega$, if $\theta^{k-1} \leq t \leq \theta^k$ and $k \geq k_0$,

$$\frac{U(t)}{\exp[t(p_2(t) + p_4(t))]\,t^{q+1}\log\log t}$$
$$\leq \frac{U(0) + \beta^{-1}\theta^{k(q+1)+1}\log k + \bar{p}(\theta^k)\theta^k}{\theta^{(k-1)(q+1)}[\log(k-1) + \log\log\theta]} \exp\left(\beta\theta^{-k(q+1)}\int_0^{\theta^k} p_3(s)ds\right).$$

If $q_3 \geq q_1 \vee q_2$, then $q = q_3$ and

$$\exp\left(\beta\theta^{-k(q+1)}\int_0^{\theta^k} p_3(s)ds\right) \to e^{\beta K_3/(q_3+1)} \quad \text{as } k \to \infty.$$

We therefore derive that

$$\limsup_{t\to\infty} \frac{U(t)}{\exp[t(p_2(t) + p_4(t))]\,t^{q+1}\log\log t} \leq \beta^{-1}\theta^{q+2}e^{\beta K_3/(q_3+1)} \quad a.s.$$

Letting $\theta \to 1$ and then choosing $\beta = (q_3+1)/K_3$ we obtain

$$\limsup_{t\to\infty} \frac{U(t)}{\exp[t(p_2(t) + p_4(t))]\,t^{q+1}\log\log t} \leq \frac{K_3 e}{q_3+1} \quad a.s.$$

On the other hand, if $q_3 < q_1 \vee q_2$, then $q > q_3$ and

$$\exp\left(\beta\theta^{-k(q+1)}\int_0^{\theta^k} p_3(s)ds\right) \to 1 \quad \text{as } k \to \infty.$$

Therefore
$$\limsup_{t\to\infty} \frac{U(t)}{\exp[t(p_2(t)+p_4(t))]\,t^{q+1}\log\log t} \le \beta^{-1}\theta^{q+2} \quad a.s.$$

Letting $\beta \to \infty$ yields
$$\limsup_{t\to\infty} \frac{U(t)}{\exp[t(p_2(t)+p_4(t))]\,t^{q+1}\log\log t} \le 0 \quad a.s.$$

The proof is complete.

If all the $p_i(t)$'s reduce to constants, namely $p_i(t) \equiv K_i$ and $q_i = 0$, we obtain the following useful result.

Corollary 5.7 *Let K_i, $1 \le i \le 4$, be positive constants. Assume that*
$$y^T b(x,y,t) \ge -K_1 - K_2\left(\frac{1}{2}|y|^2 + G(x)\right),$$
$$|\sigma(x,y,t)|^2 \le K_3,$$
$$|h(t)| \le K_4$$

for all $x,y \in R^d$ and $t \ge 0$. Then the energy of system (5.16) has the property that
$$\limsup_{t\to\infty} \frac{U(t)}{\exp[t(K_2+K_4)]\,t\log\log t} \le K_3 e \quad a.s.$$

Theorem 5.8 *Let $p_i(t)$, $1 \le i \le 4$, be real polynomials of degrees q_i on $t \ge 0$ with the leading coefficients $K_i > 0$ and the other coefficients nonnegative. Let $\rho > 0$. Assume that*
$$y^T b(x,y,t) \ge -p_1(t) - p_2(t)\left(\frac{1}{2}|y|^2 + G(x)\right),$$
$$|\sigma(x,y,t)|^2 \le p_3(t)e^{\rho t},$$
$$|h(t)| \le p_4(t)$$

for all $x,y \in R^d$ and $t \ge 0$. Then the energy of system (5.16) has the property that
$$\limsup_{t\to\infty} \frac{U(t)}{\exp[t(\rho+p_2(t)+p_4(t))]\,t^{q_3}\log t} \le \frac{K_3 e}{\rho} \quad a.s. \qquad (5.19)$$

Proof. By Itô's formula and the hypotheses we can show that
$$U(t) \le U(0) + \int_0^t \left[p_1(s) + p_4(s) + \frac{1}{2}p_3(s)e^{\rho s}\right]ds$$
$$+ \int_0^t [p_2(s) + p_4(s)]U(s)ds + \int_0^t y^T(s)\sigma(x(s),y(s),s)dB(s). \qquad (5.20)$$

Assign $\theta > 1$ and $\beta, \xi > 0$ arbitrarily. In the same way as in the proof of Theorem 5.5 we can show that for almost all $\omega \in \Omega$, there exists a random integer $k_0 = k_0(\omega)$ such that

$$\int_0^t y^T \sigma dB(s) \leq \frac{\theta}{\beta}(k\xi)^{q_3} e^{\rho k\xi} \log k + \beta(k\xi)^{-q_3} e^{-\rho k\xi} \int_0^t p_3(s) e^{\rho s} U(s) ds$$

for all $0 \leq t \leq k\xi$ and $k \geq k_0$. Substituting this into (5.20) gives that

$$U(t) \leq U(0) + \int_0^t \left[p_1(s) + p_4(s) + \frac{1}{2} p_3(s) e^{\rho s} \right] ds$$

$$+ \int_0^t [p_2(s) + p_4(s)] U(s) ds + \frac{\theta}{\beta}(k\xi)^{q_3} e^{\rho k\xi} \log k$$

$$+ \beta(k\xi)^{-q_3} e^{-\rho k\xi} \int_0^t p_3(s) e^{\rho s} U(s) ds$$

$$\leq U(0) + k\xi[p_1(k\xi) + p_4(k\xi)] + \frac{1}{2\rho} p_3(k\xi) e^{\rho k\xi} + \frac{\theta}{\beta}(k\xi)^{q_3} e^{\rho k\xi} \log k$$

$$+ \int_0^t \left[p_2(s) + p_4(s) + \beta(k\xi)^{-q_3} e^{-\rho k\xi} p_3(s) e^{\rho s} \right] U(s) ds$$

for all $0 \leq t \leq k\xi$ and $k \geq k_0$ almost surely, where we have also used the following estimate

$$\int_0^{k\xi} p_3(s) e^{\rho s} ds \leq p_3(k\xi) \int_0^{k\xi} e^{\rho s} ds \leq \frac{1}{\rho} p_3(k\xi) e^{\rho k\xi}.$$

An application of the Gronwall inequality implies that

$$U(t) \leq \left(U(0) + k\xi[p_1(k\xi) + p_4(k\xi)] + \frac{1}{2\rho} p_3(k\xi) e^{\rho k\xi} + \frac{\theta}{\beta}(k\xi)^{q_3} e^{\rho k\xi} \log k \right)$$

$$\times \exp\left(\int_0^t \left[p_2(s) + p_4(s) + \beta(k\xi)^{-q_3} e^{-\rho k\xi} p_3(s) e^{\rho s} \right] ds \right)$$

$$\leq \left(U(0) + k\xi[p_1(k\xi) + p_4(k\xi)] + \frac{1}{2\rho} p_3(k\xi) e^{\rho k\xi} + \frac{\theta}{\beta}(k\xi)^{q_3} e^{\rho k\xi} \log k \right)$$

$$\times \exp\left(t[p_2(t) + p_4(t)] + \frac{\beta}{\rho}(k\xi)^{-q_3} p_3(k\xi) \right)$$

for all $0 \leq t \leq k\xi$ and $k \geq k_0$ almost surely. Therefore, for almost all $\omega \in \Omega$, if $(k-1)\xi \leq t \leq k\xi$, then

$$\frac{U(t)}{\exp[t(\rho + p_2(t) + p_4(t))] t^{q_3} \log t}$$

$$\leq \left(e^{\rho(k-1)\xi} [(k-1)\xi]^{q_3} [\log(k-1) + \log \xi] \right)^{-1} \exp\left(\frac{\beta}{\rho}(k\xi)^{-q_3} p_3(k\xi) \right)$$

$$\times \left(U(0) + k\xi[p_1(k\xi) + p_4(k\xi)] + \frac{1}{2\rho} p_3(k\xi) e^{\rho k\xi} + \frac{\theta}{\beta}(k\xi)^{q_3} e^{\rho k\xi} \log k \right).$$

Bearing in mind that

$$(k\xi)^{-q_3} p_3(k\xi) \to K_3 \quad \text{as } k \to \infty,$$

we see immediately that

$$\limsup_{t\to\infty} \frac{U(t)}{\exp[t(\rho + p_2(t) + p_4(t))] t^{q_3} \log t} \leq \frac{\theta}{\beta} \exp\left(\rho\xi + \frac{\beta K_3}{\rho}\right) \quad a.s.$$

Letting $\theta \to 1$ and $\xi \to 0$ and then choosing $\beta = \rho/K_3$, we obtain

$$\limsup_{t\to\infty} \frac{U(t)}{\exp[t(\rho + p_2(t) + p_4(t))] t^{q_3} \log t} \leq \frac{K_3 e}{\rho} \quad a.s.$$

which is the required assertion. The proof is complete.

To close this chapter, let us state one immediate corollary.

Corollary 5.9 *Let ρ and $K_i (1 \leq i \leq 4)$ be positive constants. Assume that*

$$y^T b(x, y, t) \geq -K_1 - K_2 \left(\frac{1}{2}|y|^2 + G(x)\right),$$

$$|\sigma(x, y, t)|^2 \leq K_3 e^{\rho t},$$

$$|h(t)| \leq K_4$$

for all $x, y \in R^d$ and $t \geq 0$. Then the energy of system (5.16) has the property that

$$\limsup_{t\to\infty} \frac{U(t)}{\exp[t(\rho + K_2 + K_4)] \log t} \leq \frac{K_3 e}{\rho} \quad a.s.$$

9

Applications to Economics and Finance

9.1 INTRODUCTION

Pricing models for financial derivatives require, by their very nature, the utilization of continuous-time stochastic processes, especially Itô's stochastic calculus. For example, the Black–Scholes model (Black & Scholes (1973)) used the method of arbitrage-free pricing. But the paper was also influential because of the technical steps introduced in obtaining a closed-form formula for options prices. For an approach that used abstract notions such as the Itô calculus, the formula was accurate enough to win the attention of market participants. In brief, stochastic modelling has become more and more popular in financial economics. In this chapter we shall apply the theory of stochastic differential equations to the related problems in finance. As usual, we shall work on the given probability space $(\Omega, \mathcal{F}, \{\mathcal{F}_t\}, P)$ throughout this chapter.

9.2 STOCHASTIC MODELLING IN ASSET PRICES

One of the important problems in finance is the specification of the stochastic process governing the behaviour of an asset. In this section we shall describe a number of stochastic differential equations which have often been used in modelling asset prices.

(i) <u>Geometric Brownian Motion</u>

In the early studies people assumed that the price of an asset followed a

Gaussian process described by the Itô differential

$$dS(t) = \lambda dt + \sigma dB(t) \tag{2.1}$$

on $t \geq 0$. Here, and throughout this section, $S(t)$ is the price of the asset at time t, both λ and σ are positive constants, $B(t)$ is a one-dimensional Brownian motion and, moreover, the initial price is a positive constant S_0, i.e. $S(0) = S_0 > 0$. Clearly

$$S(t) = S_0 + \lambda t + \sigma B(t)$$

which is normally distributed with mean $S_0 + \lambda t$ and variance $\sigma^2 t$. So the price may become negative but this violates the condition of limited liability. To overcome this weakness, several people, e.g. Samuelson (1965), Black & Scholes (1973), suggested the idea of modelling by geometric Brownian motions. Influenced by their work, nowadays many economists and management scientists assume that the price of an asset follows the geometric Brownian motion, that is the price is governed by the linear Itô equation

$$dS(t) = \lambda S(t)dt + \sigma S(t)dB(t). \tag{2.2}$$

By the theory established in Chapter 3, equation (2.2) has the explicit solution

$$S(t) = S_0 \exp\left[\left(\lambda - \frac{\sigma^2}{2}\right)t + \sigma B(t)\right]. \tag{2.3}$$

Hence the price $S(t)$ is lognormally distributed. Note that for any constant λ, $\exp[-(\lambda^2/2)t + \lambda B(t)]$ is an exponential martingale on $t \geq 0$ and, therefore, $E \exp[-(\lambda^2/2)t + \lambda B(t)] = 1$. Making use of this fact, we can compute the nth moment

$$ES^n(t) = S_0^n E \exp\left[\left(\lambda - \frac{\sigma^2}{2}\right)nt + n\sigma B(t)\right]$$

$$= S_0^n \exp\left[\left(\lambda - \frac{\sigma^2}{2}\right)nt + \frac{n^2\sigma^2}{2}t\right] E \exp\left[-\frac{n^2\sigma^2}{2}t + n\sigma B(t)\right]$$

$$= S_0^n \exp\left[n\lambda t + \frac{\sigma^2}{2}n(n-1)t\right].$$

In particular, the price $S(t)$ has the mean

$$ES(t) = S_0 e^{\lambda t}$$

and variance

$$Var(S(t)) = ES^2(t) - S_0^2 e^{2\lambda t} = S_0^2 e^{2\lambda t}\left[e^{\sigma^2 t} - 1\right].$$

Therefore, the average of the price increases exponentially and is independent of parameter σ. Let us now look at the individual price, i.e. the sample properties of $S(t)$. Recall the law of the iterated logarithm

$$\limsup_{t \to \infty} \frac{B(t)}{\sqrt{2t \log \log t}} = 1 \quad \text{and} \quad \liminf_{t \to \infty} \frac{B(t)}{\sqrt{2t \log \log t}} = -1$$

almost surely. We can then easily show that

$$\lim_{t\to\infty} \frac{1}{t} \log S(t) = \lambda - \frac{\sigma^2}{2}$$

almost surely if $\lambda \neq \sigma^2/2$, while

$$\limsup_{t\to\infty} \frac{\log S(t)}{\sqrt{2t \log \log t}} = \sigma \quad \text{and} \quad \liminf_{t\to\infty} \frac{\log S(t)}{\sqrt{2t \log \log t}} = -\sigma$$

almost surely if $\lambda = \sigma^2/2$. We hence conclude:

(a) $S(t) \to \infty$ almost surely exponentially if $\lambda > \sigma^2/2$;
(b) $S(t) \to 0$ almost surely exponentially if $\lambda < \sigma^2/2$;
(c) $\limsup_{t\to\infty} S(t) = \infty$ while $\liminf_{t\to\infty} S(t) = 0$ almost surely if $\lambda = \sigma^2/2$.

Especially, it is interesting to observe that an individual who holds the asset long enough would be "almost certainly ruined" if $\lambda < \sigma^2/2$, even though in this case the average of the price is increasing.

(ii) *Mean Reverting Process*

A stochastic differential equation that has been found useful in modelling asset prices is the mean reverting model:

$$dS(t) = \lambda(\mu - S(t))dt + \sigma S(t)dB(t). \tag{2.4}$$

Especially, this is often used to model interest rate dynamics. According to the model, as $S(t)$ increases above some "mean value" μ (> 0), the drift term $\lambda(\mu - S(t))$ will become negative. This makes $dS(t)$ more likely be negative and $S(t)$ will decrease. On the other hand, as $S(t)$ falls below the value μ, $\lambda(\mu - S(t))$ will become positive. This makes $dS(t)$ more likely be positive and $S(t)$ will increase. Hence, we may expect that $S(t)$ will eventually move towards the value μ and, indeed, we shall see $ES(t) \to \mu$ as $t \to \infty$. In view of the theory of Chapter 3, we know that equation (2.4) has the explicit solution

$$S(t) = S_0 \exp\left[-(\lambda + \sigma^2/2)t + \sigma B(t)\right]$$
$$+ \lambda\mu \int_0^t \exp\left[-(\lambda + \sigma^2/2)(t-s) + \sigma(B(t) - B(s))\right] ds. \tag{2.5}$$

This formula shows clearly that $S(t)$ remains positive as long as $S_0 > 0$. Noting

$$E \exp\left[-\frac{\sigma^2}{2}(t-s) + \sigma(B(t) - B(s))\right] = 1 \quad \text{for } 0 \leq s \leq t < \infty,$$

we can compute the mean

$$ES(t) = S_0 e^{-\lambda t} + \lambda\mu \int_0^t e^{-\lambda(t-s)} ds$$
$$= S_0 e^{-\lambda t} + \mu\left[1 - e^{-\lambda t}\right] = \mu + (S_0 - \mu)e^{-\lambda t}.$$

This implies
$$\lim_{t \to \infty} ES(t) = \mu$$
as expected.

(iii) *Mean Reverting Ornstein–Uhlenbeck Process*

A model close to the one just discussed is the mean reverting Ornstein–Uhlenbeck process:
$$dS(t) = \lambda(\mu - S(t))dt + \sigma dB(t). \tag{2.6}$$

In this model, the diffusion term does not depend on $S(t)$. As a result, we shall see that $S(t)$ may become negative. Equation (2.6) has the explicit solution
$$S(t) = e^{-\lambda t}\left(S_0 + \lambda\mu \int_0^t e^{\lambda s} ds + \sigma \int_0^t e^{\lambda s} dB(s)\right)$$
$$= \mu + e^{-\lambda t}(S_0 - \mu) + \sigma e^{-\lambda t} \int_0^t e^{\lambda s} dB(s). \tag{2.7}$$

Clearly, $S(t)$ is normally distributed with mean
$$ES(t) = \mu + e^{-\lambda t}(S_0 - \mu) \to \mu \quad \text{as } t \to \infty$$
and variance
$$Var(S(t)) = \frac{\sigma^2}{2\lambda}(1 - e^{-2\lambda t}) \to \frac{\sigma^2}{2\lambda} \quad \text{as } t \to \infty.$$

We therefore observe that the distribution of $S(t)$ always approaches the normal distribution $N(\mu, \sigma^2/2\lambda)$ as $t \to \infty$ for arbitrary S_0. We also observe that $S(t)$ may become negative.

(iv) *Square Root Process*

A model close to the geometric Brownian motion is the square root process:
$$dS(t) = \lambda S(t)dt + \sigma\sqrt{S(t)}dB(t). \tag{2.8}$$

Here the mean is made to follow an exponential trend as before, while the standard deviation is made a function of the square root of $S(t)$, rather than $S(t)$ itself. This makes the "variance" of the error term proportional to $S(t)$. Hence, if the asset price volatility does not increase "too much" when $S(t)$ increases (greater than 1, of course), this model may be more appropriate. For equation (2.8), one may ask whether $S(t)$ will become negative. If so, $\sqrt{S(t)}$ would become a complex number and this would not make sense in modelling an asset price. We shall now show this is impossible. This nonnegative property is clearly equivalent to the solution of equation
$$dS(t) = \lambda S(t)dt + \sigma\sqrt{|S(t)|}dB(t) \tag{2.8'}$$

never becoming negative as long as the initial value $S_0 \geq 0$. To prove this, let $a_0 = 1$ and $a_k = e^{-k(k+1)/2}$ for every integer $k \geq 1$. Note that

$$\int_{a_k}^{a_{k-1}} \frac{du}{u} = k.$$

Let $\psi_k(u)$ be a continuous function such that its support is contained in the interval (a_k, a_{k-1}) where $0 \leq \psi_k(u) \leq 2/ku$ and, moreover,

$$\int_{a_k}^{a_{k-1}} \psi_k(u)du = 1.$$

Such a function exists obviously. Define $\varphi_k(x) = 0$ for $x \geq 0$ and

$$\varphi_k(x) = \int_0^{-x} dy \int_0^y \psi_k(u)du \quad \text{for } x < 0.$$

It is easy to see that $\varphi \in C^2(R; R)$;

$$-1 \leq \varphi'_k(x) \leq 0 \text{ if } -a_{k-1} < x < -a_k \text{ or otherwise } \varphi'_k(x) = 0;$$

$$|\varphi''_k(x)| \leq \frac{2}{k|x|} \text{ if } -a_{k-1} < x < -a_k \text{ or otherwise } \varphi''_k(x) = 0;$$

moreover,

$$x^- - a_{k-1} \leq \varphi_k(x) \leq x^- \quad \text{for all } x \in R,$$

where $x^- = -x$ if $x < 0$ or otherwise $x^- = 0$. Now for any $t \geq 0$, by Itô's formula we can derive that

$$\varphi_k(S(t)) = \varphi_k(S_0) + \int_0^t \left[\lambda S(r)\varphi'_k(S(r)) + \frac{\sigma^2}{2}|S(r)|\varphi''_k(S(r))\right] dr$$

$$+ \sigma \int_0^t \varphi'_k(S(r))\sqrt{|S(r)|}dB(r)$$

$$\leq [\lambda a_{k-1} + \sigma^2/k]t + \sigma \int_0^t \varphi'_k(S(r))\sqrt{|S(r)|}dB(r).$$

Hence

$$ES^-(t) - a_{k-1} \leq E\varphi_k(S(t)) \leq [\lambda a_{k-1} + \sigma^2/k]t.$$

Letting $k \to \infty$ we get that $ES^-(t) \leq 0$ and hence we must have

$$ES^-(t) = 0 \quad \text{for all } t \geq 0.$$

This implies

$$P\{S(t) < 0\} = 0 \quad \text{for all } t \geq 0.$$

Since $S(t)$ is continuous we must have $S(t) \geq 0$ for all $t \geq 0$ almost surely. This proves the nonnegative property of the solution of equation (2.8)', and due to this property we can certainly write equation (2.8)' as equation (2.8).

Combining the square root idea with the mean reverting one gives us the model of the mean reverting square root process:

$$dS(t) = \lambda(\mu - S(t))dt + \sigma\sqrt{S(t)}dB(t). \tag{2.9}$$

Again this process will never be negative. In fact, applying Itô's formula we have that

$$E\varphi_k(S(t)) \leq \varphi_k(S_0)$$
$$+ E\int_0^t \left[\lambda(\mu - S(t))\varphi_k'(S(r)) + \frac{\sigma^2}{2}|S(r)||\varphi_k''(S(r))|\right]dr$$
$$\leq \frac{\sigma^2 t}{k}.$$

Hence

$$-a_{k-1} \leq ES^-(t) - a_{k-1} \leq \frac{\sigma^2 t}{k}.$$

Letting $k \to \infty$ we get that $ES^-(t) = 0$ for all $t \geq 0$. This implies that $S(t) \geq 0$ for all $t \geq 0$ almost surely. Moreover, the solution of equation (2.9) still has the mean reverting trend

$$ES(t) = \mu + e^{-\lambda t}(S_0 - \mu) \to \mu \quad \text{as } t \to \infty.$$

It is particularly interesting to observe that when the parameters λ, σ and μ have the relation

$$\lambda\mu = \frac{\sigma^2}{4},$$

the square root $\sqrt{S(t)}$ is an Ornstein–Uhlenbeck process:

$$d\sqrt{S(t)} = -\frac{\lambda}{2}\sqrt{S(t)}dt + \frac{\sigma}{2}dB(t) \tag{2.10}$$

whose solution is

$$\sqrt{S(t)} = \sqrt{S_0}e^{-\lambda t/2} + \frac{\sigma}{2}\int_0^t e^{-\lambda(t-s)/2}dB(s).$$

(v) *Stochastic Volatility*

In all the previous models, the drift and diffusion parameters are constants. However, much more general models can be obtained by making these parameters random. Such models may have useful applications, since they allow us to consider the volatility not only time-varying but also random given the $S(t)$. For example, consider an asset price described by the stochastic differential equation

$$dS(t) = \lambda S(t)dt + \sigma(t)S(t)dB(t), \tag{2.11}$$

where λ is a positive constant as before, while the volatility $\sigma(t)$ is assumed to change over time. More specifically, $\sigma(t)$ is assumed to change according to an Ornstein–Uhlenbeck process

$$d\sigma(t) = -\beta\sigma(t)dt + \delta d\tilde{B}(t), \qquad (2.12)$$

with initial value $\sigma(0) = \sigma_0$, where β, δ are positive constants and $\tilde{B}(t)$ is another Brownian motion independent of $B(t)$ (it is also possible to discuss the dependent case). We can solve the equations explicitly:

$$S(t) = S_0 \exp\left[\lambda t - \frac{1}{2}\int_0^t \sigma^2(s)ds + \int_0^t \sigma(s)dB(s)\right]$$

where

$$\sigma(t) = \sigma_0 e^{-\beta t} + \delta \int_0^t e^{-\beta(t-s)}d\tilde{B}(s).$$

We see clearly that $\sigma(t)$ is normally distributed with mean $\sigma_0 e^{-\beta t}$ and variance $(\delta^2/2\beta)(1 - \varepsilon^{-2\beta t})$. Hence, in a long-run, $\sigma(t)$ will follow the normal distribution $N(0, \delta^2/2\beta)$. Recalling the model of geometric Brownian motion, we can reasonably guess that with probability

$$P\left\{\frac{\sigma^2(t)}{2} > \lambda\right\} = 2P\{\sigma(t) > \sqrt{2\lambda}\} \approx 1 - 2\mathrm{Erf}\left(\frac{2\sqrt{\lambda\beta}}{\delta}\right)$$

the $S(t)$ will tend to zero. Alternatively, we may assume that the volatility $\sigma(t)$ follows a mean reverting process

$$d\sigma(t) = \beta(\sigma_\infty - \sigma(t))dt + \delta\sigma(t)d\tilde{B}(t), \qquad (2.13)$$

where σ_∞ is a positive constant. In this case, the volatility of the asset has a long-run mean of σ_∞. We might therefore guess that if $\sigma_\infty^2/2 > \lambda$, the asset price would be most likely ruined.

Clearly, using such layers of stochastic differential equations we can obtain more and more general models for representing the financial phenomena in real life.

9.3 OPTIMAL STOPPING PROBLEMS

Suppose that a person has an asset or resource which changes according to a time-homogeneous d-dimensional stochastic differential equation

$$d\xi(t) = F(\xi(t))dt + G(\xi(t))dB(t) \quad \text{on } t \geq 0. \qquad (3.1)$$

Here $B(t)$ is an m-dimensional Brownian motion and, as a standing hypothesis, we assume that

$$F: R^d \to R^d \text{ and } G: R^d \to R^{d\times m} \text{ are uniformly Lipschitz continuous.}$$

Suppose that the person wishes to sell his asset and the price at time t is of course a function of $\xi(t)$, say $\phi(\xi(t))$. Here ϕ is a continuous nonnegative function defined on R^d and is called a *reward function*. Assume that he is given the reward function ϕ and knows the behaviour of $\xi(t)$ up to the present time t, but because of the noise in the system he is not sure at the time of sale whether his choice of time will turn out to be the best. The optimal stopping problem is to look for a stopping strategy that gives the best result in the sense that the strategy maximizes the expected profit in the long run. To formulate this problem mathematically, let us recall the notation

$$E_x \phi(\xi(t)) = \int_{R^d} \phi(y) P(x; dy, t)$$

which was introduced in Section 2.9, where $P(x; A, t)$ is the transition probability of the Markov solution $\xi(t)$. As shown in Section 2.9, this is equivalent to

$$E_x \phi(\xi(t)) = E\phi(\xi_x(t)),$$

where $\xi_x(t)$ is the unique solution of the equation

$$\xi_x(t) = x + \int_0^t F(\xi_x(s))ds + \int_0^t G(\xi_x(s))dB(s). \qquad (3.2)$$

In other words, if we denote by P_x the probability law of $\xi_x(t)$, then E_x is the expectation with respect to P_x. We also denote by $\mathcal{T}$ the family of all $\mathcal{F}_t$-stopping times (may take value ∞). Now the optimal stopping problem is to look for a stopping time $\tau^* = \tau^*(x, \omega)$ such that

$$E_x \phi(\xi(\tau^*)) = \sup_{\tau \in \mathcal{T}} E_x \phi(\xi(\tau)) \quad \text{for all } x \in R^d, \qquad (3.3)$$

where $\phi(\xi(\tau))$ is set to be 0 at the points $\omega \in \Omega$ where $\tau(\omega) = \infty$. Moreover, we also wish to find the corresponding optimal expected reward

$$\phi^*(x) := \sup_{\tau \in \mathcal{T}} E_x \phi(\xi(\tau)). \qquad (3.4).$$

To solve this problem we need to introduce some basic concepts.

Definition 3.1 *A Borel measurable function* $f : R^d \to [0, \infty]$ *is said to be* supermeanvalued *with respect to the Markov solution $\xi(t)$ of equation (3.1) if*

$$f(x) \geq E_x f(\xi(\tau))$$

for all $\tau \in \mathcal{T}$ and $x \in R^d$. The function f is said to be lower semicontinuous *if*

$$f(x) \leq \liminf_{y \to x} f(y)$$

for all $x \in R^d$. If f is not only supermeanvalued but also lower semicontinuous, then f is said to be l.s.c. superharmonic or simply **superharmonic**.

The following lemma lists a number of useful properties of supermeanvalued and superharmonic functions.

Lemma 3.2

a) If f, g are supermeanvalued (superharmonic) and $\alpha, \beta \geq 0$, then $\alpha f + \beta g$ is supermeanvalued (resp. superharmonic).

b) If $\{f_i\}_{i \in I}$ is a family of supermeanvalued functions, then $f := \inf_{i \in I} f_i$ is supermeanvalued.

c) If $\{f_i\}_{i \geq 1}$ is a sequence of supermeanvalued (superharmonic) functions and $f_i \uparrow f$ pointwise, then f is supermeanvalued (resp. superharmonic).

d) If f is supermeanvalued and $\tau_1, \tau_2 \in \mathcal{T}$ with $\tau_1 \leq \tau_2$, then $E_x f(\xi(\tau_1)) \geq E_x f(\xi(\tau_2))$.

e) If f is supermeanvalued and D is an open subset of R^d, then $f_D(x) := E_x f(\xi(\tau_D))$ is supermeanvalued, where τ_D is the first exit time of $\xi(t)$ from D, i.e. $\tau_D = \inf\{t \geq 0 : \xi(t) \notin D\}$.

f) If f is superharmonic and $\{\tau_i\}$ is any sequence of stopping times such that $\tau_i \to 0$ a.s., then

$$f(x) = \lim_{i \to \infty} E_x f(\xi(\tau_i)) \quad \text{for all } x.$$

Proof. a) is straightforward.

b) Let $\tau \in \mathcal{T}$ and $x \in R^d$ be arbitrary. Note that for every $i \in I$,

$$f_i(x) \geq E_x f_i(\xi(\tau)) \geq E_x f(\xi(\tau)).$$

Hence

$$f(x) = \inf_{i \in I} f_i(x) \geq E_x f(\xi(\tau))$$

as required.

c) First suppose that $\{f_i\}_{i \geq 1}$ is a sequence of supermeanvalued functions and $f_i \uparrow f$ pointwise. Then

$$f(x) \geq f_i(x) \geq E_x f_i(\xi(\tau)) \quad \text{for all } i.$$

So by the monotone convergence theorem,

$$f(x) \geq \lim_{i \to \infty} E_x f_i(\xi(\tau)) = E_x f(\xi(\tau))$$

which means that f is supermeanvalued. Next, if all f_i's are superharmonic, then they are lower semicontinuous and

$$f_i(x) \leq \liminf_{y \to x} f_i(y) \leq \liminf_{y \to x} f(y).$$

In consequence
$$f(x) = \lim_{i\to\infty} f_i(x) \leq \liminf_{y\to x} f(y).$$
This proves that f is lower semicontinuous and therefore is superharmonic.

d) In Chapter 2 we showed that the solution $\xi(t)$ of equation (3.1) is a homogeneous strong Markov process. Hence by the Markov property and the supermeanvalued property of f, we have
$$E[f(\xi_x(t))|\mathcal{F}_s] = E_{\xi_x(s)} f(\xi(t-s)) \leq f(\xi_x(s)), \quad 0 \leq s \leq t < \infty.$$
That is, $f(\xi_x(t))$ is a supermartingale. Therefore, by Doob's stopping theorem (see Section 1.3), we have
$$E[f(\xi_x(\tau_2))|\mathcal{F}_s] \leq f(\xi_x(\tau_1)).$$
Taking expectation on both sides yields
$$Ef(\xi_x(\tau_2)) \leq Ef(\xi_x(\tau_1)),$$
that is $E_x f(\xi(\tau_2)) \leq E_x f(\xi(\tau_1))$ as required.

e) Let $\rho \in \mathcal{T}$ be arbitrary and define $\tau_D^\rho = \inf\{t \geq \rho : \xi(t) \notin D\}$. By the strong Markov property we have that
$$E_x f(\xi(\tau_D^\rho)) = E_x \big[E_{\xi(\rho)} f(\xi(\tau_D)) \big] = E_x f_D(\xi(\rho)).$$
But $\tau_D^\rho \geq \tau_D$ so by property d) we have
$$E_x f(\xi(\tau_D^\rho)) \leq E_x f(\xi(\tau_D)) = f_D(x).$$
Therefore
$$f_D(x) \geq E_x f_D(\xi(\rho))$$
and f_D is supermeanvalued.

f) By the lower semicontinuity and the well-known Fatou lemma we have
$$f(x) \leq E_x \Big(\liminf_{i\to\infty} f(\xi(\tau_i)) \Big) \leq \liminf_{i\to\infty} E_x f(\xi(\tau_i)).$$
On the other hand, by the supermeanvalued property,
$$f(x) \geq \limsup_{i\to\infty} E_x f(\xi(\tau_i)).$$
So we must have the equality
$$f(x) = \lim_{i\to\infty} E_x f(\xi(\tau_i))$$
as desired. The proof is complete.

The following is a useful criterion (cf. Dynkin (1965)) for superharmonic functions.

Lemma 3.3 *If $f \in C^2(R^d, R_+)$, then f is superharmonic if and only if*

$$Lf(x) \leq 0 \quad \text{for all } x \in R^d, \tag{3.6}$$

where L is the diffusion operator associated with equation (3.1), that is

$$Lf(x) = f_x(x)F(x) + \frac{1}{2}\text{trace}\big[G^T(x)f_{xx}(x)G(x)\big].$$

Proof. Let (3.6) hold and $\tau \in \mathcal{T}$. For any $t \geq 0$, Itô's formula implies

$$E_x f(\xi(\tau \wedge t)) \leq f(x).$$

Letting $t \to \infty$ we obtain by the Fatou lemma that

$$E_x f(\xi(\tau)) \leq f(x).$$

So f is supermeanvalued and hence superharmonic. Conversely, assume that (3.6) is false. So there is some $\bar{x} \in R^d$ such that $Lf(\bar{x}) > 0$. Due to the continuity of $Lf(\cdot)$ we can find an open neighbourhood U of $\bar{x}$ such that

$$\theta := \sup_{x \in U} Lf(x) > 0.$$

Define the stopping time $\tau_U = \inf\{t \geq 0 : \xi_{\bar{x}}(t) \notin U\}$. Clearly, $1 \wedge \tau_U$ is also a stopping time and $1 \wedge \tau_U > 0$ a.s. Now, by Itô's formula,

$$E_{\bar{x}} f(\xi(1 \wedge \tau_U)) = Ef(\xi_{\bar{x}}(1 \wedge \tau_U))$$
$$= f(\bar{x}) + E \int_0^{1 \wedge \tau_U} Lf(\xi_{\bar{x}}(s))ds \geq f(\bar{x}) + \theta E(1 \wedge \tau_U) > f(\bar{x}).$$

This means f is not supermeanvalued and of course not superharmonic. The proof is complete.

However, it is too restrictive to require f be of C^2. Fortunately, Dynkin (1965) supplies us with another necessary and sufficient condition. To state, let us give a new definition.

Definition 3.4 *A lower semicontinuous function $f : R^d \to [0, \infty]$ is said to be* excessive *with respect to $\xi(t)$ if*

$$f(x) \geq E_x f(\xi(t)) \quad \text{for all } t \geq 0, \ x \in R^d.$$

Obviously a superharmonic function is excessive, but we now show that the converse holds as well.

Lemma 3.5 *A function f is superharmonic if and only if it is excessive.*

Proof. We only need to show the "if" part so we let f be excessive. First, we assume that f is of C^2. For any $x \in R^d$, by Itô's formula we have

$$\int_0^t E[Lf(\xi_x(s))]ds = E_x(f(\xi(t))) - f(x) \leq 0 \quad \text{for all } t \geq 0.$$

Since $E[Lf(\xi_x(s))]$ is continuous in s, we must have $E[Lf(\xi_x(0))] = Lf(x) \leq 0$ for all $x \in R^d$. By Lemma 3.3, f is therefore superharmonic. The general case can be proved by the standard approximation procedure (and the details can be found in Dynkin (1965)).

Before we state our main results in this section, we still need to introduce a few more new concepts.

Definition 3.6 *Let g be a Borel measurable real-valued function on R^d. If f is a supermeanvalued (superharmonic) function and $f \geq g$, we call f a supermeanvalued (resp. superharmonic) majorant of g. If $\bar{g}$ is a supermeanvalued majorant of g and $\bar{g} \leq f$ for any other supermeanvalued majorant f of g, then $\bar{g}$ is called the* least supermeanvalued majorant *of g. Similarly, if $\hat{g}$ is a superharmonic majorant of g and $\hat{g} \leq f$ for any other superharmonic majorant f of g, then $\hat{g}$ is called the* least superharmonic majorant *of g.*

Lemma 3.7 *The least supermeanvalued majorant $\bar{g}$ of g always exists and is given by*

$$\bar{g}(x) = \inf_f f(x) \quad \text{for } x \in R^d,$$

where inf takes over all supermeanvalued majorants f of g.

Proof. By Lemma 3.2 b), the function $\inf_f f(x)$ is again supermeanvalued and is therefore clearly the least supermeanvalued majorant of g.

The least superharmonic majorant $\hat{g}$ of g does not always exist. However, we can see clearly from Lemma 3.7 that if $\hat{g}$ exists, then $\hat{g} \geq \bar{g}$. Moreover, if $\bar{g}$ is lower semicontinuous, then $\bar{g}$ is a superharmonic majorant of g and $\bar{g} \leq f$ for any superharmonic majorant f of g and therefore, by definition, $\hat{g}$ exists and coincides with $\bar{g}$. The following theorem not only shows that $\hat{g}$ exists as long as g is nonnegative and lower semicontinuous but also gives the iterative procedure to construct $\hat{g}$.

Theorem 3.8 *Let g be a lower semicontinuous nonnegative function on R^d. Then the least superharmonic majorant $\hat{g}$ of g exists and coincides with the supermeanvalued majorant $\bar{g}$ of g, that is $\hat{g} = \bar{g}$. Moreover, let $g_0 = g$ and define iteratively*

$$g_n(x) = \sup_{t \in J_n} E_x g_{n-1}(\xi(t)) \tag{3.7}$$

for $n = 1, 2, \cdots$, where $J_n = \{k/2^n : 0 \le k \le n2^n\}$. Then $g_n \uparrow \hat{g}$.

Proof. We first claim that for any $t \ge 0$, the function

$$h(x) := E_x g_0(\xi(t))$$

is lower semicontinuous. If not, then there is some $z \in R^d$ and a sequence $\{z_k\}$ such that $z_k \to z$ and

$$h(z) > \lim_{k \to \infty} h(z_k). \tag{3.8}$$

On the other hand, by the standing hypothesis of the uniform Lipschitz continuity, we can easily show that

$$E|\xi_z(t) - \xi_{z_k}(t)|^2 \le C|z - z_k|^2,$$

where C is a positive number independent of z and z_k. Hence, there is a subsequence $\{y_k\}$ of $\{z_k\}$ such that

$$\xi_{y_k}(t) \to \xi_z(t) \quad a.s.$$

Using the lower semicontinuity of g_0 and applying the Fatou lemma we can then derive that

$$h(z) = E g_0(\xi_z(t)) \le E\left[\liminf_{k \to \infty} g_0(\xi_{y_k}(t))\right]$$

$$\le \liminf_{k \to \infty} \left[E g_0(\xi_{y_k}(t))\right] = \liminf_{k \to \infty} h(y_k) = \lim_{k \to \infty} h(z_k).$$

But this contradicts with (3.8) and hence $h(x)$ must be lower semicontinuous. Note that the supremum of any lower semicontinuous functions is lower semicontinuous. We then easily see that g_1 is lower semicontinuous and so are g_n's by induction. Moreover, g_n is clearly increasing so

$$\hat{g}(x) := \lim_{n \to \infty} g_n(x) = \sup_{n \ge 1} g_n(x)$$

is again lower semicontinuous. Noting

$$\hat{g}(x) \ge g_n(x) \ge E_x g_{n-1}(\xi(t)) \quad \text{for all } n \text{ and all } t \in J_n,$$

we have

$$\hat{g}(x) \ge \lim_{n \to \infty} E_x g_{n-1}(\xi(t)) = E_x \hat{g}(\xi(t)) \tag{3.9}$$

for all $t \in J = \bigcup_{n=1}^\infty J_n$. Since J is dense in R_+, for any $t \ge 0$ we can choose a sequence $\{t_k\}$ in J such that $t_k \to t$. Using (3.9), we then derive that

$$\hat{g}(x) \ge \liminf_{k \to \infty} E_x \hat{g}(\xi(t_k)) \ge E_x \left(\liminf_{k \to \infty} \hat{g}(\xi(t_k))\right) \ge E_x \hat{g}(\xi(t)).$$

This means that $\hat{g}$ is excessive. By Lemma 3.5, $\hat{g}$ is superharmonic and is therefore a superharmonic majorant of g. On the other hand, if f is any supermeanvalued majorant of g, we can easily show by induction that

$$f(x) \geq g_n(x) \quad \text{for all } n,$$

which implies that $f(x) \geq \hat{g}(x)$. This proves that $\hat{g}$ is the least supermeanvalued majorant $\bar{g}$ of g. But $\hat{g}$ is superharmonic so it must be the least superharmonic majorant of g as well. The proof is complete.

It should be pointed out that J_n in (3.7) can be replaced by R_+ and the proof will even become slightly easier. But, (3.7) with J_n is much easier to be used in practice.

After so many preparations we can now return to the optimal stopping problem (3.3)–(3.4). Let us first have a quick look at how the least superharmonic majorant connects with the problem. Let ϕ be the reward function so it is nonnegative and continuous. By Theorem 3.8, its least superharmonic majorant $\hat{\phi}$ exists. If $\tau \in \mathcal{T}$, then

$$\hat{\phi}(x) \geq E_x \hat{\phi}(\xi(\tau)) \geq E_x \phi(\xi(\tau)),$$

which implies

$$\hat{\phi}(x) \geq \sup_{\tau \in \mathcal{T}} E_x \phi(\xi(\tau)) = \phi^*(x). \tag{3.10}$$

What is not so obvious is that the converse inequality holds as well. In other words, we always have $\hat{\phi} = \phi^*$ and we shall now begin to prove this main result which is due to Dynkin (1963).

Theorem 3.9 *Let ϕ be a reward function (so continuous and nonnegative) and ϕ^* be the optimal reward defined by (3.4). Let $\hat{\phi}$ be the least superharmonic majorant of ϕ. Then*

$$\phi^* = \hat{\phi}. \tag{3.11}$$

Proof. First we assume that ϕ is bounded. For any $\varepsilon > 0$, set

$$D_\varepsilon = \{x \in R^d : \phi(x) < \hat{\phi}(x) - \varepsilon\}. \tag{3.12}$$

Since ϕ is continuous while $\hat{\phi}$ is lower semicontinuous, D_ε is open. Let τ_ε be the first exit time of $\xi(t)$ from D_ε, i.e.

$$\tau_\varepsilon = \inf\{t \geq 0 : \xi(t) \notin D_\varepsilon\}.$$

Clearly, τ_ε is a stopping time. Define

$$\phi_\varepsilon(x) = E_x \hat{\phi}(\xi(\tau_\varepsilon)) \quad \text{for } x \in R^d. \tag{3.13}$$

By Lemma 3.2 e), ϕ_ε is supermeanvalued. We now claim that

$$\phi(x) \leq \phi_\varepsilon(x) + \varepsilon \quad \text{for all } x \in R^d. \tag{3.14}$$

If this is false, we must have

$$\beta := \sup_{x \in R^d} [\phi(x) - \phi_\varepsilon(x)] > \varepsilon.$$

So we can find some x_0 such that

$$\phi(x_0) - \phi_\varepsilon(x_0) \geq \beta - \frac{\varepsilon}{2} > 0. \tag{3.15}$$

Note that either $x_0 \in D_\varepsilon$ or $x_0 \notin D_\varepsilon$. If the latter is true, $\tau_\varepsilon = 0$ P_{x_0}-a.s. Then $\phi_\varepsilon(x_0) = \hat\phi(x_0) \geq \phi(x_0)$ which contradicts (3.15). Therefore, we must have $x_0 \in D_\varepsilon$ and, by the continuity of the solution, $\tau_\varepsilon > 0$ P_{x_0}-a.s. Noting that $\phi_\varepsilon + \beta$ is a supermeanvalued majorant of ϕ, we have that

$$\hat\phi(x_0) \leq \phi_\varepsilon(x_0) + \beta.$$

This, together with (3.15), yields

$$\hat\phi(x_0) \leq \phi(x_0) + \frac{\varepsilon}{2}. \tag{3.16}$$

On the other hand, for any $t > 0$, by the superharmonic property of $\hat\phi$ and the definition of τ_ε we have that

$$\hat\phi(x_0) \geq E_{x_0} \hat\phi(\xi(t \wedge \tau_\varepsilon)) \geq E_{x_0} \Big([\phi(\xi(t)) + \varepsilon] I_{\{t < \tau_\varepsilon\}} \Big).$$

Applying the Fatou lemma we obtain that

$$\hat\phi(x_0) \geq \liminf_{t \to 0} E_{x_0} \Big([\phi(\xi(t)) + \varepsilon] I_{\{t < \tau_\varepsilon\}} \Big)$$
$$\geq E_{x_0} \Big(\liminf_{t \to 0} [\phi(\xi(t)) + \varepsilon] I_{\{t < \tau_\varepsilon\}} \Big) = \phi(x_0) + \varepsilon.$$

But this contradicts with (3.16) so (3.14) must hold. In consequence, $\phi_\varepsilon + \varepsilon$ is a supermeanvalued majorant of ϕ. This, together with the definition of τ_ε, implies that

$$\hat\phi(x) \leq \phi_\varepsilon(x) + \varepsilon = E_x \hat\phi(\xi(\tau_\varepsilon)) + \varepsilon$$
$$\leq E_x[\phi(\xi(\tau_\varepsilon)) + \varepsilon] + \varepsilon \leq \phi^*(x) + 2\varepsilon. \tag{3.17}$$

Since ε is arbitrary, we have $\hat\phi \leq \phi^*$. By (3.10), we must have $\phi^* = \hat\phi$. In other words, we have proved that (3.11) holds if ϕ is bounded. If ϕ is unbounded, let $\phi_n = n \wedge \phi$ for $n = 1, 2, \cdots$. Then

$$\phi^* \geq \phi_n^* = \hat\phi_n \uparrow f \quad \text{as } n \to \infty.$$

Clearly, $f \geq \phi$ and by Lemma 3.2 c), f is superharmonic. So f is a superharmonic majorant of ϕ. Thus $\phi^* \geq f \geq \hat\phi$ which, together with (3.10), implies $\phi^* = \hat\phi$ again. The proof is complete.

From the proof above, we obtain the following useful approximation result.

Corollary 3.10 *If the reward function ϕ is bounded, then τ_ε defined in the proof of Theorem 3.9 is close to the optimal stopping time in the sense that*

$$0 < \phi^*(x) - E_x\phi(\xi(\tau_\varepsilon)) \leq 2\varepsilon \qquad (3.18)$$

for all $x \in R^d$.

This corollary follows from (3.17) and (3.11) directly. We now establish two useful criteria on the optimal stopping time.

Corollary 3.11 *Let $\phi, \hat\phi$ and ϕ^* be the same as defined in Theorem 3.9. Suppose there is a stopping time $\tau_0 \in \mathcal{T}$ such that*

$$\phi_0(x) := E_x\phi(\xi(\tau_0))$$

is a supermeanvalued majorant of ϕ. Then

$$\phi^*(x) = \phi_0(x)$$

and hence $\tau^* = \tau_0$ is an optimal stopping time for problem (3.3).

Proof. Since ϕ_0 is a supermeanvalued majorant of ϕ, we have

$$\bar\phi(x) \leq \phi_0(x).$$

On the other hand, we always have that

$$\phi_0(x) \leq \sup_{\tau \in \mathcal{T}} E_x\phi(\xi(\tau)) = \phi^*(x).$$

By Theorems 3.8 and 3.9, we have $\phi^*(x) = \phi_0(x)$ and hence $\tau^* = \tau_0$ is an optimal stopping time.

Corollary 3.12 *Let $\phi, \hat\phi$ and ϕ^* be the same as defined in Theorem 3.9. Let*

$$D = \{x \in R^d : \phi(x) < \hat\phi(x)\} \quad \text{and} \quad \tau_D = \inf\{t \geq 0 : \xi(t) \notin D\}.$$

Define

$$\phi_D(x) = E_x\phi(\xi(\tau_D)).$$

If $\phi_D \geq \phi$, then $\phi^* = \phi_D$ and τ_D is an optimal stopping time.

Proof. Noting $\xi(\tau_D) \notin D$, we have $\phi(\xi(\tau_D)) \geq \hat\phi(\xi(\tau_D))$ and hence we must have $\phi(\xi(\tau_D)) = \hat\phi(\xi(\tau_D))$. By Lemma 3.2 e), $\phi_D(x) = E_x\hat\phi(\xi(\tau_D))$ is supermeanvalued. The assertions now follow from Corollary 3.11.

An optimal stopping time τ^* for problem (3.3) may not always exist. The following theorem not only gives a sufficient condition for the existence of an optimal stopping time but also characterizes it.

Theorem 3.13 *Let $\phi, \hat{\phi}$ and ϕ^* be the same as defined in Theorem 3.9. Let*

$$D = \{x \in R^d : \phi(x) < \hat{\phi}(x)\} \quad \text{and} \quad \tau_D = \inf\{t \geq 0 : \xi(t) \notin D\}.$$

For every $n = 1, 2, \cdots$, let $\phi_n = n \wedge \phi$ and define

$$D_n = \{x \in R^d : \phi_n(x) < \hat{\phi}_n(x)\} \quad \text{and} \quad \tau_n = \inf\{t \geq 0 : \xi(t) \notin D_n\}.$$

If $P_x\{\tau_n < \infty\} = 1$ for all $x \in R^d$ and $n \geq 1$, then

$$\phi^*(x) = \lim_{n \to \infty} E_x \phi(\xi(\tau_n)). \tag{3.19}$$

In particular, if for each $x \in R^d$, $P_x\{\tau_D < \infty\} = 1$ and the family $\{\phi(\xi(\tau_n))\}_{n \geq 1}$ is uniformly integrable with respect to P_x, that is

$$\lim_{K \to \infty} \left(\sup_{n \geq 1} E_x \left[\phi(\xi(\tau_n)) I_{\{\phi(\xi(\tau_n)) \geq K\}} \right] \right) = 0,$$

then

$$\phi^*(x) = E_x \phi(\xi(\tau_D)). \tag{3.20}$$

In other words, $\tau^* = \tau_D$ is an optimal stopping time for problem (3.3).

Proof. We first claim that if ϕ is bounded and $P_x\{\tau_D < \infty\} = 1$ for all $x \in R^d$, then

$$\phi^*(x) = E_x \phi(\xi(\tau_D)). \tag{3.21}$$

To show this, let τ_ε be the same as defined in the proof of Theorem 3.9. Clearly, $\tau_\varepsilon \uparrow \tau_D$ a.s when $\varepsilon \downarrow 0$. By the bounded convergence theorem, we have

$$E_x \phi(\xi(\tau_\varepsilon)) \to E_x \phi(\xi(\tau_D)) \quad \text{as } \varepsilon \to 0.$$

This, together with Corollary 3.10, yields (3.21).

We now begin to prove (3.19). By what we have just shown, we have

$$\phi_n^*(x) = E_x \phi_n(\xi(\tau_n)) \quad \text{for all } n \geq 1. \tag{3.22}$$

Since $\hat{\phi}_n$ is increasing, we can define

$$f = \lim_{n \to \infty} \hat{\phi}_n.$$

By Lemma 3.2 c), f is superharmonic. Since $\phi_n \leq \hat{\phi}_n \leq f$ for all n, we have $\phi \leq f$. Hence f is a superharmonic majorant of ϕ and so $f \geq \hat{\phi}$. On the other hand, noting that $\hat{\phi}_n \leq \hat{\phi}$ for all n, we see that $f \leq \hat{\phi}$. Therefore, we must have

$$\hat{\phi} = \lim_{n \to \infty} \hat{\phi}_n. \tag{3.23}$$

Using Theorem 3.9 and equalities (3.22)–(3.23) we then derive that

$$\phi^*(x) = \lim_{n\to\infty} \phi_n^*(x) = \lim_{n\to\infty} E_x \phi_n(\xi(\tau_n))$$
$$\leq \liminf_{n\to\infty} E_x \phi(\xi(\tau_n)) \leq \limsup_{n\to\infty} E_x \phi(\xi(\tau_n)) \leq \phi^*(x)$$

and the required assertion (3.19) follows.

We now show (3.20). Clearly, $\hat{\phi}_n \leq n$. So if $x \in D_n$, then $\phi_n(x) < n$. In consequence, $\phi(x) < n$, $\phi(x) = \phi_n(x) < \hat{\phi}_n(x) \leq \phi(x)$ and $\phi_{n+1}(x) = \phi_n(x) < \hat{\phi}_n(x) \leq \hat{\phi}_{n+1}(x)$. In other words, we have shown that

$$D_n \subset D_{n+1} \quad \text{and} \quad D_n \subset D \cap \{x : \phi(x) < n\}.$$

Recalling (3.23), we then see that D is the increasing union of D_n's and

$$\tau_D = \lim_{n\to\infty} \tau_n.$$

Thus, $\xi(\tau_n) \to \xi(\tau_D)$ P_x-a.s. and, by the the uniform integrability, this convergence is in L^1 as well. Therefore, we obtain from (3.19) that

$$\phi^*(x) = \lim_{n\to\infty} E_x \phi(\xi(\tau_n)) = E_x \phi(\xi(\tau_D))$$

which is the required (3.20). The proof is complete.

Theorem 3.13 shows that under certain conditions τ_D is an optimal stopping time. The following theorem shows the "uniqueness" in the sense that if an optimal stopping time τ^* exists, then τ_D must be an optimal stopping time (but may not be the same as τ^*).

Theorem 3.14 *Let τ_D be the same as defined in Theorem 3.13. If there exists an optimal stopping time τ^* for problem (3.3), then*

$$P_x\{\tau^* \geq \tau_D\} = 1 \quad \text{for all } x \in \mathbb{R}^d \tag{3.24}$$

and τ_D is also an optimal stopping time.

Proof. If (3.24) is not true, there is some $x_0 \in \mathbb{R}^d$ such that $P_{x_0}\{\tau^* < \tau_D\} > 0$. For $\omega \in \{\tau^* < \tau_D\}$, by the definition of τ_D and Theorem 3.9, we have that $\phi(\xi(\tau^*)) < \hat{\phi}(\xi(\tau^*)) = \phi^*(\xi(\tau^*))$. Moreover, we always have $\phi \leq \phi^*$. Thus we have a contradiction:

$$\phi^*(x_0) = E_{x_0} \phi(\xi(\tau^*)) = E_{x_0}\left[\phi(\xi(\tau^*)) I_{\{\tau^* < \tau_D\}}\right] + E_{x_0}\left[\phi(\xi(\tau^*)) I_{\{\tau^* \geq \tau_D\}}\right]$$
$$< E_{x_0}\left[\phi^*(\xi(\tau^*)) I_{\{\tau^* < \tau_D\}}\right] + E_{x_0}\left[\phi^*(\xi(\tau^*)) I_{\{\tau^* \geq \tau_D\}}\right]$$
$$= E_{x_0} \phi^*(\xi(\tau^*)) \leq \phi^*(x_0),$$

where the last inequality holds because ϕ^* is superharmonic. So (3.24) must hold. Now by Lemma 3.2 d) etc, we derive that

$$\phi^*(x) = E_x\phi(\xi(\tau^*)) \le E_x\hat{\phi}(\xi(\tau^*))$$
$$\le E_x\hat{\phi}(\xi(\tau_D)) \le E_x\phi(\xi(\tau_D)) \le \phi^*(x).$$

This proves that τ_D is an optimal stopping time.

The following two important remarks explain how the theory discussed above can be used to cope with more general problems.

Remark 3.15 In many situations the reward function ϕ not only depends on the space but also the time. That is, $\phi = \phi(x,t)$ is a continuous nonnegative function on $R^d \times R_+$. The optimal stopping problem becomes to find the optimal expected value

$$\phi_0(x) = \sup_{\tau \in \mathcal{T}} E_x \phi(\xi(\tau), \tau) \qquad (3.25)$$

and the corresponding optimal stopping time τ^*, if there is any, such that

$$\phi_0(x) = E_x\phi(\xi(\tau^*), \tau^*). \qquad (3.26)$$

Clearly, this looks more general than problem (3.3)–(3.4). However, we can use the theory established above to solve this problem. Extend ϕ to the whole $d+1$-dimensional Euclidean space $R^d \times R$ by defining

$$\phi(x,t) = \phi(x,0) \quad \text{for } x \in R^d, \ t < 0.$$

Then $\phi(x,t)$ is continuous on $R^d \times R$. Introduce the $d+1$-dimensional stochastic differential equation

$$d\eta(t) = d\begin{bmatrix}\xi(t)\\\tilde{\eta}(t)\end{bmatrix} = \begin{bmatrix}F(\xi(t))\\1\end{bmatrix}dt + \begin{bmatrix}G(\xi(t))\\0\end{bmatrix}dB(t).$$

The solution with initial value $(x,s) \in R^d \times R$ is denoted by $\eta_{x,s}(t)$ and define $E_{x,s}\phi(\eta(t)) = E\phi(\eta_{x,s}(t))$. Using the theory established above we can find the optimal mean reward

$$\phi^*(x,s) = \sup_{\tau \in \mathcal{T}} E_{x,s}\phi(\eta(\tau))$$

and, if there is one, an optimal stopping time τ^* such that

$$\phi^*(x,s) = E_{x,s}\phi(\eta(\tau^*)).$$

In particular,

$$\phi_0(x) = \phi^*(x,0) = E_{x,0}\phi(\eta(\tau^*)) = E_x\phi(\xi(\tau^*), \tau^*)$$

which solves problem (3.25)–(3.26).

Remark 3.16 Sometimes the reward at the sale time t will not only depend on the present state $x(t)$ but also the whole history $\{x(s) : 0 \leq s \leq t\}$. For example, let ϕ_1 and ϕ_2 be two continuous nonnegative functions on R^d and consider the following optimal stopping problem: Determine the optimal mean reward

$$\phi_0(x) := \sup_{\tau \in \mathcal{T}} E_x \left[\int_0^\tau \phi_1(\xi(t))dt + \phi_2(\xi(\tau)) \right], \tag{3.27}$$

and find, if there is one, an optimal stopping time τ^* such that

$$\phi_0(x) = E_x \left[\int_0^{\tau^*} \phi_1(\xi(t))dt + \phi_2(\xi(\tau^*)) \right]. \tag{3.28}$$

This again looks more general than problem (3.3)–(3.4) but we can still use the theory established above to solve it. Define

$$\phi(x,y) = \phi_2(x) + 0 \vee y \quad \text{for } (x,y) \in R^d \times R.$$

So $\phi(x,y)$ is continuous and nonnegative. Introduce the $d+1$-dimensional stochastic differential equation

$$d\eta(t) = d \begin{bmatrix} \xi(t) \\ \tilde{\eta}(t) \end{bmatrix} = \begin{bmatrix} F(\xi(t)) \\ \phi_1(\xi(t)) \end{bmatrix} dt + \begin{bmatrix} G(\xi(t)) \\ 0 \end{bmatrix} dB(t).$$

The solution with initial value $(x,y) \in R^d \times R$ is denoted by $\eta_{x,y}(t)$ and we define $E_{x,y}\phi(\eta(t)) = E\phi(\eta_{x,y}(t))$. Moreover, at the points $\omega \in \Omega$ where $\tau(\omega) = \infty$, $E_{x,y}\phi(\eta(\tau))$ is interpreted as

$$E_{x,y}\left[0 \vee \tilde{\eta}(\infty)\right] = E_x\left(0 \vee \left[y + \int_0^\infty \phi_1(\xi(t))dt\right]\right)$$

instead of it being set to 0, and this will not affect the theory discussed before. We can therefore find the optimal mean value

$$\phi^*(x,y) = \sup_{\tau \in \mathcal{T}} E_{x,y}\phi(\eta(\tau))$$

and an optimal stopping time τ^*, if there is one, such that

$$\phi^*(x,y) = E_{x,y}\phi(\eta(\tau^*)).$$

In particular,

$$\phi_0(x) = \phi^*(x,0) = E_{x,0}\phi(\eta(\tau^*)) = E_x\left[\int_0^{\tau^*} \phi_1(\xi(t))dt + \phi_2(\xi(\tau^*))\right]$$

which solves problem (3.27)–(3.28).

9.4 STOCHASTIC GAMES

Consider a d-dimensional stochastic differential equation

$$d\xi(t) = F(\xi(t))dt + G(\xi(t))dB(t) \quad \text{on } t \geq 0. \tag{4.1}$$

Here $B(t)$ is an m-dimensional Brownian motion and we assume that

(H1) $F : R^d \to R^d$ and $G : R^d \to R^{d \times m}$ are uniformly Lipschitz continuous.

Given the initial value $\xi(0) = x$, the solution of equation (4.1) is denoted by $\xi_x(t)$ and the corresponding E_x and P_x are defined as before.

For any nonempty closed subset U of R^d, denote by h_U the first hitting time of the set U by $\xi(t)$, i.e.

$$h_U = \inf\{t \geq 0 : \xi(t) \in U\}.$$

For any set $H \subset R^d$, denote by H^c the complement of H in R^d. Let D be a given nonempty open set in R^d. (In particular, one may take $D = R^d$.) Denote by ∂D the boundary of D and let $\bar{D} = D \cup \partial D$. Let A, B be two given subsets of $\bar{D}$ such that $\partial D \subset A \cap B$. For each $x \in \bar{D}$, denote by $\mathcal{A}_x$ the family of all finite stopping times σ such that $\sigma \leq h_{D^c}$ and $\xi_x(\sigma) \in A$. Similarly, denote by $\mathcal{B}_x$ the family of all finite stopping times τ such that $\tau \leq h_{D^c}$ and $\xi_x(\tau) \in B$. Note that $\sigma \equiv 0$ is in $\mathcal{A}_x$ if and only if $x \in A$. If $D = R^d$, then $\sigma \in \mathcal{A}_x$ if and only if $P_x\{\sigma < \infty, \xi(\sigma) \in A\} = 1$. If $A = B$, then $\mathcal{A}_x = \mathcal{B}_x$. Let f, φ, ϕ_1 and ϕ_2 be continuous functions defined on $\bar{D}$ with φ being nonnegative. For $x \in \bar{D}$, $\sigma \in \mathcal{A}_x$ and $\tau \in \mathcal{B}_x$, define

$$\begin{aligned}J_x(\sigma, \tau) = & E_x \int_0^{\sigma \wedge \tau} \exp\left[-\int_0^t \varphi(\xi(s))ds\right] f(\xi(t))dt \\ & + E_x\left(\exp\left[-\int_0^\sigma \varphi(\xi(s))ds\right] \phi_1(\xi(\sigma))I_{\{\sigma < \tau\}}\right) \\ & + E_x\left(\exp\left[-\int_0^\tau \varphi(\xi(s))ds\right] \phi_2(\xi(\tau))I_{\{\sigma \geq \tau\}}\right). \end{aligned} \tag{4.2}$$

This will be called the *payoff functional*.

We consider a scheme whereby, for a given $x \in \bar{D}$, player (a) chooses any stopping time $\sigma \in \mathcal{A}_x$ and player (b) chooses any stopping time $\tau \in \mathcal{B}_x$, and the resulting payoff is $J_x(\sigma, \tau)$ that player (a) pays to player (b) (Of course, if $J_x(\sigma, \tau)$ is negative, this should be interpreted as player (b) pays to player (a)). Thus, the aim of player (a) is to minimize $J_x(\sigma, \tau)$ while the aim of player (b) is to maximize $J_x(\sigma, \tau)$. We shall call this scheme the *stochastic game* associated with (4.1)–(4.2) and denote it by $\mathcal{G}_x$. We shall denote the collection $\{\mathcal{G}_x : x \in \bar{D}\}$ by $\mathcal{G}$, and call it the stochastic game associated with (4.1)–(4.2) in $\bar{D}$. If

$$\inf_{\sigma \in \mathcal{A}_x} \sup_{\tau \in \mathcal{B}_x} J_x(\sigma, \tau) = \sup_{\tau \in \mathcal{B}_x} \inf_{\sigma \in \mathcal{A}_x} J_x(\sigma, \tau), \tag{4.3}$$

then we say that the stochastic game $\mathcal{G}_x$ has *value*, and the common number in (4.3) is called the *value of game* $\mathcal{G}_x$ and is denoted by $V(x)$. If there exist stopping times σ_x^* and τ_x^* in $\mathcal{A}_x$ and $\mathcal{B}_x$, respectively, such that

$$J_x(\sigma_x^*, \tau) \leq J_x(\sigma_x^*, \tau_x^*) \leq J_x(\sigma, \tau_x^*) \tag{4.4}$$

for all $\sigma \in \mathcal{A}_x$ and $\tau \in \mathcal{B}_x$, we call (σ_x^*, τ_x^*) a *saddle point* of $\mathcal{G}_x$. If (4.4) holds, we have

$$\inf_{\sigma \in \mathcal{A}_x} \sup_{\tau \in \mathcal{B}_x} J_x(\sigma, \tau) \leq \sup_{\tau \in \mathcal{B}_x} J_x(\sigma_x^*, \tau)$$
$$\leq J_x(\sigma_x^*, \tau_x^*) \leq \inf_{\sigma \in \mathcal{A}_x} J_x(\sigma, \tau_x^*) \leq \sup_{\tau \in \mathcal{B}_x} \inf_{\sigma \in \mathcal{A}_x} J_x(\sigma, \tau).$$

On the other hand, we always have

$$\inf_{\sigma \in \mathcal{A}_x} \sup_{\tau \in \mathcal{B}_x} J_x(\sigma, \tau) \geq \sup_{\tau \in \mathcal{B}_x} \inf_{\sigma \in \mathcal{A}_x} J_x(\sigma, \tau).$$

We therefore see that if (σ_x^*, τ_x^*) is a saddle point of $\mathcal{G}_x$, then the game has its value

$$V(x) = J_x(\sigma_x^*, \tau_x^*). \tag{4.5}$$

If there exist closed sets $A^* \subset A$ and $B^* \subset B$ such that for every $x \in \bar{D}$, the pair of

$$\sigma_x^* = h_{A^*} \quad \text{and} \quad \tau_x^* = h_{B^*}$$

forms a saddle point of $\mathcal{G}_x$, then we say that the pair (h_{A^*}, h_{B^*}) is a saddle point for $\mathcal{G}$ and we call the pair (A^*, B^*) a *saddle point of sets* for $\mathcal{G}$.

To characterise the saddle points, we shall need the following conditions:

(H2) For any $x \in \bar{D}$, $\mathcal{A}_x$ and $\mathcal{B}_x$ are nonempty.

(H3) The functions f, φ, ϕ_1 and ϕ_2 are bounded and continuous with $\varphi \geq 0$ and, moreover,

$$E_x \int_0^{\sigma \wedge \tau} \exp\left[-\int_0^t \varphi(\xi(s))ds\right] f(\xi(t))dt < \infty \tag{4.6}$$

for all $x \in \bar{D}$, $\sigma \in \mathcal{A}_x$ and $\tau \in \mathcal{B}_x$.

Conditions (H2) and (H3) are irrestrictive. For example, if $P_x\{h_{D^c} < \infty\} = 1$ for every $x \in \bar{D}$, then $\mathcal{A}_x$ and $\mathcal{B}_x$ contain at least one element, namely h_{D^c}, since $\partial D \subset A \cap B$. If $E_x h_{D^c} < \infty$, then (4.6) is satisfied. Let us now establish two simple but useful lemmas that give the criteria for $E_x h_{D^c} < \infty$. Let L be the diffusion operator associated with equation (4.1), that is

$$Lu(x) = u_x(x)F(x) + \frac{1}{2}trace\left[G^T(x)u_{xx}(x)G(x)\right]$$

for a C^2-function u.

Lemma 4.1 *Suppose that there exists a function $u \in C(\bar{D}; R) \cap C^2(D; R)$ and a positive constant K such that*

$$Lu(x) \leq -1 \quad \text{and} \quad |u(x)| \leq K \quad \text{for } x \in D.$$

Then

$$E_x h_{D^c} \leq 2K \quad \text{for all } x \in D.$$

Proof. For any $t \geq 0$, by Itô's formula and the condition we derive that

$$-K \leq E_x u(\xi(t \wedge h_{D^c})) \leq u(x) + E_x \int_0^{t \wedge h_{D^c}} Lu(\xi(s))ds \leq K - E_x(t \wedge h_{D^c}).$$

That is

$$E_x(t \wedge h_{D^c}) \leq 2K.$$

Letting $t \to \infty$ we obtain the asserted conclusion.

Lemma 4.2 *Let $\Phi = (\Phi_{ij})_{d \times d} = GG^T$ and write $F = (F_1, \cdots, F_d)^T$. Let D be a domain contained in a strip $|x_1| \leq \gamma$ for some positive constant γ. Suppose that there exists a constant λ such that*

$$\lambda F_1(x) + \frac{\lambda^2}{2}\Phi_{11}(x) \geq 1 \quad \text{for all } x \in \bar{D}.$$

Then

$$E_x h_{D^c} \leq 2e^{2|\lambda|\gamma} \quad \text{for all } x \in D.$$

Proof. Let $\mu = e^{|\lambda|\gamma}$ and define

$$u(x) = -\mu e^{\lambda x_1} \quad \text{for } x \in \bar{D}.$$

Then $|u(x)| \leq e^{2|\lambda|\gamma}$ and, moreover,

$$Lu(x) = -\mu e^{\lambda x_1}\left(\lambda F_1(x) + \frac{\lambda^2}{2}\Phi_{11}(x)\right) \leq -\mu e^{\lambda x_1} \leq -\mu e^{-|\lambda|\gamma} = -1.$$

So the required conclusion follows from Lemma 4.1.

The following theorem describes the properties of the value function $V(x)$ corresponding to a saddle point of sets.

Theorem 4.3 *Let (H1)–(H3) hold and assume that (A^*, B^*) is a saddle point of sets for the stochastic game $\mathcal{G}$. Then the value function $V(x)$ has the following properties:*

$$V(x) \leq \phi_1(x) \quad \text{if } x \in A - B^*, \tag{4.7}$$
$$V(x) \geq \phi_2(x) \quad \text{if } x \in B, \tag{4.8}$$
$$V(x) = \phi_1(x) \quad \text{if } x \in A^* - B^*, \tag{4.9}$$
$$V(x) = \phi_2(x) \quad \text{if } x \in B^*, \tag{4.10}$$

also

$$V(x) \leq E_x \int_0^\alpha \exp\left[-\int_0^t \varphi(\xi(s))ds\right] f(\xi(t))dt$$
$$+ E_x \left(\exp\left[-\int_0^\alpha \varphi(\xi(s))ds\right] V(\xi(\alpha))\right) \quad (4.11)$$

if α is a stopping time such that $\alpha \leq h_{B^*}$ and, moreover,

$$V(x) \geq E_x \int_0^\beta \exp\left[-\int_0^t \varphi(\xi(s))ds\right] f(\xi(t))dt$$
$$+ E_x \left(\exp\left[-\int_0^\beta \varphi(\xi(s))ds\right] V(\xi(\beta))\right) \quad (4.12)$$

if β is a stopping time such that $\beta \leq h_{A^*}$.

Proof. By definition, we have that

$$J_x(h_{A^*}, \tau) \leq V(x) \leq J_x(\sigma, h_{B^*}) \quad \text{for all } \sigma \in \mathcal{A}_x,\ \tau \in \mathcal{B}_x. \quad (4.13)$$

If $x \in A - B^*$, the $\sigma \equiv 0$ belongs to $\mathcal{A}_x$ and $h_{B^*} > 0$ P_x-a.s. Hence

$$J_x(0, h_{B^*}) = \phi_1(x).$$

This, together with the second inequality in (4.13), yields (4.7). If $x \in B$, then $\tau \equiv 0$ belongs to $\mathcal{B}_x$ and

$$J_x(h_{A^*}, 0) = \phi_2(x).$$

This, together with the first inequality in (4.13), yields (4.8). To prove (4.9), note that if $x \in A^* - B^*$, then $h_{A^*} = 0 < h_{B^*}$ P_x-a.s. Thus

$$V(x) = J_x(h_{A^*}, h_{B^*}) = \phi_1(x).$$

Next, if $x \in B^*$, then $h_{B^*} = 0 \leq h_{A^*}$ P_x-a.s. and so

$$V(x) = J_x(h_{A^*}, h_{B^*}) = \phi_2(x)$$

which is (4.10). We proceed to prove (4.11). Let α be any stopping time such that $\alpha \leq h_{B^*}$. Note that

$$V(x) = \inf_{\sigma \in \mathcal{A}_x} J_x(\sigma, h_{B^*}) \leq \inf_{\sigma \in \mathcal{A}_x, \sigma \geq \alpha} J_x(\sigma, h_{B^*})$$
$$= \inf_{\sigma \in \mathcal{A}_x, \sigma \geq \alpha} E_x \left\{ E_x \left(\int_0^{\sigma \wedge h_{B^*}} \exp\left[-\int_0^t \varphi(\xi(s))ds\right] f(\xi(t))dt \right. \right.$$
$$\left. \left. + \exp\left[-\int_0^\sigma \varphi(\xi(s))ds\right] \phi_1(\xi(\sigma)) I_{\{\sigma < h_{B^*}\}} \right. \right.$$

$$+ \exp\left[-\int_0^{h_{B^*}} \varphi(\xi(s))ds\right]\phi_2(\xi(h_{B^*}))I_{\{\sigma \geq h_{B^*}\}}\bigg|\mathcal{F}_\alpha\bigg)\bigg\}.$$

$$= E_x \int_0^\alpha \exp\left[-\int_0^t \varphi(\xi(s))ds\right] f(\xi(t))dt$$

$$+ \inf_{\sigma \in \mathcal{A}_x, \sigma \geq \alpha} E_x \bigg\{ \exp\left[-\int_0^\alpha \varphi(\xi(s))ds\right]$$

$$\times E_x \bigg(\int_\alpha^{\sigma \wedge h_{B^*}} \exp\left[-\int_\alpha^t \varphi(\xi(s))ds\right] f(\xi(t))dt$$

$$+ \exp\left[-\int_\alpha^\sigma \varphi(\xi(s))ds\right]\phi_1(\xi(\sigma))I_{\{\sigma < h_{B^*}\}}$$

$$+ \exp\left[-\int_\alpha^{h_{B^*}} \varphi(\xi(s))ds\right]\phi_2(\xi(h_{B^*}))I_{\{\sigma \geq h_{B^*}\}}\bigg|\mathcal{F}_\alpha\bigg)\bigg\}.$$

By the strong Markov property, the right-hand side is equal to

$$E_x \int_0^\alpha \exp\left[-\int_0^t \varphi(\xi(s))ds\right] f(\xi(t))dt$$

$$+ E_x \bigg\{ \exp\left[-\int_0^\alpha \varphi(\xi(s))ds\right] \inf_{\sigma \in \mathcal{A}_{\xi(\alpha)}} J_{\xi(\alpha)}(\sigma, h_{B^*}) \bigg\}$$

$$= E_x \int_0^\alpha \exp\left[-\int_0^t \varphi(\xi(s))ds\right] f(\xi(t))dt$$

$$+ E_x \bigg\{ \exp\left[-\int_0^\alpha \varphi(\xi(s))ds\right] V(\xi(\alpha)) \bigg\}.$$

This proves (4.11). The proof of (4.12) is similar and therefore the proof of this theorem is complete.

Note that it follows from inequalities (4.7) and (4.8) that

$$\phi_1(x) \geq \phi_2(x) \quad \text{if } x \in A \cap B - B^*. \tag{4.14}$$

Thus, for the existence of a saddle point of sets (A^*, B^*), it is necessary that (4.14) holds. We shall now show the converse of Theorem 4.3.

Theorem 4.4 *Let (H1)–(H3) hold. Assume that there exists a Borel measurable function $V(x)$ defined on $\bar{D}$ and closed sets $A^* \subset A$, $B^* \subset B$ such that (4.7)–(4.12) are satisfied,*

$$h_{A^*} \in \mathcal{A}_x, \quad h_{B^*} \in \mathcal{B}_x \quad \text{for } x \in D \tag{4.15}$$

and, moreover,

$$\phi_1(x) = \phi_2(x) \quad \text{on } x \in A^* \cap B^*. \tag{4.16}$$

Then (A^*, B^*) is a saddle point of sets for the stochastic game $\mathcal{G}$ and $V(x)$ is the value of the game.

Before the proof, let us point out that if $h_{D^c} < \infty$ P_x-a.s. for every $x \in D$ and $\partial D \in A^* \cap B^*$, then condition (4.15) is satisfied. Moreover, condition (4.16) means that the game is "fair." Indeed, from the definition of $J_x(\sigma, \tau)$ we see that player (b) has a "slight" advantage for he controls ϕ_2 on the set $\{\sigma = \tau\}$, but condition (4.16) abolishes this advantage on the set $A^* \cap B^*$ while in the complement of $A^* \cap B^*$ this advantage is irrelevant.

Proof. What we have to show is that

$$J_x(h_{A^*}, \tau) \leq V(x) \leq J_x(\sigma, h_{B^*}) \tag{4.17}$$

for $\sigma \in \mathcal{A}_x$ and $\tau \in \mathcal{B}_x$. Note that we always have $\xi(h_{A^*}) \in A^*$. If $\xi(h_{A^*}) \notin B^*$, then by (4.9), $V(\xi(h_{A^*})) = \phi_1(\xi(h_{A^*}))$ while if $\xi(h_{A^*}) \in B^*$, then by (4.10) and (4.16), $V(\xi(h_{A^*})) = \phi_1(\xi(h_{A^*})) = \phi_2(\xi(h_{A^*}))$. Therefore, we have

$$V(\xi(h_{A^*})) = \phi_1(\xi(h_{A^*})). \tag{4.18}$$

Moreover, for any $\tau \in \mathcal{B}_x$, $\xi(\tau) \in B$ and hence, by (4.8),

$$V(\xi(\tau)) \geq \phi_2(\xi(\tau)). \tag{4.19}$$

Making use of (4.18)–(4.19), we then derive that

$$\begin{aligned}
J_x(h_{A^*}, \tau) &= E_x \int_0^{h_{A^*} \wedge \tau} \exp\left[-\int_0^t \varphi(\xi(s))ds\right] f(\xi(t))dt \\
&+ E_x \left(\exp\left[-\int_0^{h_{A^*}} \varphi(\xi(s))ds\right] \phi_1(\xi(h_{A^*})) I_{\{h_{A^*} < \tau\}}\right) \\
&+ E_x \left(\exp\left[-\int_0^{\tau} \varphi(\xi(s))ds\right] \phi_2(\xi(\tau)) I_{\{h_{A^*} \geq \tau\}}\right) \\
&\leq E_x \int_0^{h_{A^*} \wedge \tau} \exp\left[-\int_0^t \varphi(\xi(s))ds\right] f(\xi(t))dt \\
&+ E_x \left(\exp\left[-\int_0^{h_{A^*} \wedge \tau} \varphi(\xi(s))ds\right] V(\xi(h_{A^*} \wedge \tau))\right).
\end{aligned}$$

Using (4.12) with $\beta = h_{A^*} \wedge \tau$, we then obtain that

$$J_x(h_{A^*}, \tau) \leq V(x)$$

which is the first inequality in (4.17). The second inequality in (4.17) can be proved similarly.

The following theorem shows that the problem of finding a saddle point for the stochastic game $\mathcal{G}$ can be reduced to a problem of solving elliptic variational inequalities.

Theorem 4.5 *Let* $V \in C(\bar{D}; R) \cap C^2(D; R)$ *and set*

$$A^* = \{x \in A : V(x) = \phi_1(x)\} \quad \text{and} \quad B^* = \{x \in A : V(x) = \phi_2(x)\}.$$

Assume that

$$LV(x) - \varphi(x)V(x) + f(x) \leq 0 \quad \text{if } x \in D - A^*, \tag{4.20}$$
$$LV(x) - \varphi(x)V(x) + f(x) \geq 0 \quad \text{if } x \in D - B^*, \tag{4.21}$$
$$V(x) \leq \phi_1(x) \quad \text{if } x \in A - B^*, \tag{4.22}$$
$$V(x) \geq \phi_2(x) \quad \text{if } x \in B, \tag{4.23}$$
$$V(x) = \phi_1(x) = \phi_2(x) \quad \text{if } x \in \partial D. \tag{4.24}$$

Assume also that $h_{D^c} < \infty$ P_x-a.s. for all $x \in D$. Then (A^*, B^*) is a saddle point of sets for the stochastic game $\mathcal{G}$ and $V(x)$ is the value of the game.

Proof. Clearly, A^* and B^* are closed subsets of A and B, respectively. By (4.24), $\partial D \subset A^* \cap B^*$ and so $h_{A^*} \vee h_{B^*} \leq h_{D^c}$. This, together with the condition that $h_{D^c} < \infty$ P_x-a.s., implies that

$$h_{A^*} \in \mathcal{A}_x, \quad h_{B^*} \in \mathcal{B}_x \quad \text{for } x \in D.$$

For any $\tau \in \mathcal{B}_x$, we can easily apply the Itô formula to show that

$$E_x\left(\exp\left[-\int_0^{h_{A^*} \wedge \tau} \varphi(\xi(s))ds\right] V(\xi(h_{A^*} \wedge \tau))\right) - V(x)$$

$$= E_x \int_0^{h_{A^*} \wedge \tau} \exp\left[-\int_0^t \varphi(\xi(s))ds\right] \Big(LV(\xi(t)) - \varphi(\xi(t))V(\xi(t))\Big) dt.$$

Using conditions (4.20) and (4.23), we then see that

$$V(x) \geq E_x \int_0^{h_{A^*} \wedge \tau} \exp\left[-\int_0^t \varphi(\xi(s))ds\right] f(\xi(t)) dt$$

$$+ E_x\left(\exp\left[-\int_0^{h_{A^*}} \varphi(\xi(s))ds\right] \phi_1(\xi(h_{A^*})) I_{\{h_{A^*} < \tau\}}\right)$$

$$+ E_x\left(\exp\left[-\int_0^{\tau} \varphi(\xi(s))ds\right] \phi_2(\xi(\tau)) I_{\{h_{A^*} \geq \tau\}}\right)$$

$$= J_x(h_{A^*}, \tau). \tag{4.25}$$

On the other hand, for any $\sigma \in \mathcal{A}_x$, we have that

$$V(x) = -E_x \int_0^{\sigma \wedge h_{B^*}} \exp\left[-\int_0^t \varphi(\xi(s))ds\right] \Big(LV(\xi(t)) - \varphi(\xi(t))V(\xi(t))\Big) dt$$

$$+ E_x\left(\exp\left[-\int_0^{\sigma \wedge h_{B^*}} \varphi(\xi(s))ds\right] V(\xi(\sigma \wedge h_{B^*}))\right)$$

$$\leq E_x \int_0^{\sigma \wedge h_{B^*}} \exp\left[-\int_0^t \varphi(\xi(s))ds\right] f(\xi(t)) dt$$

$$+ E_x\left(\exp\left[-\int_0^{\sigma} \varphi(\xi(s))ds\right] \phi_1(\xi(\sigma)) I_{\{\sigma < h_{B^*}\}}\right)$$

$$+ E_x\left(\exp\left[-\int_0^{h_{B^*}} \varphi(\xi(s))ds\right] \phi_2(\xi(h_{B^*})) I_{\{\sigma \geq h_{B^*}\}}\right)$$

$$= J_x(\sigma, h_{B^*}). \tag{4.26}$$

In other words, we have proved that

$$J_x(h_{A^*}, \tau) \leq V(x) \leq J_x(\sigma, h_{B^*})$$

for all $\sigma \in \mathcal{A}_x$ and $\tau \in \mathcal{B}_x$ and therefore the desired conclusions follow. The proof is complete.

Due to the page limit we will not discuss the solution to the elliptic variational inequalities (4.20)–(4.24). The reader can find the details in Friedman (1975) or Wu & Mao (1988).

10

Stochastic Neural Networks

10.1 INTRODUCTION

Since Hopfield (1982) initiated the study of neural networks, theoretical understanding of neural network dynamics has advanced greatly and we here mention Hopfield (1984), Hopfield & Tank (1986) and Denker (1986) among the others. Much of the current interest in artificial networks stems not only from their richness as a theoretical model of collective dynamics but also from the promise they have shown as a practical tool for performing parallel computation. In performing the computation, there are various stochastic perturbations to the networks and it is important to understand how these perturbations affect the networks. Especially, it is very critical to know whether the networks are stable or not under the perturbations. Although the stability of neural networks has been studied to a great deal, the stochastic effects to the stability problem have not been investigated until Liao & Mao (1996a, b) and the main aim of this chapter is to introduce the study in this new direction.

10.2 STOCHASTIC NEURAL NETWORKS

The neural network proposed by Hopfield (1982) can be described by an ordinary differential equation of the form

$$C_i \dot{u}_i(t) = -\frac{1}{R_i} u_i(t) + \sum_{j=1}^{d} T_{ij} g_j(u_j(t)), \quad 1 \leq i \leq d, \qquad (2.1)$$

on $t \geq 0$. The variable $u_i(t)$ represents the voltage on the input of the ith neuron. Each neuron is characterized by an input capacitance C_i and a transfer function $g_i(u)$. The connection matrix element T_{ij} has a value either $+1/R_{ij}$ or $-1/R_{ij}$ depending on whether the noninverting or inverting output of the jth neuron is connected to the input of the ith neuron through a resistance R_{ij}. The parallel resistance at the input of the ith neuron is

$$R_i = \frac{1}{\sum_{j=1}^{d} |T_{ij}|}.$$

The nonlinear transfer function $g_i(u)$ is sigmoidal, saturating at ± 1 with maximum slope at $u = 0$. In terms of mathematics, $g_i(u)$ is a nondecreasing Lipschitz continuous function with properties that

$$ug_i(u) \geq 0 \quad \text{and} \quad |g_i(u)| \leq 1 \wedge \beta_i |u| \quad \text{on} \quad -\infty < u < \infty, \tag{2.2}$$

where β_i is the slope of $g_i(u)$ at $u = 0$ and is supposed to be positive and finite. By defining

$$b_i = \frac{1}{C_i R_i} \quad \text{and} \quad a_{ij} = \frac{T_{ij}}{C_i}$$

equation (2.1) can be re-written as

$$\dot{u}_i(t) = -b_i u_i(t) + \sum_{j=1}^{d} a_{ij} g_j(u_j(t)), \quad 1 \leq i \leq n, \tag{2.3}$$

or equivalently

$$\dot{u}(t) = -\bar{B} u(t) + A g(u(t)), \tag{2.4}$$

where

$$u(t) = (u_1(t), \cdots, u_d(t))^T, \quad \bar{B} = \text{diag.}(b_1, \cdots, b_d),$$
$$A = (a_{ij})_{d \times d}, \quad g(u) = (g_1(u_1), \cdots, g_d(u_d))^T.$$

It is useful to note that

$$b_i = \sum_{j=1}^{d} |a_{ij}|, \quad 1 \leq i \leq d. \tag{2.5}$$

Suppose there exists a stochastic perturbation to the neural network and the stochastically perturbed network is described by a stochastic differential equation

$$dx(t) = [-\bar{B} x(t) + A g(x(t))] dt + \sigma(x(t)) dB(t) \quad \text{on } t \geq 0. \tag{2.6}$$

Here $B(t)$ is an m-dimensional Brownian motion defined on the given complete probability space $(\Omega, \mathcal{F}, \{\mathcal{F}_t\}, P)$ and $\sigma(x) = (\sigma_{ij}(x))_{d \times m}$ is a $d \times m$-matrix valued function defined on R^d. We always assume that $\sigma(x)$ is locally Lipschitz continuous and satisfies the linear growth condition as well. By the theory of Chapter 2, we know that given any initial value $x(0) = x_0 \in R^d$, equation

(2.6) has a unique global solution on $t \geq 0$ and we shall denote the solution by $x(t; x_0)$. Moreover, we also assume that $\sigma(0) = 0$ for the stability purpose of this chapter. So equation (2.6) admits a trivial solution $x(t; 0) \equiv 0$. Moreover, by Lemma 4.3.2, we know that if the initial value $x_0 \neq 0$, the solution will never be zero with probability one, that is $x(t; x_0) \neq 0$ for all $t \geq 0$ a.s.

Now that equation (2.6) is a stochastically perturbed system of equation (2.4), it is interesting to know how the stochastic perturbation affects the stability property of equation (2.4). More precisely speaking, when equation (2.4) is stable, it is useful to know whether the perturbed equation (2.6) remains stable or becomes unstable; but when equation (2.4) is unstable, it is then useful to know whether the perturbed equation (2.6) becomes stable or remains unstable. In the sequel of this section we shall discuss these problems in detail.

(i) *Exponential Stability*

Let us first state a useful lemma.

Lemma 2.1 *Assume that there exists a symmetric positive definite matrix $Q = (q_{ij})_{d \times d}$ and two numbers $\mu \in R$ and $\rho \geq 0$ such that*

$$2x^T Q[-\bar{B}x + Ag(x)] + trace[\sigma^T(x)Q\sigma(x)] \leq \mu x^T Q x \tag{2.7}$$

and

$$|x^T Q \sigma(x)|^2 \geq \rho (x^T Q x)^2 \tag{2.8}$$

for all $x \in R^d$. Then the solution of equation (2.6) has the property that

$$\limsup_{t \to \infty} \frac{1}{t} \log(|x(t; x_0)|) \leq -\left(\rho - \frac{\mu}{2}\right) \quad a.s. \tag{2.9}$$

whenever $x_0 \neq 0$. In particular, if $\rho > \mu/2$, then the stochastic neural network (2.6) is almost surely exponentially stable.

This lemma follows directly from Theorem 4.3.3 by letting $V(x) = x^T Q x$. We next employ this Lemma to establish a number of useful theorems on the almost sure exponential stability for the stochastic neural network (2.6).

Theorem 2.2 *Let (2.2) hold. Assume that there exists a positive definite diagonal matrix $Q = \mathrm{diag.}(q_1, q_2, \cdots, q_d)$ and two real numbers $\mu > 0$ and $\rho \geq 0$ such that*

$$trace[\sigma^T(x)Q\sigma(x)] \leq \mu x^T Q x$$

and

$$|x^T Q \sigma(x)|^2 \geq \rho (x^T Q x)^2$$

for all $x \in R^d$. Let $H = (h_{ij})_{d \times d}$ be the symmetric matrix defined by

$$h_{ij} = \begin{cases} 2q_i[-b_i + (0 \vee a_{ii})\beta_i] & \text{for } i = j, \\ q_i |a_{ij}|\beta_j + q_j |a_{ji}|\beta_i & \text{for } i \neq j. \end{cases}$$

Then the solution of equation (2.6) has the properties that

$$\limsup_{t\to\infty} \frac{1}{t}\log(|x(t;x_0)|) \leq -\left(\rho - \frac{1}{2}\left[\mu + \frac{\lambda_{\max}(H)}{\min_{1\leq i\leq n} q_i}\right]\right) \quad a.s. \qquad (2.10)$$

if $\lambda_{\max}(H) \geq 0$, or otherwise

$$\limsup_{t\to\infty} \frac{1}{t}\log(|x(t;x_0)|) \leq -\left(\rho - \frac{1}{2}\left[\mu + \frac{\lambda_{\max}(H)}{\max_{1\leq i\leq n} q_i}\right]\right) \quad a.s. \qquad (2.11)$$

whenever $x_0 \neq 0$.

Proof. Compute, by (2.2),

$$2x^T Q A g(x) = 2\sum_{i,j=1}^d x_i q_i a_{ij} g_j(x_j)$$

$$\leq 2\sum_i q_i(0 \vee a_{ii})x_i g_i(x_i) + 2\sum_{i\neq j} |x_i|\, q_i\, |a_{ij}|\, \beta_j\, |x_j|$$

$$\leq 2\sum_i q_i(0 \vee a_{ii})\beta_i x_i^2 + \sum_{i\neq j} |x_i|(q_i\, |a_{ij}|\, \beta_j + q_j\, |a_{ji}|\, \beta_i)|x_j|.$$

In the case $\lambda_{\max}(H) \geq 0$,

$$2x^T Q[-\bar{B}x + Ag(x)] \leq (|x_1|,\cdots,|x_d|)\, H\, (|x_1|,\cdots,|x_d|)^T$$

$$\leq \lambda_{\max}(H)|x|^2 \leq \frac{\lambda_{\max}(H)}{\min_{1\leq i\leq n} q_i} x^T Q x,$$

and then conclusion (2.10) follows from Lemma 2.1 easily. Similarly, in the case $\lambda_{\max}(H) < 0$,

$$2x^T Q[-\bar{B}x + Ag(x)] \leq \lambda_{\max}(H)|x|^2 \leq \frac{\lambda_{\max}(H)}{\max_{1\leq i\leq n} q_i} x^T Q x$$

and then conclusion (2.11) follows from Lemma 2.1 again.

Theorem 2.3 *Let (2.2) and (2.5) hold. Assume that there exist d positive numbers $q_1, q_2, \cdots, q_d$ such that*

$$\beta_j^2 \sum_{i=1}^d q_i\, [0 \vee \text{sign}(a_{ii})]^{\delta_{ij}}\, |a_{ij}| \leq q_j b_j, \qquad 1 \leq j \leq d,$$

where δ_{ij} is the Dirac delta function, i.e.

$$\delta_{ij} = \begin{cases} 1 & \text{for } i = j, \\ 0 & \text{for } i \neq j. \end{cases}$$

Assume moreover that

$$\text{trace}[\sigma^T(x)Q\sigma(x)] \leq \mu x^T Q x$$

and

$$|x^T Q\sigma(x)|^2 \geq \rho(x^T Q x)^2$$

for all $x \in R^d$, where $Q = \text{diag.}(q_1, q_2, \cdots, q_d)$, $\mu > 0$ and $\rho \geq 0$ are both constants. Then the solution of equation (2.6) satisfies

$$\limsup_{t\to\infty} \frac{1}{t} \log(|x(t;x_0)|) \leq -\left(\rho - \frac{\mu}{2}\right) \quad a.s.$$

whenever $x_0 \neq 0$.

Proof. Compute, by the conditions,

$$2x^T Q A g(x) = 2 \sum_{i,j=1}^{d} x_i \, q_i \, a_{ij} \, g_j(x_j)$$

$$\leq 2 \sum_{i,j=1}^{d} |x_i| \, q_i \, [0 \vee \text{sign}(a_{ii})]^{\delta_{ij}} \, |a_{ij}| \, \beta_j |x_j|$$

$$\leq \sum_{i,j=1}^{d} q_i \, [0 \vee \text{sign}(a_{ii})]^{\delta_{ij}} \, |a_{ij}|(x_i^2 + \beta_j^2 x_j^2)$$

$$\leq \sum_{i=1}^{d} q_i \left(\sum_{j=1}^{d} |a_{ij}|\right) x_i^2 + \sum_{j=1}^{d} \left(\beta_j^2 \sum_{i=1}^{d} q_i \, [0 \vee \text{sign}(a_{ii})]^{\delta_{ij}} \, |a_{ij}|\right) x_j^2$$

$$\leq \sum_{i=1}^{d} q_i b_i x_i^2 + \sum_{j=1}^{d} q_j b_j x_j^2 = 2x^T Q \bar{B} x.$$

Hence

$$2x^T Q[-\bar{B}x + Ag(x)] + \text{trace}[\sigma^T(x)Q\sigma(x)] \leq \mu x^T Q x$$

and the conclusion follows from Lemma 2.1.

Theorem 2.4 *Let (2.2) and (2.5) hold. Assume that the network is symmetric in the sense*

$$|a_{ij}| = |a_{ji}| \quad \text{for all } 1 \leq i,j \leq d.$$

Assume also that there is a pair of constants $\mu > 0$ and $\rho \geq 0$ such that

$$|\sigma(x)|^2 \leq \mu |x|^2 \quad \text{and} \quad |x^T \sigma(x)|^2 \geq \rho |x|^4$$

for all $x \in R^d$. Then the solution of equation (2.6) has the properties that

$$\limsup_{t\to\infty} \frac{1}{t} \log(|x(t;x_0)|) \leq -\left[\rho + \hat{b}(1-\check{\beta}) - \frac{\mu}{2}\right] \quad a.s. \quad (2.12)$$

if $\check{\beta} \leq 1$, and

$$\limsup_{t\to\infty} \frac{1}{t} \log(|x(t;x_0)|) \leq -\left[\rho - \check{b}(\check{\beta} - 1) - \frac{\mu}{2}\right] \quad a.s. \quad (2.13)$$

if $\check{\beta} > 1$, whenever $x_0 \neq 0$, where

$$\check{\beta} = \max_{1\leq i\leq d} \beta_i, \qquad \check{b} = \max_{1\leq i\leq d} b_i, \qquad \hat{b} = \min_{1\leq i\leq d} b_i.$$

Proof. Compute

$$2x^T Ag(x) = 2\sum_{i,j=1}^{d} x_i a_{ij} g_j(x_j)$$

$$\leq 2\sum_{i,j=1}^{d} |x_i| |a_{ij}| \beta_j |x_j| \leq \check{\beta} \sum_{i,j=1}^{d} |a_{ij}|(x_i^2 + x_j^2)$$

$$= \check{\beta}\left[\sum_{i=1}^{d}\left(\sum_{j=1}^{d} |a_{ij}|\right) x_i^2 + \sum_{j=1}^{d}\left(\sum_{i=1}^{d} |a_{ji}|\right) x_j^2\right]$$

$$= \check{\beta}\left[\sum_{i=1}^{d} b_i x_i^2 + \sum_{j=1}^{d} b_j x_j^2\right] = 2\check{\beta} x^T \bar{B} x.$$

Hence

$$2x^T[-\bar{B}x + Ag(x)] \leq -2(1-\check{\beta})x^T \bar{B}x.$$

If $\check{\beta} \leq 1$, then

$$2x^T[-\bar{B}x + Ag(x)] + |\sigma(x)|^2 \leq [-2\hat{b}(1-\check{\beta}) + \mu]|x|^2$$

and conclusion (2.12) follows from Lemma 2.1 with Q the identity matrix. On the other hand, if $\check{\beta} > 1$, then

$$2x^T[-\bar{B}x + Ag(x)] + |\sigma(x)|^2 \leq [2\check{b}(\check{\beta} - 1) + \mu]|x|^2,$$

and conclusion (2.13) follows from Lemma 2.1 again.

(ii) *Exponential Instability*

We now begin to discuss the exponential instability for the stochastic neural network (2.6). The following lemma is useful

Lemma 2.5 *Assume that there exists a symmetric positive definite matrix $Q = (q_{ij})_{d\times d}$ and two real numbers $\mu \in R$ and $\rho > 0$ such that*

$$2x^T Q[-\bar{B}x + Ag(x)] + \text{trace}[\sigma^T(x)Q\sigma(x)] \geq \mu x^T Qx \quad (2.14)$$

and
$$|x^T Q\sigma(x)|^2 \leq \rho(x^T Q x)^2 \tag{2.15}$$

for all $x \in R^d$. Then the solution of equation (2.6) has the property that

$$\liminf_{t\to\infty} \frac{1}{t} \log(|x(t; x_0)|) \geq \frac{\mu}{2} - \rho \quad a.s. \tag{2.16}$$

whenever $x_0 \neq 0$. In particular, if $\rho < \mu/2$ then the stochastic neural network (2.6) is almost surely exponentially unstable.

This lemma follows directly from Theorem 4.3.5 by using $V(x) = x^T Q x$. We now apply this lemma to establish a couple of useful results.

Theorem 2.6 Let (2.2) hold. Assume that there exists a positive definite diagonal matrix $Q = \text{diag.}(q_1, q_2, \cdots, q_d)$ and two positive numbers μ and ρ such that
$$\text{trace}[\sigma^T(x) Q \sigma(x)] \geq \mu x^T Q x$$

and
$$|x^T Q\sigma(x)|^2 \leq \rho(x^T Q x)^2$$

for all $x \in R^d$. Let $S = (s_{ij})_{d\times d}$ be the symmetric matrix defined by

$$s_{ij} = \begin{cases} 2q_i[-b_i + (0 \wedge a_{ii})\beta_i] & \text{for } i = j, \\ -q_i |a_{ij}| \beta_j - q_j |a_{ji}| \beta_i & \text{for } i \neq j. \end{cases}$$

Then the solution of equation (2.6) satisfies

$$\liminf_{t\to\infty} \frac{1}{t} \log(|x(t; x_0)|) \geq \frac{1}{2}\left[\mu + \frac{\lambda_{\min}(S)}{\min_{1\leq i\leq n} q_i}\right] - \rho \quad a.s.$$

whenever $x_0 \neq 0$.

Proof. In the same way as in the proof of Theorem 2.2 one can show that

$$2x^T Q[-\bar{B}x + Ag(x)] \geq (|x_1|, \cdots, |x_d|) S (|x_1|, \cdots, |x_d|)^T \geq \lambda_{\min}(S)|x|^2.$$

Note that we must have $\lambda_{\min}(S) \leq 0$ since all the elements of S are non-positive. So
$$2x^T Q[-\bar{B}x + Ag(x)] \geq \frac{\lambda_{\min}(S)}{\min_{1\leq i\leq d} q_i} x^T Q x$$

and the assertion follows from Lemma 2.5 immediately.

Theorem 2.7 Let (2.2) and (2.5) hold. Assume that the network is symmetric in the sense
$$|a_{ij}| = |a_{ji}| \quad \text{for all } 1 \leq i, j \leq d.$$

Assume also that there are two positive constants μ and ρ such that

$$|\sigma(x)|^2 \geq \mu |x|^2 \quad \text{and} \quad |x^T \sigma(x)|^2 \leq \rho |x|^4$$

for all $x \in R^d$. Then the solution of equation (2.6) satisfies that

$$\liminf_{t \to \infty} \frac{1}{t} \log(|x(t; x_0)|) \geq \frac{\mu}{2} - \check{b}(1 + \check{\beta}) - \rho \quad a.s.$$

whenever $x_0 \neq 0$, where

$$\check{\beta} = \max_{1 \leq i \leq n} \beta_i \quad \text{and} \quad \check{b} = \max_{1 \leq i \leq n} b_i.$$

Proof. Compute

$$2x^T A g(x) = 2 \sum_{i,j=1}^{d} x_i a_{ij} g_j(x_j)$$

$$\geq -2 \sum_{i,j=1}^{d} |x_i| |a_{ij}| \beta_j |x_j| \geq -\check{\beta} \sum_{i,j=1}^{d} |a_{ij}|(x_i^2 + x_j^2)$$

$$= -\check{\beta} \left[\sum_{i=1}^{d} \left(\sum_{j=1}^{d} |a_{ij}| \right) x_i^2 + \sum_{j=1}^{d} \left(\sum_{i=1}^{d} |a_{ji}| \right) x_j^2 \right]$$

$$= -\check{\beta} \left[\sum_{i=1}^{d} b_i x_i^2 + \sum_{j=1}^{d} b_j x_j^2 \right] = -2\check{\beta} x^T \bar{B} x.$$

Hence

$$2x^T[-\bar{B}x + Ag(x)] \geq -2(1 + \check{\beta}) x^T \bar{B} x \geq -2\check{b}(1 + \check{\beta})|x|^2.$$

In consequence,

$$2x^T[-\bar{B}x + Ag(x)] + |\sigma(x)|^2 \geq [\mu - 2\check{b}(1 + \check{\beta})]|x|^2,$$

and the required conclusion follows from Lemma 2.5 with Q the identity matrix. The proof is complete.

(iii) *Robustness of Stability and Stochastic Stabilization*

The results obtained in subsection (i) can be applied to study the robustness of stability as well as to investigate the stochastic stabilization. To explain the application to the study of robustness of stability, let us suppose that there exists a symmetric positive definite matrix $Q = (q_{ij})_{d \times d}$ and a positive constant μ_1 such that

$$2x^T Q[-\bar{B}x + Ag(x)] \leq -\mu_1 x^T Q x \quad \text{for } x \in R^n. \tag{2.17}$$

It is well-known that hypothesis (2.17) guarantees the exponential stability of the neural network (2.4). We further assume that the stochastic perturbation is not too strong in the sense that there is a constant μ_2 such that

$$0 < \mu_2 < \frac{\mu_1 \lambda_{\min}(Q)}{\lambda_{\max}(Q)} \quad \text{and} \quad |\sigma(x)|^2 \leq \mu_2 |x|^2 \quad \text{for } x \in R^n. \tag{2.18}$$

It is easy to verify that

$$2x^T Q[-\bar{B}x + Ag(x)] + trace[\sigma^T(x)Q\sigma(x)] \leq -\left[\mu_1 - \frac{\mu_2 \lambda_{\max}(Q)}{\lambda_{\min}(Q)}\right] x^T Q x.$$

Thus, by Lemma 2.1 with $\rho = 0$ (since (2.8) always holds with $\rho = 0$), the stochastic neural network (2.6) is almost surely exponentially stable. In other words, the stochastic perturbation does not change the stability property of the neural network (2.4).

Let us now discuss the stochastic stabilization. We know that the neural network (2.4) may be unstable sometimes. Perhaps one might imagine that an unstable neural network should behave even worse (more unstable) if the network subjects to stochastic perturbations. However, this is not always true. As every thing has two sides, stochastic perturbations may make the given unstable network nicer (stable). Indeed, we shall show that any neural network of form (2.4) can be stabilized by stochastic perturbations. From the practical point of view we shall restrict ourselves to the linear stochastic perturbation only. In other words, we only consider the stochastic perturbation of the form

$$\sigma(x(t))dB(t) = \sum_{k=1}^{m} G_k x(t) dB_k(t),$$

i.e. $\sigma(x) = (G_1 x, G_2 x, \cdots, G_m x)$, where $G_k, 1 \leq k \leq m$ are all $d \times d$ matrices. In this case, the stochastically perturbed network (2.6) becomes

$$dx(t) = [-\bar{B}x(t) + Ag(x(t))]dt + \sum_{k=1}^{m} G_k x(t) dB_k(t) \quad \text{on } t \geq 0. \tag{2.19}$$

Note that for any symmetric $d \times d$-matrix Q,

$$trace[\sigma^T(x)Q\sigma(x)] = \sum_{k=1}^{m} x^T G_k^T Q G_k x$$

and

$$|x^T Q \sigma(x)|^2 = trace[\sigma^T(x) Q x x^T Q \sigma(x)]$$
$$= \sum_{k=1}^{m} x^T G_k^T Q x x^T Q G_k x = \sum_{k=1}^{m} (x^T Q G_k x)^2.$$

We therefore obtain the following useful result from Lemma 2.1.

Theorem 2.8 *Assume that there exists a symmetric positive definite matrix $Q = (q_{ij})_{d \times d}$ and two numbers $\mu \in R$ and $\rho \geq 0$ such that*

$$2x^T Q[-\bar{B}x + Ag(x)] + \sum_{k=1}^{m} x^T G_k^T Q G_k x \leq \mu x^T Q x$$

and

$$\sum_{k=1}^{m} (x^T Q G_k x)^2 \geq \rho (x^T Q x)^2$$

for all $x \in R^d$. Then the solution of equation (2.19) satisfies

$$\limsup_{t \to \infty} \frac{1}{t} \log(|x(t; x_0)|) \leq -\left(\rho - \frac{\mu}{2}\right) \quad a.s.$$

whenever $x_0 \neq 0$. In particular, if $\rho > \mu/2$ then the stochastic neural network (2.19) is almost surely exponentially stable.

Let us explain through examples how we can apply this theorem to stabilize a given neural network.

Example 2.9 Let

$$G_k = \theta_k I \quad \text{for } 1 \leq k \leq m,$$

where I is the identity matrix and $\theta_k, 1 \leq k \leq m$ are all real numbers. Then equation (2.19) becomes

$$dx(t) = [-\bar{B}x(t) + Ag(x(t))]dt + \sum_{k=1}^{m} \theta_k x(t) dB_k(t). \quad (2.20)$$

Clearly, we can interpret the numbers $\theta_k, 1 \leq k \leq m$ as the intensity of the stochastic perturbation. Choose Q the identity matrix. Then

$$\sum_{k=1}^{m} x^T G_k^T Q G_k x = \sum_{k=1}^{m} |G_k x|^2 = \sum_{k=1}^{m} \theta_k^2 |x|^2 \quad (2.21)$$

and

$$\sum_{k=1}^{m} (x^T Q G_k x)^2 = \sum_{k=1}^{m} (x^T \theta_k x)^2 = \sum_{k=1}^{m} \theta_k^2 |x|^4. \quad (2.22)$$

Moreover, in view of (2.2), we have

$$2x^T Q Ag(x) \leq 2|x| \, ||A|| \, |g(x)| \leq 2\check{\beta} ||A|| \, |x|^2,$$

where $\check{\beta} = \max_{1 \leq k \leq d} \beta_k$. Hence

$$2x^T Q[-\bar{B}x + Ag(x)] \leq 2(\check{\beta}||A|| - \hat{b})|x|^2, \quad (2.23)$$

where $\hat{b} = \min_{1 \le k \le d} b_k$. Combining (2.21)–(2.23) and then applying Theorem 2.8 we see that the solution of equation (2.20) satisfies

$$\limsup_{t \to \infty} \frac{1}{t} \log(|x(t; x_0)|) \le -\left[\frac{1}{2}\sum_{k=1}^{m} \theta_k^2 - (\check{\beta}\|A\| - \hat{b})\right] \quad a.s.$$

whenever $x_0 \neq 0$. In particular, if we choose θ_k's sufficiently large for

$$\sum_{k=1}^{m} \theta_k^2 > 2(\check{\beta}\|A\| - \hat{b}),$$

then the stochastic neural network (2.20) is always almost surely exponentially stable no matter whether the neural network (2.4) is unstable. Furthermore, if we let $\theta_k = 0$ for $2 \le k \le m$, then equation (2.20) becomes an even simpler one

$$dx(t) = [-\bar{B}x(t) + Ag(x(t))]dt + \theta_1 x(t) dB_1(t). \quad (2.24)$$

That is we only use a scalar Brownian motion as the source of stochastic perturbation. This stochastic network is always almost surely exponentially stable provided

$$\theta_1^2 > 2(\check{\beta}\|A\| - \hat{b}).$$

From this simple example we see that if a strong enough stochastic perturbation is added onto a neural network $\dot{u}(t) = -\bar{B}u(t) + Ag(u(t))$ in a certain way then the network can be stabilized. In other words, we have already obtained the following general result.

Theorem 2.10 *Any neural network of the form*

$$\dot{u}(t) = -\bar{B}u(t) + Ag(u(t))$$

can be stabilized by Brownian motions provided (2.2) is satisfied. Moreover, one can even use only a scalar Brownian motion to do so.

Theorem 2.8 ensures that there are many choices for the matrices G_k in order to stabilize a given network. Of course the choices in Example 2.9 are just the simplest ones. For illustration we give one more example.

Example 2.11 For each k, choose a positive definite $d \times d$ matrix D_k such that

$$x^T D_k x \ge \frac{\sqrt{3}}{2} \|D_k\| \, |x|^2,$$

Obviously, there are lots of such matrices. Let θ be a real number and set $G_k = \theta D_k$. Then equation (2.19) becomes

$$dx(t) = [-\bar{B}x(t) + Ag(x(t))]dt + \theta \sum_{k=1}^{m} D_k x(t) dB_k(t). \quad (2.25)$$

Again let Q be the identity matrix. Note that

$$\sum_{k=1}^{m} x^T G_k^T Q G_k x = \sum_{k=1}^{m} |\theta D_k x|^2 \leq \theta^2 \sum_{k=1}^{m} ||D_k||^2 |x|^2$$

and

$$\sum_{k=1}^{m} (x^T Q G_k x)^2 = \theta^2 \sum_{k=1}^{m} (x^T D_k x)^2 \geq \frac{3\theta^2}{4} \sum_{k=1}^{m} ||D_k||^2 |x|^4.$$

Combining these together with (2.23) and then applying Theorem 2.8 we obtain that the solution of equation (2.25) satisfies

$$\limsup_{t \to \infty} \frac{1}{t} \log(|x(t; x_0)|) \leq -\left[\frac{\theta^2}{4} \sum_{k=1}^{m} ||D_k||^2 - (\check{\beta}||A|| - \hat{b})\right] \quad a.s.$$

whenever $x_0 \neq 0$. Particularly, if choose θ sufficiently large for

$$\theta^2 > \frac{4(\check{\beta}||A|| - \hat{b})}{\sum_{k=1}^{m} ||D_k||^2},$$

then the stochastic network (2.25) is almost surely exponentially stable.

From the above examples we see clearly that in order to stabilize a given unstable network the stochastic perturbation should be strong enough. This is not surprising since if the stochastic perturbation is too weak it may not be able to change the instability property of the network and we shall see this more clearly in the next subsection.

(iv) *Robustness of Instability and Stochastic Destabilization*

The results established in subsection (ii) can be applied to the study of robustness of instability as well as stochastic destabilization. To explain the former, let us assume that network (2.4) is exponentially unstable, which is guaranteed by the condition that

$$2x^T Q[-\bar{B}x + Ag(x)] \geq \mu_1 x^T Q x, \quad x \in R^n \qquad (2.26)$$

for some $\mu_1 > 0$ and some symmetric positive definite matrix Q. Assume also that the stochastic perturbation is so small that

$$|\sigma(x)|^2 \leq \mu_2 |x|^2, \quad x \in R^n, \qquad (2.27)$$

where $0 < \mu_2 < \mu_1 [\lambda_{\min}(Q)]^2 / (2||Q||^2)$. Note that

$$|x^T Q \sigma(x)|^2 \leq \frac{\mu_2 ||Q||^2}{[\lambda_{\min}(Q)]^2} (x^T Q x)^2 \quad \text{and} \quad \operatorname{trace}[\sigma^T(x) Q \sigma(x)] \geq 0.$$

By Lemma 2.5, we therefore see that under conditions (2.26) and (2.27), the solution of equation (2.6) has the property that

$$\liminf_{t\to\infty} \frac{1}{t} \log(|x(t;x_o)|) \geq \frac{\mu_1}{2} - \frac{\mu_2 ||Q||^2}{[\lambda_{min}(Q)]^2} > 0 \quad a.s.$$

That is, the stochastic neural network (2.6) remains unstable. This supports once again the fact that if the stochastic perturbation is too weak it will not be able to change the instability property of the network.

In the previous subsection we have discussed the stochastic stabilization problem. Let us now turn to consider the opposite problem—stochastic destabilization. That is, we shall add stochastic perturbations onto a given stable network in the hope that the stochastically perturbed network becomes unstable. We have seen in the previous subsection that the stochastic perturbation should be strong enough or else the stability property will not be destroyed. However, the strength of the perturbation is not the only factor, since, as shown in the previous subsection, sometimes the stronger the stochastic perturbation is added the more stable the network becomes. As a matter of fact, the way how the stochastic perturbation is added onto the network is more important. From the practical point of view, we again restrict ourselves to linear stochastic perturbation only. In other words, we still assume the stochastically perturbed network is described by equation (2.19). Applying Lemma 2.5 to equation (2.19) we immediately obtain the following useful result.

Theorem 2.12 *Assume that there exists a symmetric positive definite matrix $Q = (q_{ij})_{d\times d}$ and two numbers $\mu \in R$ and $\rho > 0$ such that*

$$2x^T Q[-\bar{B}x + Ag(x)] + \sum_{k=1}^{m} x^T G_k^T Q G_k x \geq \mu x^T Qx$$

and

$$\sum_{k=1}^{m} (x^T Q G_k x)^2 \leq \rho (x^T Qx)^2$$

for all $x \in R^d$. Then the solution of equation (2.19) satisfies

$$\liminf_{t\to\infty} \frac{1}{t} \log(|x(t;x_0)|) \geq \frac{\mu}{2} - \rho \quad a.s.$$

whenever $x_0 \neq 0$. In particular, if $\rho < \mu/2$ then the stochastic neural network (2.19) is almost surely exponentially unstable.

Let us now apply this theorem to show how one can use stochastic perturbation to destabilize a given network.

Example 2.13 First of all, let the dimension of the network $d \geq 3$. Let $m = d$, that is we choose a d-dimensional Brownian motion $(B_1(t), B_2(t), \cdots, B_d(t))^T$.

Let θ be a real number. For each $k = 1, 2, \cdots, d-1$, define $G_k = (g_{ij}^k)_{d \times d}$ by $g_{ij}^k = \theta$ if $i = k$ and $j = k+1$ or otherwise $g_{ij}^k = 0$. Moreover, define $G_d = (g_{ij}^d)_{d \times d}$ by $g_{ij}^d = \theta$ if $i = d$ and $j = 1$ or otherwise $g_{ij}^d = 0$. Then the stochastic network (2.19) becomes

$$dx(t) = [-\bar{B}x(t) + Ag(x(t))]dt + \theta \begin{pmatrix} x_2(t)dB_1(t) \\ \vdots \\ x_d(t)dB_{d-1}(t) \\ x_1(t)dB_d(t) \end{pmatrix}. \quad (2.28)$$

Let Q be the identity matrix. Note that

$$\sum_{k=1}^{d} x^T G_k^T Q G_k x = \sum_{k=1}^{d} |G_k x|^2 = \sum_{k=1}^{d} |\theta x_k|^2 = \theta^2 |x|^2. \quad (2.29)$$

Also, setting $x_{d+1} = x_1$,

$$\sum_{k=1}^{d} (x^T Q G_k x)^2 = \theta^2 \sum_{k=1}^{d} x_k^2 x_{k+1}^2$$

$$\leq \frac{2\theta^2}{3} \sum_{k=1}^{d} x_k^2 x_{k+1}^2 + \frac{\theta^2}{6} \sum_{k=1}^{d} (x_k^4 + x_{k+1}^4) \leq \frac{\theta^2}{3} |x|^4. \quad (2.30)$$

Moreover, by (2.2),

$$2x^T Q[-\bar{B}x + Ag(x)] \geq -2(\check{b} + \check{\beta}\|A\|)|x|^2, \quad (2.31)$$

where $\check{b} = \max_{1 \leq k \leq d} b_k$ and $\check{\beta} = \max_{1 \leq k \leq d} \beta_k$. Combining (2.29)–(2.31) and then applying Theorem 2.12 we see that the solution of equation (2.28) satisfies

$$\liminf_{t \to \infty} \frac{1}{t} \log(|x(t; x_0)|) \geq \frac{\theta^2}{2} - (\check{b} + \check{\beta}\|A\|) - \frac{\theta^2}{3} = \frac{\theta^2}{6} - (\check{b} + \check{\beta}\|A\|) \quad a.s.$$

whenever $x_0 \neq 0$. So the stochastic neural network (2.28) is almost surely exponentially unstable if

$$\theta^2 > 6(\check{b} + \check{\beta}\|A\|).$$

Secondly, let us consider the case when the dimension of the network d is an even number. Let $m = 1$, that is choose a scalar Brownian motion $B_1(t)$. Let θ be a real number. Define

$$G_1 = \begin{pmatrix} 0 & \theta & & & 0 \\ -\theta & 0 & & & \\ & & \ddots & & \\ & & & 0 & \theta \\ 0 & & & -\theta & 0 \end{pmatrix}$$

Then equation (2.19) becomes

$$dx(t) = [-\bar{B}x(t) + Ag(x(t))]dt + \theta \begin{pmatrix} x_2(t) \\ -x_1(t) \\ \vdots \\ x_d(t) \\ -x_{d-1}(t) \end{pmatrix} dB_1(t). \qquad (2.32)$$

Let Q be the identity matrix again. Note that

$$x^T G_1^T Q G_1 x = \theta^2 |x|^2 \quad \text{and} \quad (x^T Q G_1 x)^2 = 0. \qquad (2.33)$$

Combining this with (2.31) and then applying Theorem 2.12 we see that the solution of equation (2.32) satisfies

$$\liminf_{t \to \infty} \frac{1}{t} \log(|x(t; x_0)|) \geq \frac{\theta^2}{2} - (\check{b} + \check{\beta}||A||) \quad a.s.$$

whenever $x_0 \neq 0$. So the stochastic neural network (2.32) is almost surely exponentially unstable if

$$\theta^2 > 2(\check{b} + \check{\beta}||A||).$$

This example proves the following theorem.

Theorem 2.14 *Any neural network of the form*

$$\dot{x}(t) = -\bar{B}x(t) + Ag(x(t))$$

can be destabilized by Brownian motions provided the dimension $d \geq 2$ and (2.2) is satisfied.

Naturally, one would ask what happens when the dimension $d = 1$. Although from the practical point of view one-dimensional networks are rare, the question needs to be answered for the completeness of theory. So let us consider a one-dimensional network

$$\dot{u}(t) = -bu(t) + ag(u(t)), \qquad (2.34)$$

where $b > 0$ and $a = b$ or $-b$, and $g(u)$ is a sigmoidal real-valued function such that

$$ug(u) \geq 0 \quad \text{and} \quad |g(u)| \leq 1 \wedge \beta |u| \quad \text{for all } -\infty < u < \infty.$$

Assume $\beta < 1$. Then it is easy to verify that the solution, denoted by $u(t; x_0)$, of equation (2.34) with initial value $u(0) = x_0 \neq 0$ satisfies

$$\limsup_{t \to \infty} \frac{1}{t} \log(|u(t; x_0)|) \leq -b[1 - \beta(0 \vee \text{sign}(a))] < 0.$$

In other words, network (2.34) is exponentially stable. Now perturb this network stochastically and assume the perturbed network is described by

$$dx(t) = [-bx(t) + ag(x(t))]dt + \sum_{k=1}^{m} \theta_k x(t) dB_k(t), \qquad (2.35)$$

where θ_k's are all real numbers. It is not difficult to show by Theorem 2.8 that the solution $x(t; x_0)$ of equation (2.35) with initial value $x(0) = x_0 \neq 0$ satisfies

$$\limsup_{t \to \infty} \frac{1}{t} \log(|x(t; x_0)|) \leq -b[1 - \beta(0 \vee \text{sign}(a))] - \frac{1}{2}\sum_{k=1}^{m} \theta_k^2 < 0 \quad a.s.$$

Hence the stochastic neural network (2.35) becomes even more stable. We therefore see that a one-dimensional stable network may not be destabilized by Brownian motions if the stochastic perturbation is restricted to be linear.

10.3 STOCHASTIC NEURAL NETWORKS WITH DELAYS

In many networks, time delays can not be avoided. For example, in electronic neural networks, time delays will be present due to the finite switching speed of amplifiers. In a similar way as Hopfield (1982), a model for a network with delays can be described by a differential delay equation

$$C_i \dot{u}_i(t) = -\frac{1}{R_i} u_i(t) + \sum_{j=1}^{d} T_{ij} g_j(u_j(t - \tau_j)), \quad 1 \leq i \leq d, \qquad (3.1)$$

where C_i, R_i etc. are the same as described in the previous section while τ_j's are positive constants and stand for the time delays. Let $A, \bar{B}, g$ and u be the same as defined in the previous section and define

$$u_\tau(t) = (u_1(t - \tau_1), \cdots, u_d(t - \tau_d))^T.$$

Then equation (3.1) can be re-written as

$$\dot{u}(t) = -\bar{B}u(t) + Ag(u_\tau(t)). \qquad (3.2)$$

In this section we shall discuss the stochastic effects to this delay neural network. Suppose that there exists a stochastic perturbation to the delay neural network (3.2) and the stochastically perturbed network is described by the stochastic differential delay equation

$$dx(t) = [-\bar{B}x(t) + Ag(x_\tau(t))]dt + \sigma(x(t), x_\tau(t))dB(t) \quad \text{on } t \geq 0 \qquad (3.3)$$

with initial data $x(s) = \xi(s)$ for $-\bar{\tau} \leq s \leq 0$. Here

$$x_\tau(t) = (x_1(t - \tau_1), \cdots, x_d(t - \tau_d))^T, \quad \sigma: R^d \times R^d \to R^{d \times m},$$
$$\bar{\tau} = \max_{1 \leq i \leq d} \tau_i, \quad \xi = \{\xi(s): -\bar{\tau} \leq s \leq 0\} \in C([-\bar{\tau}, 0]; R^d).$$

We assume that $\sigma(x,y)$ is locally Lipschitz continuous and satisfies the linear growth condition as well. By the theory of Chapter 5, equation (3.3) has a unique global solution on $t \geq 0$ and we denote the solution by $x(t;\xi)$. Moreover, we also assume that $\sigma(0,0) = 0$ for the stability purpose and hence equation (3.3) admits a trivial solution $x(t;0) \equiv 0$.

Theorem 3.1 *Let (2.2) hold. Assume that there exist d positive constants $\delta_i, 1 \leq i \leq d$ such that the symmetric matrix*

$$H = \begin{pmatrix} C & A \\ A^T & D \end{pmatrix}$$

is negative definite, where

$$C = \mathrm{diag.}(-2b_1 + \delta_1\beta_1^2, \cdots, -2b_d + \delta_d\beta_d^2) \quad \text{and} \quad D = \mathrm{diag.}(-\delta_1, \cdots, -\delta_d).$$

Let $-\lambda = \lambda_{\max}(H)$ (so $\lambda > 0$). Assume also that there exists an $\mu \in [0,\lambda)$ such that

$$|\sigma(x,y)|^2 \leq \mu|x|^2 + \lambda|g(y)|^2 \tag{3.4}$$

for all $(x,y) \in R^d \times R^d$. Then the solution of equation (3.3) has the property that

$$\limsup_{t \to \infty} \frac{1}{t} \log(E|x(t;\xi)|^2) \leq -\varepsilon, \tag{3.5}$$

where $\varepsilon \in (0, \lambda - \mu)$ is the unique root to the equation

$$\max_{1 \leq i \leq d} \left(\varepsilon\delta_i\tau_i\beta_i^2 e^{\varepsilon\tau_i} + \varepsilon\right) = \lambda - \mu. \tag{3.6}$$

In other words, the stochastic delay network (3.3) is exponentially stable in mean square.

Proof. Fix the initial data ξ arbitrarily and write $x(t;\xi) = x(t)$. Introduce a Lyapunov function

$$V(x,t) = |x|^2 + \sum_{i=1}^{d} \delta_i \int_{-\tau_i}^{0} g_i^2(x_i(t+s))ds$$

for $(x,t) \in R^d \times [0,\infty)$. By the Itô formula,

$$dV(x(t),t) = \left(\sum_{i=1}^{d} \delta_i[g_i^2(x_i(t)) - g_i^2(x_i(t-\tau_i))]\right.$$

$$\left. + 2x^T(t)[-\bar{B}x(t) + Ag(x_\tau(t))] + |\sigma(x(t),x_\tau(t))|^2\right)dt$$

$$+ 2x^T(t)\sigma(x(t),x_\tau(t))dB(t). \tag{3.7}$$

By (2.2) and the hypothesis on H one can derive that

$$\sum_{i=1}^{d} \delta_i[g_i^2(x_i(t)) - g_i^2(x_i(t-\tau_i))] + 2x^T(t)[-\bar{B}x(t) + Ag(x_\tau(t))]$$

$$\leq (x^T(t), g^T(x_\tau(t))) H \begin{pmatrix} x(t) \\ g(x_\tau(t)) \end{pmatrix} \leq -\lambda(|x(t)|^2 + |g(x_\tau(t))|^2).$$

Also by (3.4),

$$|\sigma(x(t), x_\tau(t))|^2 \leq \mu|x(t)|^2 + \lambda|g(x_\tau(t))|^2.$$

Substituting these into (3.7) yields

$$dV(x(t), t) \leq -(\lambda - \mu)|x(t)|^2 dt + 2x^T(t)\sigma(x(t), x_\tau(t))dB(t). \quad (3.8)$$

Let $0 < \varepsilon < \lambda - \mu$ be the root to equation (3.6). By Itô's formula again,

$$d[e^{\varepsilon t} V(x(t), t)] = e^{\varepsilon t}\left[\varepsilon V(x(t), t)dt + dV(x(t), t)\right]$$

$$\leq e^{\varepsilon t}\left[\varepsilon \sum_{i=1}^{d} \delta_i \int_{-\tau_i}^{0} g_i^2(x_i(t+s))ds - (\lambda - \mu - \varepsilon)|x(t)|^2\right]dt$$

$$+ 2e^{\varepsilon t} x^T(t)\sigma(x(t), x_\tau(t))dB(t),$$

where (3.8) has been used. Integrating both sides of this inequality from 0 to $T > 0$ and then taking the expectation one obtains that

$$e^{\varepsilon T} EV(x(T), T) \leq c_1$$

$$+ E \int_0^T e^{\varepsilon t}\left[\varepsilon \sum_{i=1}^{d} \delta_i \int_{-\tau_i}^{0} g_i^2(x_i(t+s))ds - (\lambda - \mu - \varepsilon)|x(t)|^2\right]dt, \quad (3.9)$$

where

$$c_1 = |\xi(0)|^2 + \sum_{i=1}^{d} \delta_i \tau_i.$$

Compute

$$\int_0^T e^{\varepsilon t} \int_{-\tau_i}^{0} g_i^2(x_i(t+s))dsdt = \int_0^T e^{\varepsilon t} \int_{t-\tau_i}^{t} g_i^2(x_i(s))dsdt$$

$$= \int_{-\tau_i}^{T} \left[\int_{s\vee 0}^{(s+\tau_i)\wedge T} e^{\varepsilon t} dt\right] g_i^2(x_i(s))ds \leq \int_{-\tau_i}^{T} \tau_i e^{\varepsilon(s+\tau_i)} g_i^2(x_i(s))ds$$

$$\leq \tau_i^2 e^{\varepsilon \tau_i} + \tau_i \beta_i^2 e^{\varepsilon \tau_i} \int_0^T e^{\varepsilon s} |x_i(s)|^2 ds. \quad (3.10)$$

Consequently, in light of (3.6),

$$\int_0^T e^{\varepsilon t}\left(\varepsilon \sum_{i=1}^d \delta_i \int_{-\tau_i}^0 g_i^2(x_i(t+s))ds\right)dt$$

$$\leq c_2 + (\lambda - \mu - \varepsilon)\int_0^T e^{\varepsilon t}|x(t)|^2 dt, \qquad (3.11)$$

where

$$c_2 = \varepsilon \sum_{i=1}^d \delta_i \tau_i^2 e^{\varepsilon \tau_i}.$$

Substituting (3.11) into (3.9) yields

$$e^{\varepsilon T} EV(x(T), T) \leq c_1 + c_2.$$

In particular,

$$e^{\varepsilon T} E|x(T)|^2 \leq c_1 + c_2.$$

Therefore

$$\limsup_{T\to\infty} \frac{1}{T} \log(E|x(T)|^2) \leq -\varepsilon$$

which is the required (3.5). The proof is complete.

In this theorem, condition (3.4) is a little bit restrictive although it covers some interesting cases (see examples below). The following theorem is an improvement.

Theorem 3.2 *Assume that all the conditions of Theorem 3.1 hold except (3.4) is replaced by*

$$|\sigma(x,y)|^2 \leq \mu|x|^2 + \lambda|g(y)|^2 + \rho|y|^2 \qquad (3.12)$$

for all $(x,y) \in R^d \times R^d$, *where* μ *and* ρ *are both nonnegative numbers such that*

$$\mu + \rho < \lambda.$$

Then the solution of equation (3.3) satisfies

$$\limsup_{t\to\infty} \frac{1}{t} \log(E|x(t;\xi)|^2) \leq -\varepsilon, \qquad (3.13)$$

where $\varepsilon \in (0, \lambda - \mu - \delta)$ *is the unique solution to*

$$\max_{1\leq i\leq d}\left[(\varepsilon\delta_i\tau_i\beta_i^2 + \rho)e^{\varepsilon\tau_i} + \varepsilon\right] = \lambda - \mu. \qquad (3.14)$$

Proof. Again fix ξ arbitrarily and write $x(t;\xi) = x(t)$. Let $\varepsilon \in (0, \lambda - \mu - \rho)$ be the unique solution to equation (3.14) and $T > 0$. In the same way as in the proof of Theorem 3.1 we can show that

$$e^{\varepsilon T} EV(x(T), T) \leq c_1 + E \int_0^T e^{\varepsilon t} \Big\{ \varepsilon \sum_{i=1}^d \delta_i \int_{-\tau_i}^0 g_i^2(x_i(t+s)) ds$$

$$- (\lambda - \mu - \varepsilon)|x(t)|^2 + \rho|x_\tau(t)|^2 \Big\} dt. \qquad (3.15)$$

Compute

$$\int_0^T \rho e^{\varepsilon t} |x_\tau(t)|^2 dt = \sum_{i=1}^d \int_0^T \rho e^{\varepsilon t} |x_i(t - \tau_i)|^2 dt$$

$$= \sum_{i=1}^d \int_{-\tau_i}^{T-\tau_i} \rho e^{\varepsilon(t+\tau_i)} |x_i(t)|^2 dt \leq c_3 + \sum_{i=1}^d \rho e^{\varepsilon \tau_i} \int_0^T e^{\varepsilon t} |x_i(t)|^2 dt, \qquad (3.16)$$

where

$$c_3 = \rho \bar{\tau} e^{\varepsilon \bar{\tau}} \sup_{-\bar{\tau} \leq s \leq 0} |\xi(s)|^2.$$

Combining (3.10), (3.14) and (3.16) we obtain that

$$\int_0^T e^{\varepsilon t} \Big(\varepsilon \sum_{i=1}^d \delta_i \int_{-\tau_i}^0 g_i^2(x_i(t+s)) ds + \rho|x_\tau(t)|^2 \Big) dt$$

$$\leq c_2 + c_3 + \sum_{i=1}^d (\varepsilon \delta_i \tau_i \beta_i^2 + \rho) e^{\varepsilon \tau_i} \int_0^T e^{\varepsilon t} |x_i(t)|^2 dt$$

$$\leq c_2 + c_3 + (\lambda - \mu - \varepsilon) \int_0^T e^{\varepsilon t} |x(t)|^2 dt. \qquad (3.17)$$

Substituting (3.17) into (3.15) gives

$$e^{\varepsilon T} EV(x(T), T) \leq c_1 + c_2 + c_3.$$

The remainder of the proof is the same as before. The proof is complete.

With slightly more careful arguments we can show the following even more general result.

Theorem 3.3 *Assume that all the conditions of Theorem 3.1 hold except (3.4) is replaced by*

$$|\sigma(x, y)|^2 \leq \sum_{i=1}^d (\mu_i x_i^2 + \rho_i y_i^2) + \lambda |g(y)|^2 \qquad (3.18)$$

for all $(x, y) \in R^d \times R^d$, where μ_i and ρ_i are all nonnegative numbers such that

$$\mu_i + \rho_i < \lambda \quad \text{for all } 1 \leq i \leq d. \tag{3.19}$$

Then the solution of equation (3.3) satisfies

$$\limsup_{t \to \infty} \frac{1}{t} \log(E|x(t;\xi)|^2) \leq -\varepsilon,$$

where $\varepsilon \in (0, \min_{1 \leq i \leq d}(\lambda - \mu_i - \rho_i))$ is the unique solution to

$$\max_{1 \leq i \leq d} \left[(\varepsilon \delta_i \tau_i \beta_i^2 + \rho_i) e^{\varepsilon \tau_i} + \mu_i + \varepsilon \right] = \lambda.$$

The details of the proof are left to the reader. We shall now use this theorem to establish a number of useful corollaries.

Corollary 3.4 *Let (2.2) and (2.5) hold. Assume that*

$$b_i > \beta_i^2 \sum_{j=1}^{d} |a_{ji}| \quad \text{for all } 1 \leq i \leq d.$$

Assume also that both (3.18) and (3.19) are satisfied with

$$\lambda = \min_{1 \leq i \leq d} \frac{b_i - \beta_i^2 \sum_{j=1}^{d} |a_{ji}|}{1 + \beta_i^2}.$$

Then the stochastic delay network (3.3) is exponentially stable in mean square.

Proof. Set

$$\delta_i = \frac{b_i + \sum_{j=1}^{d} |a_{ji}|}{1 + \beta_i^2} \quad \text{for all } 1 \leq i \leq d.$$

Let the symmetric matrix H be the same as defined in Theorem 3.1. We claim that $\lambda_{\max}(H) \leq -\lambda$. In fact, for any $x, y \in R^d$,

$$(x^T, y^T) H \begin{pmatrix} x \\ y \end{pmatrix} = \sum_{i=1}^{d}(-2b_i + \delta_i \beta_i^2)x_i^2 + 2\sum_{i,j=1}^{d} a_{ij} x_i y_j - \sum_{i=1}^{d} \delta_i y_i^2$$

$$\leq \sum_{i=1}^{d}(-2b_i + \delta_i \beta_i^2)x_i^2 + \sum_{i,j=1}^{d} |a_{ij}|(x_i^2 + y_j^2) - \sum_{i=1}^{d} \delta_i y_i^2$$

$$= -\sum_{i=1}^{d}(b_i - \delta_i \beta_i^2)x_i^2 - \sum_{i=1}^{d}\left(\delta_i - \sum_{j=1}^{d}|a_{ji}|\right)y_i^2,$$

where condition (2.5) has been used. But

$$b_i - \delta_i \beta_i^2 = \delta_i - \sum_{j=1}^{d} |a_{ji}| = \frac{b_i - \beta_i^2 \sum_{j=1}^{d} |a_{ji}|}{1 + \beta_i^2} \geq \lambda.$$

So

$$(x^T, y^T) H \begin{pmatrix} x \\ y \end{pmatrix} \leq -\lambda(|x|^2 + |y|^2).$$

Now the conclusion of the corollary follows from Theorem 3.3 immediately. The proof is complete.

Corollary 3.5 *Let (2.2) hold. Assume that*

$$2b_i > ||A||(1 + \beta_i^2) \quad \text{for all } 1 \leq i \leq d.$$

Assume also that both (3.18) and (3.19) are satisfied with

$$\lambda = \min_{1 \leq i \leq d} \left(\frac{2b_i}{1 + \beta_i^2} - ||A|| \right).$$

Then the stochastic delay network (3.3) is exponentially stable in mean square.

Proof. Let

$$\delta_i = \frac{2b_i}{1 + \beta_i^2} \quad \text{for all } 1 \leq i \leq d$$

and then define the symmetric matrix H the same as in Theorem 3.1. One can then show $\lambda_{\max}(H) \leq -\lambda$ in the same way as in the proof of Corollary 3.4, and hence the conclusion follows from Theorem 3.3. The proof is complete.

In practice, networks are often symmetric in the sense $|a_{ij}| = |a_{ji}|$. For such symmetric networks we have the following useful result.

Corollary 3.6 *Let (2.2) and (2.5) hold. Assume that the network is symmetric in the sense*

$$|a_{ij}| = |a_{ji}| \quad \text{for all } 1 \leq i, j \leq d.$$

Assume that

$$\beta_i < 1 \quad \text{for all } 1 \leq i \leq d.$$

Assume also that both (3.18) and (3.19) are satisfied with

$$\lambda = \min_{1 \leq i \leq d} \frac{b_i(1 - \beta_i^2)}{1 + \beta_i^2}. \tag{3.20}$$

Then the stochastic delay network (3.3) is exponentially stable in mean square.

Proof. By the assumptions,

$$\beta_i^2 \sum_{j=1}^{d} |a_{ji}| < \sum_{j=1}^{d} |a_{ij}| = b_i$$

for all $1 \le i \le d$. Also

$$\min_{1 \le i \le d} \frac{b_i - \beta_i^2 \sum_{j=1}^{d} |a_{ji}|}{1 + \beta_i^2} = \min_{1 \le i \le d} \frac{b_i(1 - \beta_i^2)}{1 + \beta_i^2}.$$

Hence the conclusion follows from Corollary 3.4 directly. The proof is complete.

Let us now turn to discuss the almost sure exponential stability for the stochastic delay network (3.3). The following lemma is useful.

Lemma 3.7 *Let (2.2) hold. Assume that there exists a constant $K > 0$ such that*

$$|\sigma(x,y)|^2 \le K(|x|^2 + |y|^2) \quad \text{for all } x,y \in R^d. \tag{3.21}$$

If

$$\limsup_{t \to \infty} \frac{1}{t} \log(E|x(t;\xi)|^2) \le -\gamma \tag{3.22}$$

for some $\gamma > 0$, then

$$\limsup_{t \to \infty} \frac{1}{t} \log(|x(t;\xi)|) \le -\frac{\gamma}{2} \quad a.s. \tag{3.23}$$

In other words, under conditions (2.2) and (3.21), the exponential stability in mean square of equation (3.3) implies the almost sure exponential stability.

Proof. Fix any ξ and write $x(t;\xi) = x(t)$ again. Let $\varepsilon \in (0, \gamma/2)$ be arbitrary. By (3.22) there is a constant $M > 0$ such that

$$E|x(t)|^2 \le M e^{-(\gamma-\varepsilon)t} \quad \text{for all } t \ge -\tau. \tag{3.24}$$

Let $k = 1, 2, \cdots$. By the Doob martingale inequality, the Hölder inequality as well as conditions (2.2) and (3.21), one can easily show that

$$E\left[\sup_{k \le t \le k+1} |x(t)|^2\right] \le 3E|x(k)|^2 + c_4 \int_k^{k+1} \left[E|x(t)|^2 + E|x_\tau(t)|^2\right] dt, \tag{3.25}$$

where c_4 and the following c_5 are both positive constants independent of k. Note that

$$E|x_\tau(t)|^2 = \sum_{i=1}^{d} E|x_i(t-\tau_i)|^2 \le \sum_{i=1}^{d} E|x(t-\tau_i)|^2.$$

Substituting this and (3.24) into (3.25) yields

$$E\left[\sup_{k\leq t\leq k+1} |x(t)|^2\right] \leq c_5 e^{-(\gamma-\varepsilon)k}. \qquad (3.26)$$

Consequently

$$P\left\{\omega: \sup_{k\leq t\leq k+1} |x(t)|^2 > e^{-(\gamma-2\varepsilon)k}\right\} \leq c_5 e^{-\varepsilon k}.$$

By the Borel–Cantelli lemma, one sees that for almost all $\omega \in \Omega$ there exists a random integer $k_0(\omega)$ such that for all $k \geq k_0$,

$$\sup_{k\leq t\leq k+1} |x(t)|^2 \leq e^{-(\gamma-2\varepsilon)k}.$$

This implies that

$$\limsup_{t\to\infty} \frac{1}{t} \log(|x(t;\xi)|) \leq -\frac{\gamma}{2} + \varepsilon \qquad a.s.$$

Finally the desired (3.23) follows by letting $\varepsilon \to 0$. The proof is complete.

The following result follows immediately from this lemma and Theorem 3.3 etc.

Theorem 3.8 *Let (2.2) and (3.21) hold. Then, under the conditions of Theorem 3.3 or one of Corollary 3.4–3.6, the stochastic delay network (3.3) is almost surely exponentially stable.*

Let us now discuss a number of examples to illustrate our theory.

Example 3.9 Consider the 2-dimensional stochastic delay network

$$d\begin{pmatrix} x_1(t) \\ x_2(t) \end{pmatrix} = \begin{pmatrix} -4 & 0 \\ 0 & -2 \end{pmatrix}\begin{pmatrix} x_1(t) \\ x_2(t) \end{pmatrix} dt + \begin{pmatrix} 2 & -2 \\ 1 & 1 \end{pmatrix}\begin{pmatrix} g_1(x_1(t-\tau_1)) \\ g_2(x_2(t-\tau_2)) \end{pmatrix} dt$$

$$+ G_1\begin{pmatrix} g_1(x_1(t-\tau_1)) \\ g_2(x_2(t-\tau_2)) \end{pmatrix} dB(t). \qquad (3.27)$$

Here τ_1, τ_2 are both positive constants, $B(t)$ a real-valued Brownian motion, G_1 a 2×2 constant matrix, and

$$g_i(u) = \frac{e^u - e^{-u}}{e^u + e^{-u}} \qquad \text{for } u \in R, \ i = 1, 2.$$

So (2.2) is satisfied with $\beta_i = 1$. To apply Theorem 3.1, let $\delta_1 = 4$ and $\delta_2 = 2$. Then

$$H = \begin{pmatrix} -4 & 0 & 2 & -2 \\ 0 & -2 & 1 & 1 \\ 2 & 1 & -4 & 0 \\ -2 & 1 & 0 & -2 \end{pmatrix}.$$

Sec.10.3]   Stochastic Neural Networks with Delays

It is not difficult to compute $\lambda_{\max}(H) = -0.2474$. Note also that

$$\sigma(x, y) = G_1 g(y) \quad \text{for } (x, y) \in R^2 \times R^2,$$

where $g(y) = (g_1(y_1), g_2(y_2))^T$. Hence

$$|\sigma(x, y)|^2 = |G_1 g(y)|^2 \le ||G_1||^2 |g(y)|^2.$$

Therefore, if

$$||G_1||^2 \le 0.2474, \tag{3.28}$$

then, by Theorem 2.1, the stochastic delay network (3.27) is exponentially stable in mean square. Moreover, let $\varepsilon > 0$ be the root to the equation

$$\max\{4\varepsilon\tau_1 e^{\varepsilon\tau_1} + \varepsilon,\ 2\varepsilon\tau_2 e^{\varepsilon\tau_2} + \varepsilon\} = 0.2474. \tag{3.29}$$

Then the second moment Lyapunov exponent should not be greater than $-\varepsilon$. For example, if $\tau_1 = 0.005$ and $\tau_2 = 0.01$, then (3.29) becomes

$$0.02\varepsilon e^{0.01\varepsilon} + \varepsilon = 0.2474,$$

which has the root $\varepsilon = 0.24253$, hence in this case the second moment Lyapunov exponent should not be greater than -0.24253. In view of Theorem 3.8, we can also conclude that the stochastic delay network (3.27) is almost surely exponentially stable as long as (3.28) is satisfied.

Example 3.10 Consider a more general stochastic delay network than (3.27) which is described by the equation

$$d\begin{pmatrix} x_1(t) \\ x_2(t) \end{pmatrix} = \begin{pmatrix} -4 & 0 \\ 0 & -2 \end{pmatrix} \begin{pmatrix} x_1(t) \\ x_2(t) \end{pmatrix} dt + \begin{pmatrix} 2 & -2 \\ 1 & 1 \end{pmatrix} \begin{pmatrix} g_1(x_1(t-\tau_1)) \\ g_2(x_2(t-\tau_2)) \end{pmatrix} dt$$

$$+ G_1 \begin{pmatrix} g_1(x_1(t-\tau_1)) \\ g_2(x_2(t-\tau_2)) \end{pmatrix} dB_1(t) + G_2 \begin{pmatrix} x_1(t) \\ x_2(t) \end{pmatrix} dB_2(t)$$

$$+ G_3 \begin{pmatrix} x_1(t-\tau_1) \\ x_2(t-\tau_2) \end{pmatrix} dB_3(t). \tag{3.30}$$

Here $(B_1(t), B_2(t), B_3(t))$ is a 3-dimensional Brownian motion, g_1 and g_2 are the same as in Example 3.9 while G_i, $1 \le i \le 3$ are all 2×2 constant matrices. Assume that (3.28) holds and also

$$||G_2||^2 \vee ||G_3||^2 \le 0.1. \tag{3.31}$$

Note in this case that

$$\sigma(x, y) = (G_1 g(y), G_2 x, G_3 y) \quad \text{for } (x, y) \in R^2 \times R^2,$$

where $g(y) = (g_1(y_1), g_2(y_2))^T$. So
$$|\sigma(x,y)|^2 = |G_1 g(y)|^2 + |G_2 x|^2 + |G_3 y|^2$$
$$\leq 0.2474|g(y)|^2 + 0.1|x|^2 + 0.1|y|^2.$$

Therefore, by Theorems 3.2 and 3.8, we conclude that the stochastic delay network (3.30) is exponentially stable in mean square and is also almost surely exponentially stable. Moreover, if $\varepsilon > 0$ is the root to the equation

$$\max\{(4\varepsilon\tau_1 + 0.1)e^{\varepsilon\tau_1} + \varepsilon, \ (2\varepsilon\tau_2 + 0.1)e^{\varepsilon\tau_2} + \varepsilon\} = 0.1474, \quad (3.32)$$

then the second moment Lyapunov exponent should not be greater than $-\varepsilon$. For example, if $\tau_1 = \tau_2 = 0.01$, then equation (3.32) becomes

$$(0.04\varepsilon + 0.1)e^{0.01\varepsilon} + \varepsilon = 0.1474$$

which has the root $\varepsilon = 0.04553$, and hence the second moment Lyapunov exponent should not be greater than -0.04553 and the sample Lyapunov exponent should not be greater than -0.022765.

Example 3.11 Finally, consider the 3-dimensional symmetric stochastic delay network

$$dx(t) = [-\bar{B}x(t) + Ag(x_\tau(t))]dt + G_1 x(t)dB_1(t)$$
$$+ G_2 \begin{bmatrix} \sin(x_1(t-\tau_1)) \\ \sin(x_2(t-\tau_2)) \\ \sin(x_3(t-\tau_3)) \end{bmatrix} dB_2(t). \quad (3.33)$$

Here $(B_1(t), B_2(t))$ is a 2-dimensional Brownian motion, G_1 and G_2 are both 3×3 constant matrices, and

$$\bar{B} = \text{diag.}(2,3,4), \quad A = \begin{pmatrix} 0 & 1 & 1 \\ 1 & 1 & 1 \\ 1 & 1 & 2 \end{pmatrix},$$

$$g_i(u_i) = (0.5u_i \wedge 1) \vee (-1), \quad g(u) = (g_1(u_1), g_2(u_2), g_3(u_3))^T.$$

Note that (2.2) and (2.5) are satisfied with $\beta_1 = \beta_2 = \beta_3 = 0.5$. To apply Corollary 3.6, we compute $\lambda = 1.2$ by (3.20). Assume that

$$||G_1||^2 < 1.2 \quad \text{and} \quad ||G_2||^2 \leq 0.3. \quad (3.34)$$

Note that
$$\sigma(x,y) = (G_1 x, G_2(\sin y_1, \sin y_2, \sin y_3)^T)$$
for $(x,y) \in R^3 \times R^3$. Note also that

$$\sin^2 z \leq [(z \wedge 1) \vee (-1)]^2 \leq 4[(0.5z \wedge 1) \vee (-1)]^2 \quad \text{for } -\infty < z < \infty.$$

Hence
$$|\sigma(x,y)|^2 \leq ||G_1||^2 |x|^2 + 4||G_2||^2 |g(y)|^2$$
$$\leq ||G_1||^2 |x|^2 + 1.2|g(y)|^2.$$

Applying Corollary 3.6 and Theorem 3.8 we can conclude that the stochastic delay network (3.33) is exponentially stable in mean square and is also almost surely exponentially stable.

Bibliographical Notes

Chapter 1: The material in this chapter is classical and we refer Arnold (1974), Doob (1953), Friedman (1975), Gihman & Skorohod (1972), Liptser & Shiryayev (1986) etc.

Chapter 2: Most of the material in this chapter is classical, but Section 2.5 is based on Mao (1992c) while Section 2.6 is based on Bell & Mohammed (1989) and Mao (1994b).

Chapter 3: Stochastic Liouville's formula is due to Mao (1983) while the variation-of-constants formula etc. are classical.

Chapter 4: Most of the material in this chapter is essentially from Mao (1991a, 1994a) while Section 4.5 is based on Mao (1994c) and Theorem 4.6.2 is new.

Chapter 5: The existence-and-uniqueness theorems and the exponential estimates are classical and we refer Kolmonovskiĭ & Nosov (1986), Mao (1994a) and Mohammed (1984). Sections 5.5, 5.6 and 5.7 are based on Mao (1991d, e), Mao (1996b) and Mao (1996c), respectively.

Chapter 6: Stochastic functional differential equations of neutral type have been studied by Kolmonovskii & Nosov (1986) but some results have only been stated without proofs. In this chapter we systematically study the neutral-type equations and our treatments are independent. Many results in this chapter are new, for example, the pathwise estimates, the L^p-continuity and the Razumikhin theorems.

Chapter 7: The martingale representation theorem is classical. Section 7.3 is based on Pardoux & Peng (1990) while Section 7.4 is based on Mao (1995a). The generalized Feynman–Kac formula is due to Pardoux & Peng (1992).

Chapter 8: The Cameron–Martin–Girsanov theorem is classical. Sections 8.3 and 8.4 are based on Markus & Weerasinghe (1988) while Section 8.5 is based on Mao & Markus (1991).

Chapter 9: Section 9.2 describes several useful and well-known stochastic models for asset prices and we refer Neftci (1996), Oksendal (1995) etc. The main results in Section 9.3 are due to Dynkin (1963, 1965) and we follow the treatment of Oksendal (1995). Section 9.4 is based on Friedman (1975) and Wu & Mao (1988).

Chapter 10: The results of this chapter are essentially due to Liao & Mao (1996a, b).

References

Arnold, L. (1974), Stochastic Differential Equations: Theory and Applications, John Wiley and Sons.

Arnold, L. and Crauel, H. (1991), Random dynamical system, Lecture Notes in Mathematics 1486, pp1–22.

Arnold, L. and Kliemann, W. (1987), On unique ergodicity for degenerate diffusions, Stochastics 21, pp41–61.

Arnold, L., Oeljeklaus, E. and Pardoux, E. (1984), Almost sure and moment stability for linear Itô equations, Lecture Note in Math. 1186, pp129–159.

Baxendale, P. (1994), A stochastic Hopf bifurcation, Probab. Theory Relat. Fields 99, pp581–616.

Baxendale, P. and Henning, E.M. (1993), Stabilization of a linear system, Random Comput. Dyn. 1(4), pp395–421.

Beckenbach, E.F. and Bellman, R. (1961), Inequalities, Springer.

Bell, D.R. and Mohammed, S.E.A. (1989), On the solution of stochastic differential equations via small delays, Stochastics and Stochastics Reports 29, pp293–299.

Bellman, R. and Cooke, K.L. (1963), Differential–Difference Equations, Academic Press.

Bensoussan, A. (1982), Lectures on stochastic control, Lecture Notes in Math. 972, Springer.

Bihari, I. (1957), A generalization of a lemma of Bellman and its application to uniqueness problem of differential equations, Acta Math. Acad. Sci. Hungar. 7, pp71–94.

Bismut, J.M. (1973), Théorie probabiliste du contrôle des diffusions, Mem. Amer. Math. Soc. No. 176.

Black, F. and Scholes, M. (1973), The prices of options and corporate liabilities, J. Political Economy 81, 637–654.

Brayton, R.(1976), Nonlinear oscillations in a distributed network, Quart. Appl.

Math. 24, pp289–301.

Bucy, H.J. (1965), Stability and positive supermartingales, J. Differential Equation 1, pp151–155.

Carmona, R. and Nulart, D. (1990), Nonlinear Stochastic Integrators, Equations and Flows, Gordon and Breach.

Chow, P. (1982), Stability of nonlinear stochastic evolution equations, J. Math. Anal. Appl. 89, pp400–419.

Coddington, R.F. and Levinson, N. (1955), Theory of Ordinary Differential Equations, McGraw–Hill.

Curtain, R.F. and Pritchard, A.J. (1978), Infinite Dimensional Linear System Theory, Lecture Notes in Control and Information Sciences 8, Springer.

Da Prato, G. and Zabczyk, J. (1992), Stochastic Equations in Infinite Dimensions, Cambridge University Press.

Davis, M. (1994), Linear Stochastic Systems, Chapman and Hall.

Dellacherie, C. and Meyer, P.A. (1980), Probabilités et Potentiels, 2e édition, Chapitres V–VIII, Hermann.

Denker, J.S.(1986), Neural Networks for Computing, Proceedings of the Conference on Neural Networks for Computing (Snowbird, UT, 1986), AIP, New York, 1986.

Doléans-Dade, C. and Meyer, P.A. (1977), Equations Différentilles Stochastiques, Sém. Probab. XI, Lect. Notes in Math. 581.

Doob, J.L. (1953), Stochastic Processes, John Wiley.

Driver, R.D. (1963), A functional differential system of neutral type arising in a two-body problem of classical electrodynamics, in "Nonlinear Differential Equations and Nonlinear Mechanics," Academic Press, pp474–484.

Dunkel, G. (1968), Single-species model for population growth depending on past history, Lecture Notes in Math. 60, pp92–99.

Dynkin,E.B. (1963), The optimum choice of the instant for stopping a Markov process, Soviet Mathematics 4, 627–629.

Dynkin, E.B. (1965), Markov Processes, Vol.1 and 2, Springer.

Elliott, R.J. (1982), Stochastic Calculus and Applications, Springer.

Elworthy, K.D. (1982), Stochastic Differential Equations on Manifolds, London Math. Society, Lect. Notes Series 70, C. U. P.

Ergen, W.K. (1954), Kinetics of the circulating fuel nuclear reaction, J. Appl. Phys. 25, pp702–711.

Freidlin, M.I. (1985), Functional Integration and Partial Differential Equations, Princeton University Press.

Freidlin, M.I. and Wentzell, A.D. (1984), Random Perturbations of Dynamical Systems, Springer.

Friedman, A. (1975), Stochastic Differential Equations and Applications, Vol.1

and 2, Academic Press.

Gihman, I.I. and Skorohod, A.V. (1972), Stochastic Differential Equations, Springer.

Girsanov, I.V. (1962), An example of nonuniqueness of the solution of K. Itô's stochastic equations, Teoriya veroyatnostey i yeye primeneniya 7, pp336–342.

Hahn, W. (1967), Stability of Motion, Springer.

Hale, J.K. and Lunel, S.M.V. (1993), Introduction to Functional Differential Equations, Springer.

Halmos, P.R. (1974), Measure Theory, Springer.

Has'minskii, R.Z. (1967), Necessary and sufficient conditions for the asymptotic stability of linear stochastic systems, Theory Probability Appl. 12, pp144–147.

Has'minskii, R.Z. (1980), Stochastic Stability of Differential Equations, Alphen: Sijtjoff and Noordhoff (translation of the Russian edition, Moscow, Nauka 1969).

Haussmann, U.G. (1986), A Stochastic Maximum Principle for Optimal Control of Diffusions, Pitman Research Notes in Math. 151, Longman.

Hopfield, J.J. (1982), Neural networks and physical systems with emergent collect computational abilities, Proc. Natl. Acad. Sci. USA 79, pp2554–2558.

Hopfield, J.J. (1984), Neurons with graded response have collective computational properties like those of two-state neurons, Proc. Natl. Acad. Sci. USA 81, pp3088–3092.

Hopfield, J.J. and Tank, D.W. (1986), Computing with neural circuits, Model Science 233, pp3088–3092.

Ichikawa, A. (1982), Stability of semilinear stochastic evolution equations, J. Math. Anal. Appl. 90, pp12–44.

Ikeda, N. and Watanabe, S. (1981), Stochastic Differential Equations and Diffusion Processes, North–Holland.

Itô, K. (1942), Differential equations determining Markov processes, Zenkoku Shijo Suguku Danwakai 1077, pp1352–1400.

Jacod, J. (1979), Calcul Stochastique et Problèmes de Martingales, Lect. Notes in Math. 714, Springer.

Kaneko, T. and Nakao, S. (1988), A note on approximation for stochastic differential equations, Séminaire de Probabilités 22, Lect. Notes Math. 1321, pp155–162.

Karatzas, I. and Shreve, S.E. (1988), Brownian Motion and Stochastic Calculus, Springer.

Kolmanovskii, V.B. and Nosov, V.R. (1986), Stability of Functional Differential Equations, Academic Press.

Kolmogorov, A., Petrovskii, I. and Piskunov, N. (1973), Etude de l'équation de la diffusion avec croissance de al matière et son application a un problème biologique, Moscow Univ. Math. Bell. 1 , 1-25.

Krasovskii, N. (1963), Stability of Motion, Standford University Press.

Krylov, N.V. (1980), Controlled Diffusion Processes, Springer.

Kunita, H. (1990), Stochastic Flows and Stochastic Differential Equations, Cambridge University Press.

Kushner, H.J. (1965), On the construction of stochastic Lyapunov functions, IEEE Trans. Automatic Control AC-10, pp477-478.

Kushner, H.J. (1967), Stochastic Stability and Control, Academic Press.

Ladde, G.S. and Lakshmikantham, V. (1980), Random Differential Inequalities, Academic Press.

Lakshmikantham, V., Leeda, S. and Martynyuk, A.A. (1989), Stability Analysis of Nonlinear Systems, Marcel Dekker.

Le Jan, Y. (1982), Flots de diffusions dans Rd, C. R. Acad. Sci., Paris, Ser. I 294 (1982), pp697-699.

Li, C.W. and Mao, X. (1986a), Delay stochastic differential equations, J. East China Inst. Chem. Tech. 6, pp779-792.

Li, C.W. and Mao, X. (1986b), Global asymptotic stability of solutions of stochastic differential equations, J. East China Inst. Chem. Tech. 4, pp531-536.

Li, C.W. and Mao, X. (1987), Stochastic boundedness of solutions of Itô's equations, J. East China Inst. Chem. Tech.3, pp383-388.

Liao, X.X. and Mao, X. (1996a), Exponential stability and instability of stochastic neural networks, Sto. Anal. Appl. 14(2), pp165-185.

Liao, X.X. and Mao, X. (1996b), Stability of stochastic neural networks, Neural, Parallel and Scientific Computations 4, pp205-224.

Liao, X.X. and Mao, X. (1997), Stability of large-scale neutral-type stochastic differential equations, Dynamics of Continuous, Discrete and Impulsive Systems 3(1), pp43-56.

Liptser, R.Sh. and Shiryayev, A.N. (1986), Theorey of Martingales, Klumer Academic Publishers.

Lyapunov, A.M. (1892), Probleme general de la stabilite du mouvement, Comm. Soc. Math. Kharkov 2, pp265-272.

Mao, X. (1983), Liouville's formula for stochastic integral equations, Journal of Fuzhou Univ. 3, pp8-19.

Mao, X. (1989a), Existence and uniqueness of the solutions of delay stochastic integral equations, Sto. Anal. Appl. 7(1), pp59-74.

Mao, X. (1990a), Eventual asymptotic stability for stochastic differential equations with respect to semimartingales, Quarterly J. Math. Oxford (2), 41, pp71-77.

Mao, X. (1990b), Lyapunov functions and almost sure exponential stability of stochastic differential equations based on semimartingales with spatial parameters, SIAM J. Control Optim. 28(6), pp1481–1490.

Mao, X. (1990c), Exponential stability for stochastic differential equations with respect to semimartingales, Sto. Proce. Their Appl. 35, pp267–277.

Mao, X. (1990d), A note on global solution to stochastic differential equations based on semimartingales with spatial parameters, Journal of Theoretical Probability 4(1), pp161–167.

Mao, X. (1991a), Stability of Stochastic Differential Equations with Respect to Semimartingales, Pitman Research Notes in Mathematics Series 251, Longman Scientific and Technical.

Mao, X. (1991b), Almost sure exponential stability for delay stochastic differential equations with respect to semimartingales, Sto. Anal. Appl. 9(2), pp177–194.

Mao, X. (1991c), A note on comparison theorems for stochastic differential equations with respect to semimartingales, Stochastics and Stochastics Reports 37, pp49–59.

Mao, X. (1991d), Approximate solutions for a class of delay stochastic differential equations, Stochastics and Stochastics Reports 35, pp111–123.

Mao, X. (1991e), Approximate solutions for a class of stochastic evolution equations with variable delays, Numer. Funct. Anal and Optimiz. 12, pp525–533.

Mao, X. (1992a), Polynomial stability for perturbed stochastic differential equations with respect to semimartingales, Stochastic Processes and Their Applications 41, pp101–116.

Mao, X. (1992b), Almost sure polynomial stability for a class of stochastic differential equations, Quarterly J. Math. Oxford (2) 43, pp339–348.

Mao, X. (1992c), Almost sure asymptotic bounds for a class of stochastic differential equations, Stochastics and Stochastics Reports 41, pp57–69.

Mao, X. (1992d), Solutions of stochastic differential–functional equations via bounded stochastic integral contractors, J. Theoretical Probability 5(3), pp487–502.

Mao, X. (1992e), Exponential stability of large–scale stochastic differential equations, Systems and Control Letters 19, pp71–81.

Mao, X. (1992f), Robustness of stability of nonlinear systems with stochastic delay perturbations, Systems and Control Letters 19, pp391–400.

Mao, X. (1993), Almost sure exponential stability for a class of stochastic differential equations with applications to stochastic flows, Sto. Anal. Appl. 11(1), pp77–95.

Mao, X. (1994a), Exponential Stability of Stochastic Differential Equations, Marcel Dekker.

Mao, X. (1994b), Approximate solutions of stochastic differential equations with pathwise uniqueness, Sto. Anal. Appl. 13(3), pp355–367.

Mao, X. (1994c), Stochastic stabilization and destabilization, Systems and Control Letters 23, pp279–290.

Mao, X. (1995a) Adapted solutions of backward stochastic differential equations with non–Lipschitz coefficients, Stochastic Processes and Their Applications 58, pp281–292.

Mao, X. (1995b) Exponential stability in mean square of neutral stochastic differential functional equations, Systems and Control Letters 26, pp245–251.

Mao, X. (1996a), Robustness of exponential stability of stochastic differential delay equations, IEEE Transactions AC–41, pp442–447.

Mao, X. (1996b), Razumikhin-type theorems on exponential stability of stochastic functional differential equations, Sto. Proc. Their Appl. 65, pp233–250.

Mao, X. (1996c), Stochastic self-stabilization, Stochastics and Stochastics Reports 57, pp57-70

Mao, X. (1997), Razumikhin-type theorems on exponential stability of neutral stochastic functional differential equations, to appear in SIAM J. Math. Anal.

Mao, X. and Markus, L. (1991), Energy bounds for nonlinear dissipative stochastic differential equations with respect to semimartingales, Stochastics and Stochastics Reports 37, pp1–14.

Mao, X. and Markus, L. (1993), Wave equations with stochastic boundary conditions, J. Math. Anal. Appl. 177 (2), pp315–341.

Mao, X. and Rodkina, A.E. (1995), Exponential stability of stochastic differential equations driven by discontinuous semimartingales, Stochastics and Stochastics Reports 55, pp207–224.

Markus, L. and Weerasinghe, A. (1988), Stochastic oscillators, J. Differential Equations 71(2), pp288–314.

McKean, H.P. (1969), Stochastic Integrals, Academic Press.

Mcshane, E.J. (1974), Stochastic Calculus and Stochastic Models, Academic Press.

Métivier, M. (1982), Semimartingales, Wslter de Gruyter.

Meyer, P.A. (1972), Martingales and stochastic integrals I, Lect. Notes in Math. 284, Springer.

Mizel, V.J. and Trutzer, V. (1984), Stochastic hereditary equations: existence and asymptotic stability, J. Integral Equations 7, pp1–72.

Mohammed, S–E. A. (1984), Stochastic Functional Differential Equations, Longman Scientific and Technical.

Mohammed, S–E. A., Scheutzow, M. and WeizsaÉcher, H. V. (1986), Hyperbolic state space decomposition for a linear stochastic delay equations, SIAM J. on Control and Optimization 24(3), pp543–551.

Natanson, I.P. (1964), Theory of Functions of Real Variables, Vol.1, Frederick Unger.

Neftci, S.N. (1996), An Introduction to the Mathematics of Financial Derivatives, Academic Press.

Oksendal, B. (1995), Stochastic Differential Equations: An Introduction with Applications, 4th Ed., Springer.

Oseledec, V.I. (1968), A multiplicative ergodic theorem: Laypunov characteristic numbers for dynamical systems, Trans. moscow Math. Soc. 19, pp197–231.

Pardoux, E. and Peng, S.G. (1990), Adapted solution of a backward stochastic differential equation, Systems and Control Letters 14, pp55–61.

Pardoux, E. and Peng, S.G. (1992), Backward stochastic differential equations and quasilinear parabolic partial differential equations, Lecture Notes in Control and Information Science 176, pp200–217.

Pardoux, E. and Wihstutz, V. (1988), Lyapunov exponent and rotation number of two–dimensional stochastic systems with small diffusion, SIAM J. Applied Math. 48, pp442–457.

Pinsky, M.A. and Wihstutz, V. (1988), Lyapunov exponents of nilponent Itô systems, Stochastics 25, pp43–57.

Rao, A.N.V. and Tsokos, C.P. (1975), On the existence, uniqueness, and stability behavior of a random solution to a nonlinear perturbed stochastic integro–differential equation, Inform. and Control 27, pp61–74.

Razumikhin, B.S. (1956), On the stability of systems with a delay, Prikl. Mat. Meh. 20, pp500–512.

Razumikhin, B.S. (1960), Application of Lyapunov's method to problems in the stability of systems with a delay, Automat. i Telemeh. 21, pp740–749.

Rogers, L.C.G. and Williams, D. (1987), Diffusions, Markov Processes and Martingales, Vol.2, John Wiley and Sons.

Rubanik, V.P. (1969), Oscillations of quasilinear systems with retardation, Nauk, Moscow.

Samuelson, P.A. (1965), Rational theory of warrant pricing (with Appendix by H.P. McKean), Industrial Management Review 6, 13–31.

Truman, A. (1986), An introduction to the stochastic mechanics of stationary states with applications, in "From Local Times to Global Geometry, Control and Physics" edited by K. D. Elworthy, Pitman Research Notes in Math. 150, pp329–344.

Tsokos, C.P. and Padgett, W.J. (1974), Random Integral Equations with Applications to Life Sciences and Engineering, Academic Press.

Uhlenbeck, G.E. and Ornstein, L.S. (1930), On the theory of Brownian motion, Pyys. Rev. 36, pp362–271.

Volterra, V. (1928), Sur la théorie mathématique des phénomènes héréditaires, J. Math. Pures Appl. 7, pp249–298.

Reference

Wang, Z. K. (1986), Theory of Stochastic Processes, China Science Press.

Watanabe, S. and Yamada, T. (1971), On the uniqueness of solutions of stochastic differential equations II, J. Math. Kyoto University 11, pp553–563.

Wright, E.M. (1961), A functional equation in the heuristic theory of primes, Mathematical Gazette 45, pp15–16.

Wu, R.Q. and Mao, X. (1983), Existence and uniqueness of the solutions of stochastic differential equations, Stochastics 11, pp19–32.

Wu, R.Q. and Mao, X. (1986), On Comparison theorems for a kind of integral equations with respect to semimartingales, J. Engineering Math. 3(1), pp1–10.

Wu, R.Q. and Mao, X. (1988), A class of stochastic games, Mathematical Statistics and Applied Probability 3(1), pp99–111.

Yamada, T. (1981), On the successive approximation of solutions of stochastic differential equations, J. Math. Kyoto University 21, pp501–515.

Yamada, T. and Watanabe, S. (1971), On the uniqueness of solutions of stochastic differential equations, J. Math. Kyoto University 11, pp155–167.

Yan, J.A. (1981), Introduction to Martingales and Stochastic Integrals, Shanghai Science and Technology Press.

Index

adapted process 10
asymptotic stability 110
augmentation 15

backward stochastic differential
   equation 239
Bihari's inequality 45
Borel
   measurable 3
   set 3
   σ-algebra 3
Borel–Cantelli's lemma 7
Brownian bridge 104
Brownian motion 15, 17
   on the unit circle 49, 103
Burkholder–Davis–Gundy's
   inequality 40

cadlag process 10
Cameron–Martin–Girsanov's
   theorem 270
Caratheodory's approximation 71, 162
Cauchy problem 82
Cauchy–Maruyama's
   approximation 76, 166
Chapman–Kolmogorov's equation 85
Chebyshev's inequality 5
coloured noise 102
complement 2
completion 4
conditional expectation 8
conditional probability 8

continuous process 10
convergence
   almost sure $\sim$ 5
   with probability one 5
   stochastic $\sim$ 6
   in probability 6
   in pth moment 6
   in distribution 6
convergence rate function 189
convergence theorem
   bounded $\sim$ 6
   dominated $\sim$ 6
   monotonic $\sim$ 6
covariance matrix 5

decrescent function 108
delay equation 155
destabilization 135, 338
differential equation 48
diffusion operator 110
Dirac delta function 17, 330
Dirichlet problem 78
dissipative force 286
distribution 5
Doob's martingale
   convergence theorem 13
   inequality 14
Doob's stopping theorem 11
Doob–Dynkin's lemma 3

energy 286
energy bound 286

equilibrium position 107, 110, 170
error function 279
excessive 309
exit time 11
expectation 4
exponential instability 123, 332
exponential martingale inequality 43
exponential stability
   almost sure $\sim$ 119, 173, 223
   moment $\sim$ 127, 171, 219
external driving force 286

Feller property 86
Feynman–Kac's formula 78, 259
   generalized $\sim$ 266
finite variation process 10
filtration 9
   natural $\sim$ 15
   right continuous $\sim$ 9
flow property 48
Fourier transform 236
functional differential equation 147
fundamental matrix 92
$\mathcal{F}$-measurable 3

geometric Brownian motion 105, 299
Girsanov's theorem 272
global solution 58, 154, 207
Gronwall's inequality 44

Has'minskii condition 58
hitting time 319
Hölder's inequality 5

increasing process 10
indicator function 3
indefinite integral 23, 30
independent increments 15
independent
   random variables 7
   sets 6
   σ-algebras 7
indistinguishable processes 11
initial-boundary value problem 81
integrable process 11
integrable random variable 4

integration by parts formula 36
Itô's formula 32, 35
Itô's stochastic integral 18
Itô's process 31, 36

Jensen's inequality 75
joint quadratic variation 12

Kolmogorov backward equation 83

Laplace operator 38
law of the iterated logarithm 16
least superharmonic majorant 310
least supermeanvalued majorant 310
left continuous process 10
Lévy's theorem 17
linear growth condition 51, 150, 202
linear stochastic differential eq.
   autonomous $\sim$ 92, 100
   homogeneous $\sim$ 92
   in the narrow sense 92, 99
Liouville's formula 92
Lipschitz condition 51, 150, 202
local Lipschitz condition 56, 153, 207
local martingale 12
lower semicontinuous 306
Lyapunov exponent
   moment $\sim$ 63, 127
   sample $\sim$ 63, 119
Lyapunov function 109
   stochastic $\sim$ 115
$L^1(R_+; R^d)$-stable 189

Markov process 84
   homogeneous $\sim$ 85
   strong $\sim$ 86
Markov property 84
   strong $\sim$ 86
martingale 11
   representation theorem 234
mean reverting process 301
measurable process 10
measurable space 3
Minkovskii's inequality 5
modification 10
moment 4

monotone condition 58

negative definite function 108
neural network 327
  with delay 342
neutral stochastic
  functional differential eq. 201
  differential delay eq. 207

optimal
  expected reward 306
  stopping problem 306
  stopping time 306
optional process 10
Ornstein–Uhlenbeck position
  process 102
Ornstein–Uhlenbeck process 101
  mean reverting $\sim$ 103, 302

partial differential equation 78
pathwise uniqueness 78
payoff functional 319
Picard's iterations 53, 205, 248
positive-definite function 108
predictable process 10
probability measure 4
probability space 4
  complete $\sim$ 4
progressive process 10
progressively measurable 10

quadratic variation 12
quasilinear partial differential
  equation 265

radially unbounded 108
Radon–Nikodym's theorem 8
random variable 3
Razumikhin theorem 169, 219
reachability problem 233
reflection principle 279
restoring force 286
reward function 306
right continuous process 10
robustness of
  instability 338
  stability 334

saddle point 320
  of sets 320
sample path 10
semigroup property 48
simple process 19
solution 48, 149, 201, 239
square integrable martingale 12
square root process 302
  mean reverting $\sim$ 304
stability in probability 110
stabilization 135, 334
state space 9
stationary increments 15
step function 8
step process 18
stochastic asymptotic stability 110
  in the large
stochastic differential 31, 36
stochastic game 319
stochastic instability 110
stochastic interval 11
stochastic oscillator 269
  linear $\sim$ 276
  nonlinear $\sim$ 272
stochastic process 9
stochastic self-stabilization 188
stochastic stability 110
stopped process 11
stopping time 11
strong law of large numbers 12, 16
strong solution 78
submartingale 13
  inequality 14
superharmonic 307
  majorant 310
supermartingale 13
  inequality 13
  convergence theorem 13
supermeanvalued 306
  majorant 310
symmetric neural network 331, 333
σ-algebra 3
  generated 3

transformation formula 5
transition probability 84
trivial solution 107, 110, 170

uncorrelated 4
uniformly integrable 13, 315
uniformly L^2-continuous 167
unique solution 48, 149, 210, 239
upper limit of sets 7
usual conditions 9

variance 4
variational inequality 324
variation-of-constants formula 96
version 10
volatility 304

weak solution 78
weak uniqueness 78
Wronskian determinant 92

Horwood books on Mathematics and Astronomy

FUNDAMENTALS OF UNIVERSITY MATHEMATICS

COLIN McGREGOR, JOHN NIMMO and WILSON W. STOTHERS, Department of Mathematics, University of Glasgow

ISBN: 1-898563-10-1   540 pages

A unified course for first year mathematics, bridging the school/university gap, suitable for pure and applied mathematics courses, and those leading to degrees in physics, chemical physics, computing science, or statistics.

Contents: Preliminaries, Functions & Inverse Functions, Polynomials & Rational Functions, Induction & the Binomial Theorem, Trigonometry, Complex Numbers, Limits & Continuity, Differentiation - Fundamentals, Differentiation - Applications, Curve Sketching, Matrices & Linear Equations, Vectors & Three Dimensional Geometry, Products of Vectors, Integration - Fundamentals, Logarithms & Exponentials, Integration - Methods & Applications, Ordinary Differential Equations, Sequences & Series, Numerical Methods, Appendices.

DECISION AND DISCRETE MATHEMATICS: Maths for Decision-making in Business and Industry

THE SPODE GROUP, Truro School, Cornwall

ISBN: 1-898563-27-6   240 pages

The coverage in the Decision Mathematics Module (Discrete Mathematics) of the syllabuses of English A level examination boards is suitable for foundation first year and undergraduate courses in qualitative studies and operational research, or for access courses for students needing strengthening in mathematics or those moving into mathematics from another subject.

Contents: An Introduction to Networks; Recursion; Shortest Route; Dynamic Programming; Flows in Networks; Critical Path Analysis; Linear Programming (graphical) Linear Programming: Simplex Method; The Transportation Problem; Matching and Assignment Problems; Game Theory; Recurrence Relations; Simulation; Iterative Processes; Sorting and Packing; Algorithms.

MATHEMATICAL ANALYSIS AND PROOF

DAVID S.G. STIRLING, Senior Lecturer in Mathematics, University of Reading

ISBN: 1-898563-36-5 **250 pages**

This fundamental and straightforward text for first and second year degree courses in the UK addresses a weakness observed among students, namely a lack of familiarity with formal proof. Dr Stirling begins with the need for mathematical proof, developing associated technical and logical skills. This lucid analysis and development reads naturally and in straightforward progression by indicating how proofs are constructed. The text emphasises (1) the need for familiarity with long mathematical arguments and manipulation, (2) the importance of the ability to construct proofs in analysis.

Contents: Setting the Scene; Logic and Deduction; Mathematical Induction; Sets and Numbers; Order and Inequalities; Decimals; Limits; Infinite Series; Structure of Real Number System; Continuity; Differentiation; Functions Defined by Power Series; Integration; Functions of Several Variables; Appendix: Decimal Expansion of Integers.

SIGNAL PROCESSING IN ELECTRONIC COMMUNICATIONS FOR ENGINEERS AND MATHEMATICIANS

MICHAEL J. CHAPMAN, DAVID P. GOODALL, and NIGEL C. STEELE, School of Mathematics and Information Sciences, University of Coventry

ISBN: 1-898563-23-3 **288 pages**

This text for advanced undergraduate and postgraduate courses in electronic engineering, applied mathematics, and computer science, deals with signal processing as an important aspect of electronic communications in its role of transmitting information, and the language of its expression. It develops the required mathematics in an interesting and informative way, leading to confidence on the part of the reader.

Contents: Signal and Linear System Fundamentals; System Responses; Fourier Methods; Analogue Filters; Discrete-time Signals and Systems; Discrete-time System Responses; Discrete-time Fourier Analysis; The Design of Digital Filters; Aspects of Speech Processing; Appendices: The Complex Exponential; Linear Predictive Coding Algorithms; Answers.

CALCULUS: Introduction to Theory and Applications in Physical and Life Science

R.M. JOHNSON, Senior Lecturer, Department of Mathematics and Statistics, University of Paisley, Paisley, Scotland.

ISBN: 1-898563-06-3 **336 pages**

A lucid text conveying clear understanding of the fundamentals and applications for first year undergraduates in applied mathematics, computing, physics, electrical, mechanical and civil engineering, chemical science, biology and life science.

Contents: Prerequisites from Algebra; Geometry and Trigonometry; Limits and Differentiation; Differentiation of Products and Quotients; Higher-order Derivatives; Integration; Definite Integrals; Stationary Points of Inflexion; Applications of the Function of a Function Rule; The Exponential, Logarithmic and Hyperbolic Functions; Methods of Integration; Further Applications of Integration; Approximate Integration; Infinite Series; Differential Equations.

LINEAR DIFFERENTIAL AND DIFFERENCE EQUATIONS
A systems approach for mathematicians and engineers

R.M.JOHNSON, Senior Lecturer, Department of Mathematics and Statistics, University of Paisley, Paisley, Scotland.

ISBN: 1-898563-12-8 **125 pages**

An advanced text for senior undergraduates and graduates, and professional workers in applied mathematics, and electrical and mechanical engineering.

"Should find wide application by undergraduate students in engineering and computer science ... the author is to be congratulated on the importance that he attaches to conveying the parallelism of continuous and discrete systems" - *Institute of Electrical Engineers (IEE) Proceedings*

Contents: Part I CONTINUOUS SYSTEMS: Approach to Laplace Transforms; Solution of Linear Differential Equations; Steady State Oscillations; Piece-wise Continuous Functions; Part II DISCRETE SYSTEMS: From the Ideal Sampler to the z-transform; Solution of Linear Differential Equations; Digital Filters; TABLES: Laplace Transforms of $f(t)$; z-transforms of $\{xn\}$; z-transforms of $f\{t\}$; continuous and discrete systems compared.

TEACHING AND LEARNING MATHEMATICAL MODELLING

Editors: S.K. HOUSTON, University of Ulster, Northern Ireland; W.BLUM, University of Kassel, Germany; IAN HUNTLEY, University of Bristol; N.T. NEILL, University of Ulster, Northern Ireland

ISBN: 1-898563-29-2   **320 pages**

Mathematicians from 10 countries contribute to mathematical modelling. Interdisciplinary topics reflect applications in mechanics and engineering, computing science, traffic control, business studies, and mathematics (fractals and analysis).

Contents: Section A: Reflections and Investigations; Section B: Assessment at Tertiary Level; Section C: Secondary Courses and Case Studies; Section D: Tertiary Case Studies.

PRACTICAL ASTRONOMY

H. ROBERT MILLS, OBE, MSc, FRAS, former Science and Engineering Director, The British Council, London
Foreword by Heather Couper, Gresham Professor of Astronomy, London

ISBN: 1-898563-00-4   **235 pages**

"Excellent basic material ... assembling the wealth of practical and experimental projects, easily conducted at home ... many of the ideas are original and well presented. Simple instructions, perfectly clear illustrations" - *Spaceflight*
"A unique compendium of "do it yourself " projects' and a vast store of general astronomical information ... get this book, and revel in the good things in it" Commander H.R. Hatfield R.N., *British Astronomical Association Journal*
"Useful work for teachers of astronomy as well as stargazers ... glossary, reading list, and compendium of useful astronomical data" - *Choice, American Library Assn.*

THE GREAT ASTRONOMICAL REVOLUTION:
1534-1687 and the Space Age Epilogue

PATRICK MOORE, CBE, DSc, FRAS,
Past President, British Astronomical Association

ISBN: 1-898563-19-5   **256 pages**

"Summarizes lives, achievements of Copernicus, Brahe, Kepler, Galileo, Newton and methods of scientific enquiry. Relates their discoveries to modern cosmology and space exploration" - *College and Research Libraries News*, **USA**
"An epic and fascinating story about a great area of scientific history ... an enormous enhancement to our knowledge of the Universe" - *Spaceflight*
"Describes this revolutionary change with clarity and insight ... up to date with a telling Epilogue ... an invigorating account of events that determined the future of the human race "highly recommended" - *Journal of the British Astronomical Association*